Space Law

SCIENCES

Space Exploration and Technology,
Field Director – Jean-Luc Lefebvre

Space and Humankind, Subject Head – Jean-Luc Lefebvre

Space Law

Legal Framework for Space Activities

Coordinated by
Thomas Leclerc

ISTE

WILEY

First published 2023 in Great Britain and the United States by ISTE Ltd and John Wiley & Sons, Inc.

ISTE Ltd
27-37 St George's Road
London SW19 4EU
UK

www.iste.co.uk

John Wiley & Sons, Inc.
111 River Street
Hoboken, NJ 07030
USA

www.wiley.com

Library of Congress Control Number: 2023933067

British Library Cataloguing-in-Publication Data
A CIP record for this book is available from the British Library
ISBN 978-1-78945-152-8

ERC code:
PE9 Universe Sciences
PE9_15 Space Sciences
SH2 Institutions, Values, Environment and Space
SH2_5 International relations, global and transnational governance
SH2_6 Sustainability sciences, environment and resources

Contents

Introduction

Thomas LECLERC
Université de Bretagne Occidentale, Brest, France

Outer space, as it is known and exploited by humans, is far from being a lawless area. The tiny portion of outer space, whose access is within the reach of human technology, is indeed subject to a legal framework. There is a set of rules, both from public and private law, specifically dedicated to outer space and to the activities carried out there.

Nevertheless, many steps separate us from the context that surrounded the launch, on October 4, 1957, of the first artificial Earth satellite, a launch followed by the adoption by the General Assembly of the United Nations, on December 20, 1961, of a principle that has become the normative and legal cornerstone of space activities: outer space shall be free for exploration and use by all States.

To understand the current state of this legal corpus, it seems necessary to analyze the changes that have affected the technology used and the actors involved, as well as the domains concerned by the exploration and the use, by humanity, of outer space.

This volume thus proposes to bear witness to the progressive densification of the legal framework applicable to outer space and to the activities that are carried out there.

Without claiming to be exhaustive, the structure of this volume has been designed to facilitate the discovery of this legal environment by presenting the main sources of space law (Part 1), its main principles (Part 2), the diversity of its fields of application (Part 3) and the challenges and issues that the development of space activities inevitably raises (Part 4).

First of all, space law can only be understood as a composite universe whose sources are to be found both in international law (Chapter 1) and in the diversity of national legal orders (Chapter 2).

The common framework for all these rules is a treaty drawn up under the aegis of the United Nations: the Treaty on Principles Governing the Activities of States in the Exploration and Use of Outer Space, including the Moon and Other Celestial Bodies, opened for signature on January 27, 1967. Among the main principles that it lays down – and whose understanding and analysis will enable us to understand the normative architecture of space law – are the principle of freedom of use and exploration of outer space (Chapter 3), the principle of exploration and use of outer space for the benefit and in the interest of all countries (Chapter 4), the non-appropriation principle of outer space (Chapter 5), the principle of the peaceful use of outer space (Chapter 6), or the widely discussed principle of State responsibility and liability for space activities (Chapter 7).

All of these principles have made it possible to provide a framework for the emergence, development and deployment of space technologies that are largely dedicated to the placement in orbit and operation of artificial Earth satellites. The legal framework established by the major United Nations treaties dedicated to outer space could not, however, meet the growing need for specific rules governing space activities. These activities, and the challenges and legal issues relating to their development, explained the development of multiple branches of space law that are explored in this volume.

The circulation of objects placed in orbit appeared to be one of the first challenges linked to the occupation of the Earth's orbital environment. Today, analyzed from the point of view of the law of space traffic management (Chapter 8), the control of this circulation implies an allocation and a coordinated management of the spectrum of frequencies and associated orbits (Chapter 9). The allocation of these frequencies is indispensable to the numerous fields of application that space technology offers human society today. This volume could not do without a legal examination, in the current context of *New Space*, of some of these fields of application: satellite radio communication (Chapter 10), navigation satellite systems (Chapter 11) and space teledetection (Chapter 12). The legal issues related to the development of small satellites (Chapter 13) and to projects for the exploitation of material resources of celestial bodies (Chapter 14) also needed to be analyzed, especially in view of the commercial interest of an ever-increasing number of private actors.

Behind the development of technologies and activities also lies a series of legal issues and challenges. Projects for the exploration of the universe near and far from Earth raise a certain number of questions, which the law must apprehend and sometimes anticipate (Chapter 15), including the legal framework dedicated to the risks of contamination of the explored planets (Chapter 16). Our growing dependence on space technologies, in both civil and military domains, finally explains a new perception of the space environment and of the activities carried out in it, sometimes far from an outer space that some people hoped would be exclusively reserved for peace and science. A volume dedicated to space law should therefore address the issues of conflict and the responses that international law provides today to the development of a new type of threat to international security (Chapters 17 and 18). As for the increasingly pressing issue of space debris, the choice was made not to devote a specific chapter to it. Raised indirectly in different chapters of this volume, the technicality of this issue will certainly justify, in the future, the publication of a specialized volume.

This volume of space law is intended for a wide audience, from students to specialists and experts, as well as citizens who are curious to learn from reliable sources. It owes its existence to the enthusiastic investment of a large number of people, whom we wish to thank.

First of all, it benefited from the precious supervision of Jean-Luc Lefebvre, director of the editorial domain “Spatiology” of the “Sciences” encyclopedia and coordinator of the spatiology collection at ISTE editions. My deepest gratitude also goes to Professor Armel Kerrest, to whom the inspiration for this project belongs and who was its first architect. Armel thought up this volume and brought together a large number of its contributors before entrusting me with its direction. He has never ceased to participate in this project providing his advice and expertise.

I also would like to thank all the contributors to this volume, academics and practitioners belonging to various French and international institutions all recognized for their expertise in the field of space law. The enthusiastic welcome they have given to this project has been the driving force behind it. I would like to thank them for having agreed to make this effort to transmit and make their expertise accessible, thus providing everyone with the knowledge needed to understand the many issues raised by the development of space activities.

I would also like to thank the editorial team of ISTE Group for their competence and efficiency in the design and material development of this volume.

This volume would not have seen the light of day without the friendly support of my colleagues and friends at the Université de Bretagne Occidentale, starting with Professor Valère Ndior, whose advice and experience were an invaluable support throughout this project. I address to them my most sincere and warm thanks.

Part 1

The Main Sources of Space Law

1

History, the Treaties, the Resolutions

Sergio MARCHISIO
Sapienza University of Rome, Italy

1.1. Notion and evolution of international space law

International space law can be described as the special branch of international law that governs human activities in outer space, including the Moon and other celestial bodies (Kerrest 2007). It comprises, much like international law at large, a variety of international agreements, treaties, conventions and United Nations (UN) General Assembly (GA) resolutions, as well as rules of customary international law. The term "international space law" is most often associated with the rules and principles contained in the five international treaties and five sets of declarations of principles on space matters, which have been developed since the 1960s under the auspices of the UN.

Moreover, the notion of space law also encompasses the rules of other international organizations of universal or regional character, which carry on space-related activities, such as the International Telecommunication Union (ITU), the European Space Agency (ESA) and the European Union (EU). According to Article 2.1 (*j*) of the Convention on the Law of Treaties between States and International Organizations or between International Organizations of March 21, 1986, not yet in force, these rules mean "the constituent instruments, decisions and resolutions adopted in accordance with them, and the established practice of the organizations" (Marchisio 1986).

In addition to these international instruments, many States have adopted national legislation governing space activities, mainly to implement their international obligations. A well-established rule of general international law, codified in Articles 26 and 27 of the Vienna Convention on the Law of Treaties of 23 May 1969, entered into force on 27 January 1980 (VCLT), establishes that States must perform in good faith treaties in force binding upon them and that they may not invoke the provisions of their internal law as justification for their failure to perform them. Although the way in which international law applies within a State is a matter regulated by the law of that State, the outcome affects the State's position in international law. International law requires that States fulfill their obligations and they will be held responsible if they do not. Furthermore, often international treaties are not fully self-executing and they may require implementing national legislation.

Thus, in a broader sense, the notion of "space law" means a specialized body of law, both of international and national nature, which is aimed at maintaining order and co-ordinating relations among the subjects involved in space activities, States and private persons. Every entity carrying out activities in outer space must generally behave in a fashion that does not breach legal rules or hamper the rights of other subjects. Transgressions of legal rules would provoke social disruption, reactions and disputes to be solved in accordance with applicable legal norms.

Space law is a relatively new branch of international law. This body of law has grown from the necessity of creating norms to govern the expanding uses of outer space science and technology in improving functions and providing new services on the Earth. When the space age began in 1957 with the Soviet launch of Sputnik 1, the first man-made satellite, the international community immediately realized that it was essential to formulate international norms for the conduct of human activities in outer space. Then, space law has developed over time and will continue to develop as new challenges arise.

At the beginning, it was natural that the responsibility to regulate the activities of States in outer space would fall upon the UN that had been established after the Second World War to maintain international peace and security and charged with the task of encouraging the codification of international law and its progressive development.

The process began in 1958, in a climate of intense rivalry between the United States and the Soviet Union (USSR) within the Cold War. Shortly after the launching of the first artificial satellite, the Permanent Representative of the United States to the UN requested the Secretary General (SG) that an item called "Programme for International Cooperation in the Field of Outer Space" be placed on the agenda of the GA. This letter called for the Assembly to establish an ad hoc committee "to make the necessary detailed studies and recommendations as to what specific steps the Assembly might take to further man's progress" in outer space and "to assure that outer space (would) be used solely for the benefit of all mankind".

At the beginning, in the context of the Cold War, the concern of the UN was in preventing an extension of the arms race into outer space. Between 1959 and 1962, the major space faring nations made a series of proposals for banning the weaponization of outer space. An important step was reached with the conclusion in Moscow of the Treaty Banning Nuclear Weapon Tests in the Atmosphere, in Outer Space and Under Water of August 5, 1963, entered into force on October 10, 1963, by the United States, United Kingdom and the USSR.

By an imaginative effort at international legislation within the UN, and through the arduous work painstakingly carried out over a relatively short period of time, the UNGA elaborated a set of multilateral treaties and legal principles, which provide the framework of international law that governs space activities (Kopal 2011). The UN become the focal point for international cooperation in outer space and for the development of international space law. The UNGA resolution 1348 (XIII) of 1958 established the Committee on the Peaceful Uses of Outer Space (COPUOS), first as an ad hoc body with 18 Member States; one year later, on December 12, 1959, UNGA resolution 1472 (XIV) gave it the status of permanent body and reaffirmed its mandate. From the legal point of view, the COPUOS was established as a subsidiary organ of the GA, based on Articles 7, paragraph 2, and 22 of the UN Charter, following which "the General Assembly may establish such subsidiary organs as it deems necessary for the performance of its functions". Thus, the Committee was not established as an independent intergovernmental organization founded on a multilateral treaty.

According to the founding resolutions, the main task of the Committee was to facilitate international cooperation in the field of outer space within the framework of the UN. Moreover, letter (*d*) of UNGA resolution 1472 (XIV) of 1959 opened also the way for consideration of the "legal problems which might arise in programs to explore outer space". In 1961, UNGA resolution 1721(XVI) mandated the COPUOS to assist in the study of measures for the promotion of international cooperation in outer space activities and requested the SG to maintain a public registry based on information supplied by States launching objects into orbit or beyond. It called also upon launching States to "furnish information promptly" to the COPUOS, through the SG, for the registration of launchings. This recommendation has not been superseded by the subsequent Convention on Registration of Objects Launched into Outer Space of January 14, 1975, entered into force on September 15, 1976 (Registration Convention) and is still utilized by some States that have not yet ratified the Convention for registering on a voluntary basis their objects launched in outer space.

Like many other subsidiary organs of the UN, the COPUOS has its own internal structure, composed by two Subcommittees: the Scientific and Technical Subcommittee (STSC) and the Legal Subcommittee (LSC), created at the second session of COPUOS in 1962. Each subcommittee is composed of the same member States that comprise the parent body and is mandated to assist the COPUOS in the study of the specific proposals concerning, on the one hand, the scientific and technical aspects of space activities, and, on the other hand, the legal matters raised by member States for the development of international cooperation in space exploration for peaceful purposes. The STSC held its first session from May 28 to June 13, 1962, and the LSC first convened in Geneva on May 28, 1962. This latter date may be considered as the starting point of the evolutionary stages of the COPUOS.

1.2. Space law as a fruit of the United Nations

By examining the accomplishments of the COPUOS and its LSC in the field of international space law, it is possible to identify three main evolutionary phases (Marchisio 2005). The first corresponds to the law-making phase, when the five UN Space Treaties were concluded, which began just after LSC's creation, and ended in the 1980s. The second phase was the soft law phase, characterized by the adoption of four sets of non-legally binding principles until the middle half of the 1990s. During the third and current phase, efforts have been made in order to broaden the acceptance of the UN Space Treaties and improve their application through the adoption of UNGA resolutions, technical norms and international standard called "guidelines". Each of these stages presents its specific features and results.

Right in the beginning of COPUOS deliberations, an important decision was made which since then has defined the working methods of this part: the conclusions to be adopted by the Committee and both its subcommittees should be subject to agreement without need for voting. Thus, the COPUOS became the first UN body that started applying in its proceedings a decision-making principle later known as the rule of consensus and expanded in the practice of the UN and other international organizations.

The application of this procedure went on together with the method of a progressive elaboration of appropriate normative instruments. The rule of law in outer space was thus established not by a single, all embracing international treaty, but step-by-step, by several legal instruments dealing with the most urgent problems of space activities. In the first stage, the UNGA felt it necessary to give urgently some guidance to member States conducting space activities in order to avoid the development of practices dictated exclusively by national interests. Moreover, the initial debate in the LSC led to the conclusion that the basis for space activities should be conceived rather in principles than in detailed norms in order to reach the necessary agreement relatively soon.

This was realized thanks to a declaration of principles, belonging to the genus of Assembly recommendations, which are endowed, in legal terms, with a merely hortatory value, as the UNGA does not have a legislative function. However, the Assembly's "Declaration of Principles" or "Principles" tout court, are considered important tools in the process of evolving international law. In this sense, the adoption of a *corpus* of general principles, to be translated later into a binding treaty, was the best way to dictate the rules of the road for the emerging space activities of the space faring nations.

The Declaration of Legal Principles Governing the Activities of States in the Exploration and Use of Outer Space, as adopted by UNGA resolution 1962 (XVIII) of 1963 (Legal Principles Declaration), had in origin only a recommendatory value, but afterwards some of these principles acquired binding legal nature when they were restated in the Treaty on Principles Governing the Activities of States in the Exploration and Use of Outer Space, including the Moon and Other Celestial Bodies of January 27, 1967, entered into force on October 10, 1967 (OST). Furthermore, the universal acceptance of several of these principles has consolidated their customary value, which can hardly be questioned even by the stricter test of legal effectiveness. International custom is generally considered to be the product of two constitutive elements: *diuturnitas* and *opinio iuris*. The first element refers to general and consistent conduct by States, while the second element means that the practice stems from a belief of legal obligation. This definition helps us immediately underline the importance, in establishing the customary value of the principles, the conduct of States, international organizations and private entities acting under the States' control and supervision according to international space law.

In this regard, the practice of States has consolidated the customary value of several principles contained in the declaration, such as the exploration and use of outer space for the benefit and in the interests of all mankind; the freedom of exploration and use of outer space and celestial bodies by all States on a basis of equality and in accordance with international law; the prohibition of national appropriation of outer space and celestial bodies; the applicability of international law, including the UN Charter, to the exploration and use of outer space; the international responsibility for national activities in space and the authorization of the private entities activities by the State concerned; the principles of co-operation and mutual assistance, as well as of due regard for the corresponding interests of other States; the avoidance of harmful interference; the protection of the astronauts as envoys of mankind.

1.3. The outer space treaty of 1967: Legal past, legal future

While the adoption of an instrument not legally binding was the first step toward a new legal regime for outer space, soon after the time seemed mature for entering multilateral treaties to clarify and progressively develop the rules applicable to space activities.

The LSC become the most appropriate forum for reaching consensus on the major issues involved and transforming the principles on mandatory norms of international law. On June 16, 1966, both the United States and the USSR submitted draft treaties. The US draft dealt only with celestial bodies; the Soviet draft covered the whole outer space environment. The United States accepted the Soviet position on the scope of the Treaty, and by September agreement had been reached on most Treaty provisions. Differences on the few remaining issues – chiefly involving access to facilities on celestial bodies, reporting on space activities and the use of military equipment and personnel in space exploration – were satisfactorily resolved in private consultations during the UNGA session by December.

These were the origins of the OST which became one of the outstanding law-making treaties of contemporary international law (Lachs 2010). It significantly contributed to the progressive development and codification of international law in the meaning of Article 13 of the UN Charter. By the OST, an attempt was made at finding a balanced compromise between the common interests of all nations, the aims of humankind and the interests of individual States as subjects of international law.

Article I of the OST solemnly declares that "[t]he exploration and use of outer space, including the Moon and other celestial bodies, shall be carried out for the benefit and in the interests of all countries, irrespective of their degree of economic or scientific development, and shall be the province of all mankind". The Treaty develops the principles of the freedom in the exploration and use of outer space for all; the freedom of scientific investigation in outer space through international cooperation; and the prohibition of appropriation of outer space, no exception being admitted. Thus, space and celestial bodies belong to the category of *res communes omnium*, free for exploration and use by all States without discrimination of any kind, on a basis of equality and in accordance with international law.

The OST is a foundational instrument, a law-making Treaty of universal character. It lies at the top of the chain of normative instruments containing the rules governing space activities, whatever their nature and source, international, regional or national. The Treaty is also one of the most significant law-making treaties concluded in the second half of the 20th century. Law-making treaties are international multilateral agreements concluded with the intent of establishing, in the general interest, a set of rules universally valid, representing the only international regime applicable to a certain situation. The OST sets a legal regime of permanent character that make rules applicable to space activities relevant not only for the States parties, but also for States that are not parties to it. Evidently, not all the principles of the OST have acquired customary nature in international law, but some of them have certainly acquired a status that go beyond the conventional nature of the instrument in which they are contained. Some are even of peremptory nature and cannot be derogated through subsequent agreements. An agreement between two States aimed to appropriate a celestial body in contrast with Article II of the OST would be null and void in line with Article 53 of the 1969 VCLT.

The OST has enjoyed the widest acceptance by the international community from among all the UN space treaties. It has received, as of January 1, 2022, 112 ratifications. Although the number of States parties to the Treaty has now been increasing rather slowly, the fact that its status has reached more than 100 States parties demonstrates that it belongs to a category of international instruments that have been endorsed by a great majority of the international community.

As said, the main achievement of the OST has been the translation into treaty language of a series of legal principles governing the activities of States in the exploration and use of space. These principles are normative prescriptions of a general character, fundamental for the sector object of regulation and open-ended for further implementation.

Furthermore, the link between the Treaty and the objective of preserving peace in outer space is another structural feature that cannot be altered without disrupting the orderly development of outer space activities of public and private actors. From the outset, the very nature of the OST was to establish a legal regime to maintain peace and security in outer space. While freedom of access, exploration and use of outer space are recognized only for peaceful purposes, Article IV of the Treaty confirms the undertaking not to place in orbit around the Earth any objects carrying nuclear weapons or any other kinds of weapons of mass destruction, install such weapons on celestial bodies, or station them in outer space in any other manner. Yet, the use of the Moon and other celestial bodies is allowed exclusively for peaceful purposes, while the establishment of military bases, installations or fortifications; the test of weapons of any kind or the conduct of military maneuvers are prohibited.

At the same time, the content of the OST per principles has allowed a certain flexibility and permitted the adaptation of its legal framework to the evolution of space activities. Moreover, the Treaty also sets out restraints on States in two different ways: by requiring compliance with its provisions and imposing conditions on their outer space activities. Thanks to the OST, the conduct of States and private entities in space is "ruled by law", that is, by a level of normativity sufficiently certain and predictable. The development of national space legislation to implement Article VI of the Treaty is the most eloquent factor of stability of the legal regime governing activities in outer space.

1.3.1. *The relevance of the OST for private actors*

A special significance presents the principle that States parties shall bear international responsibility for national activities in outer space whether such activities are carried on by governmental agencies or by non-governmental entities, and for assuring that national activities are carried out in conformity with the provisions of the Treaty. The plain language of Article VI of the OST requires authorization and supervision of the activities of a country's commercial space actors in order to assure their conformity with the provisions of the Treaty.

The first sentence providing that States Parties bear international responsibility for their private entities' activities is quite unique in international law. Normally, a government is not responsible for purely private conduct in the absence of a strong link such as the government exercising direction or effective control over the private activity. This provision was part of the trade-off in the negotiation of the OST in which the original Soviet proposal was to ban private actors from space altogether (Dembling and Arons 1967). The OST clearly allows for and anticipates commercial space activity but makes State parties internationally responsible for such activity. The last clause of the first sentence of Article VI of the OST also provides that States parties must assure that national activities (including those by its commercial actors) are carried out in conformity with the OST. The second sentence then requires the appropriate State to undertake authorization and give official permission and continuing supervision of its non-governmental activities. In the current age of expanding private activity in space, Article VI of the OST preserves law and order in space by requiring countries to take steps to ensure that their nationals act in accordance with international law. Now that private operators outnumber government operators, Article VI has become critical to prevent a chaotic environment in space.

After having listed the many merits of the OST, we should also be aware that the Treaty would and could not regulate all existing and foreseeable aspects of space activities. Already on the 30th anniversary of the OST, it was noted that the Treaty provides only rudimentary protection of the space environment, in a single sentence contained in Article IX (Lafferranderie and Crowther 1997). There are certainly aspects of the OST that need clarification. But, from a general point of view, the Treaty seems to be sufficiently adaptable to absorb new challenges and face new issues.

The Treaty is an open-ended instrument, so that the lack of some specific provisions can be filled referring to international law at large. It is a leaving instrument because of its Article III, which establishes that States parties to the Treaty shall carry out on activities in the exploration and use of outer space including the Moon and other celestial bodies in accordance with international law, including the UN Charter, in the interest of maintaining international peace and security and promoting international cooperation and understanding.

Along the last decades, space activities have rapidly evolved. New emerging space nations have appeared on the one side and new private actors are engaged in space with innovative, flexible organizations focused on new technologies. New projects are developing, such as those regarding the mega constellations of small satellites that want to facilitate access to space through the reduction of costs and the acceleration of production. Similarly, there are initiatives for rendering repair services to satellites in-orbit and proximity operations.

But, as the other face of the same coin, there are also growing concerns relating to emerging challenges, such as the handling of space debris, the possible effects of large constellations deployments on the current and future orbital debris environment, the possible risks imposed on space missions by the new applications, the emerging threats to the security and resiliency of orbital infrastructures.

Considering these two related aspects, the question has been raised whether the Treaty could provide an adequate framework to address the complex relations resulting from the rapid growth of commercial activities in outer space and the increasing number of concerns related to the safety, security and sustainability of space activities. As we shall see later, the international community is trying to face these challenges mainly through the adoption of non-legally binding instruments.

1.4. The other UN treaties on space activities

1.4.1. *The Rescue and Return Agreement*

The general legal framework set up by the OST has been complemented by four other treaties, all negotiated within the LSC during the 1960s/1970s, following a method of progressive elaboration of appropriate space law instruments. They are in chronological order: the Agreement on the Rescue of Astronauts, the Return of Astronauts and the Return of Objects Launched into Outer Space of April 22, 1968, entered into force on December 3, 1968 (Rescue and Return Agreement); the Convention on International Liability for Damage Caused by Space Objects of March 29, 1972, entered into force on September 1, 1972 (Liability Convention); the Registration Convention and, finally, the Agreement Governing the Activities of States on the Moon and Other Celestial Bodies of December 18, 1979, entered into force on July 11, 1984 (Moon Agreement). Under the legal framework of these treaties, space exploration by nations, international organizations and private entities has flourished. As a result, space technology and services might better contribute to economic growth and improvements in the quality of life around the world.

Building upon Article V of the OST, which qualifies astronauts as "envoys of mankind" the 1968 Agreement was the first in the series of UN space treaties that followed the 1967 Treaty. The negotiating process initiated on May 1962, when two proposals were presented for discussion by the USSR and the United States. Later, a draft compromise text was submitted by Australia and Canada. The main outstanding issues were whether a general obligation existed to assist and return astronauts and space objects and whether this obligation applied to the high seas and beyond the national jurisdiction of States (Kopal 1969). A compromise text was after all submitted to the GA, while the successful entry into force of the OST hastened the negotiations of the Rescue and Return Agreement. On December 19, 1967, the UNGA adopted unanimously its resolution 2345 (XXII) commending the Agreement, which entered into force on December 3, 1968, after the deposit of the fifth instrument of ratification.

The shortest among the UN space treaties, the Rescue and Return Agreement contains on the one side provisions, based on humanitarian considerations, on the assistance to astronauts in distress and their return in case of emergency or unintended landing, and, on the other side, on recovery and return of space objects. The Agreement stipulated that States should render to astronauts all possible assistance in the event of accident, distress or emergency landing on the territory of another State Party or on the high seas. In the event of a landing, astronauts should be safely and promptly returned to the State of registry of their space vehicles.

Furthermore, the Rescue and Return Agreement included a provision concerning the return of objects launched into outer space or their component parts found beyond the limits of the Party on whose registry they would be carried on. The duty to return a recovered space object was a logical consequence of the principle inserted in Article VIII of the OST. According to it, ownership of objects launched into outer space and of their component parts is not affected by their presence in outer space or on a celestial body or by their return to Earth.

The legal framework of the Rescue and Return Agreement has been complemented by other international agreements, regulating more specific aspects through a detailed discipline. These agreements have been completed by regulations enacted by national space agencies. Among the international instruments, mention should be made of the Intergovernmental Agreement (IGA) between Canada, the States members of the ESA, Japan, the Russian Federation, and the United States concerning cooperation on the civil International Space Station (ISS), and of the Memorandum of Understanding between ESA and NASA (MOU), signed in Washington on January 29, 1998 (von der Dunk and Brus 2006; Marchisio 2011).

A Code of Conduct for the ISS crew was approved on September 15, 2000 by the Multilateral Coordination Board (MCB), the highest level cooperative body established by the MOU. It defines the crew as one integrated team with one Commander, responsible for the mission program implementation and crew safety guarantee aboard the Space Station. It specifies that ISS crewmember means "any person approved for flight to the ISS, including both ISS expedition crew and visiting crew, beginning upon assignment to the crew for a specific and ending upon completion of the post flight activities related to the mission". After the conclusion of the IGA, the ESA's Council adopted, on March 25, 1998, a resolution on the development of a single European astronaut corps, which recalls the provisions of various legal instruments relevant to astronaut activity adopted under the auspices of the UN and ratified by the ESA Member States, and in particular the Astronauts Agreement.

It is still an open issue whether the expressions "astronauts" – with its variants the Russian "cosmonauts" and the Chinese "taikonauts" – remain appropriate to cover new categories of individuals, such as flight participants, scientists members of multinational crews, temporary visitors and tourists, flying in outer space. Other legal texts have offered a clearer distinction between professional astronauts, space flight participants and visiting crew members, or space tourists. On December 23, 2004, the Commercial Space Launch Amendments Act (CSLAA) was signed as law in the United States. It establishes, inter alia, an "experimental permit" regime to allow reusable suborbital (passenger carrying) launch vehicle developers to build and test their vehicles without undue regulatory constraints.

1.4.2. *The Liability Convention*

Discussions on liability for damage caused by space objects continued as from 1962 within the LSC with a focus on three drafts submitted by the United States, Belgium and Hungary, in 1962, 1963 and 1964, respectively. A later proposal from France was submitted as of 1968. The *travaux préparatoires* compiled over the period of 6 years during which it took to achieve agreement on the Liability Convention of 1972, adopted by the UNGA in its resolution 2777 (XXVI).

The Convention was drafted with a view to achieving a victim orientated and unlimited system of liability (Kerrest 2017). The preamble to the Liability Convention reflects the overriding objective "to establish a uniform rule of liability and a simple and expeditious procedure governing financial compensation for damage". The Convention clearly builds upon Article VII of the OST, which establishes that a State that launches or procures the launching of an object into outer space, including the Moon and other celestial bodies, and from whose territory or facility an object is launched, is internationally liable for damage to another State or to its natural or juridical persons by such object or its component parts on the Earth, in air space or in outer space, including the Moon and other celestial bodies.

Accordingly, the requirements for international liability are damage, launching State and space object (Kerrest 2005). The definition of damage within the Convention is very broad. The type of damage for which a launching State can become liable to provide compensation are "loss of life, personal injury or other impairment of health; or loss of or damage to property of States or of persons, natural or juridical, or property of international governmental organisations". It includes environmental damage. The same provision confirms that the launching States that can become liable for damages made by their space objects are those mentioned in Article VII of the OST. They include a State which launches the object, as well a State which procure the launch (a State placing an order for a launch) and a State from whose territory or facility a space object is launched. Finally, the term space object includes component parts of a space object as well as its launch vehicle and parts thereof.

The Convention establishes two different regimes of liability depending on the place where the damage occurs. Following Article II, a launching State shall be absolutely liable to pay compensation for damage caused by its space object on the surface of the Earth or to aircraft in flight. This non-fault based, absolute liability system providing compensation for victims better corresponds to the ultra-hazardous nature of space activities and to the immense risk posed by space technology.

Liability for collisions in outer space was to remain with the launching State, but only on proof of fault. In the event of damage being caused elsewhere than on the surface of the Earth to a space object of one launching State or to persons or property on board such a space object by a space object of another launching State, the latter shall be liable only if the damage is due to its fault or the fault of persons for whom it is responsible (Liability Convention, Art. III). In case of joint launches or cooperation programmes, States should consider the conclusion of agreements in accordance with the Liability Convention.

A State which suffers damage, or whose natural or juridical persons suffer damage, may present to a launching State a claim for compensation for such damage. The Convention does not impose an obligation, but a discretionary faculty and, for that, if the State of nationality has not presented a claim, another State may, in respect of damage sustained in its territory by any natural or juridical person, present a claim to a launching State. The claim shall be presented to a launching State through diplomatic channels or the SG of the UN, provided the claimant State and the launching State are both Members of the Organization.

The compensation which the launching State shall be liable to pay for damage under the Convention shall be determined in accordance with international law and the principles of justice and equity, in order to provide such reparation in respect of the damage as will restore the person, natural or juridical, State or international organization on whose behalf the claim is presented to the condition which would have existed if the damage had not occurred. If no settlement of a claim is arrived at through diplomatic negotiations, within 1 year from the date on which the claimant State notifies the launching State that it has submitted the documentation of its claim, the parties concerned shall establish a Claims Commission at the request of either party. This Commission shall decide the merits of the claim for compensation and determine the amount of compensation payable, if any. The decision of the Commission shall be final and binding if the parties have so agreed; otherwise, the Commission shall render a final and recommendatory award, which the parties shall consider in good faith.

1.4.3. *The Registration Convention*

The Registration Convention was annexed to UNGA resolution 3235 (XXIX), adopted on November 12, 1974 and opened for signature on January 14, 1975. As already mentioned, UNGA resolution 1721B (XVI) of 1961 asked the SG to establish a public registry of objects launched into outer space to promote international cooperation in the exploration and use of outer space. Later in 1967, Article VIII of the OST codified the principle that a State of registry shall "retain jurisdiction and control over such object". Then, after the Registration Convention, UNGA resolution 62/101 of 2007 recommended means for States to improve registration practices.

Thus, there are at the UN levels, two complementary registers of objects launched into outer space maintained by the SG: Resolution Register and Convention Register. After the entry into force of the latter, States began providing information on space objects under the Registration Convention instead of resolution 1721B (XVI). Some States, like France, have re-registered all their space objects under the Convention.

The Registration Convention aims at providing a mandatory system of registering objects launched into outer space and additional means and procedures to assist States parties in the identification of space objects. When a space object is launched into Earth orbit or beyond, the launching State shall register the space object by means of an entry in an appropriate registry which it shall maintain. Each launching State shall inform the SG of the establishment of such a registry. Where there are two or more launching States in respect of any such space object, they shall jointly determine which one of them shall register the object. There shall be full and open access to the information in the Register maintained by the SG on the basis of the information furnished by States.

The Convention establishes the types of information to be provided "as soon as practicable", including the date of the launch, the territory or location of launch, the basic orbital parameters and the general function. States parties may submit, if they wish, any additional information, such as the re-entry of a space objects, the notification when a space object is no longer in Earth orbit, or when a space object is marked with the designator or registration number prior to launch. States, particularly those with tracking capabilities, are required to respond to the greatest extent feasible to requests in the identification of a space object which has caused damage to it or to any of its natural or juridical persons, or which may be of a hazardous or deleterious nature.

According to the UN Office for Outer Space Affairs (OOSA), since the establishment of the first UN Register in 1961, 89% of functional space objects that are presently in Earth orbit or beyond have been registered; 96% of functional space objects that were in Earth orbit were registered; 87% of functional space objects that are/were in geostationary orbit (GSO) have been registered; and 90% of functional space objects that are in low and medium orbits (LEO/MEO) have been registered. Space objects on deep space/planetary missions have been registered, as well as space objects carrying nuclear power sources and crewed spacecraft. Space station flight elements (including modules and robotic arms) are registered. Only 8% of functional space objects have not been registered since 1957 to present (Di Pippo 2016).

With the adoption of UNGA resolution 62/101 "Recommendations on enhancing the practice of States and international intergovernmental organizations in registering space objects", there have been substantial changes to registration practices of States of registry. A growing number of States have begun using the model registration form developed in accordance with the resolution. More States are providing information on transfer of ownership/supervision, mission termination, in-orbit disposal and re-entry of space objects. Only 23% of all space objects presently being tracked in Earth orbit are satellites, manned spacecraft or space station flight elements. Approximately 2,100 objects are rocket-stages and related. The remaining approximately 11,400 objects are non-functional objects: "space debris".

1.4.4. *The Moon Agreement*

The 1979 Moon Agreement is a case apart. It has been accepted but by 18 States, failing to collect support, notwithstanding that, like the other UN space treaties, it was adopted in the UNGA by consensus. There are many reasons for the hesitation shown by a great number of States to adhere to this Agreement, but the most evident is perhaps the contradiction between the legal qualification of the Moon and other celestial bodies as *res communes omnium* under the OST, and the legal regime of the Moon and its resources provided for by the 1979 Treaty. The latter utilizes in Article 11 the concept of common heritage of humankind, which in principle excludes any other type of exploitation but collective through an international authority. The notion of the common heritage of humankind has been adopted by the UN Convention on the Law of the Sea of December 10, 1982, entered into force on 16 November 1994 (Montego Bay Convention) for qualifying the sea bed and ocean floor and subsoil thereof beyond national jurisdiction (the Area) and for setting up the International Sea Bed Authority, the body through which States parties are to organize and control the activities concerned with sea-bed minerals. The Moon Agreement requires also the exploitation of the natural resources of the Moon to be governed by a future "international legal regime", and its full establishment has been postponed until such exploitation is about to become feasible (Tronchetti 2009).

In this sense, it is a matter of fact that at the end of the 1970s the LSC concluded its law-making era with one of the most controversial legal regime of international space law. As many non-space faring States have not yet accepted the core UN treaties, including some Members of COPUOS, one of the main functions of the LSC consists presently in acting for broadening their universal acceptance, inviting States to consider the reasons why their ratification should be considered highly beneficial. At the same time, the LSC should also encourage States that have accepted these conventions to look at the sufficiency of their national laws to implement them.

Furthermore, there is still a noticeable variable geometry among the UN five space treaties in terms of ratifications. Each one of the five UN treaties is indeed an independent multilateral treaty, with different content and a variable sphere of application, as shown by the range of State parties in 2022: OST 112, Rescue and Return Agreement 99, Liability Convention 98, Registration Convention 72, Moon Agreement 18.

This situation leads to a great degree of divergence, first depending on the fact that each State is bound only by those treaties that has accepted and that a treaty does not create either obligations or rights for a third State without its consent. Second, in international law, the rights and obligations of States parties to treaties successive in time relating to the same subject matter are normally governed by the principle *lex posterior derogat priori*, unless the earlier treaty contains peremptory rules, which is normally not the case in space law, or unless the treaty specifies that it is subject to, or that it is not to be considered as incompatible with an earlier or later treaty. In this case, the provisions of that other treaty prevail (so-called priority clause). Now, no one of the UN treaties regulates the issue of their reciprocal relations, apart the Liability Convention which in its Article XXIII establishes that its provisions shall not affect other international agreements in force insofar as relations between the States parties to such agreements are concerned. This clause makes the OST prevail in case of inconsistency.

Furthermore, as the UN space treaties were negotiated and concluded in different times and among different States, the only subjects competent to interpret them are the respective States parties in the exercise of their sovereignty. However, there is no institutionalized body created by these treaties that can interpret, nor does this competence fall within the powers of the COPUOS, as delegations rather firmly restate at every opportunity during the COPUOS sessions.

1.5. The declarations of principles

We should turn now to the instruments of soft law which have been adopted after the end of the 1970s. These instruments, while non-legally binding, can be considered as subsequent practice to the treaties developed to address new issues and challenges without recurring to legally binding instruments. During the second and third evolutionary phases of the COPUOS, international space law has developed mainly through such instruments, whose utility is without doubt (Marboe 2012).

Though the elaboration of further UN space treaties was discontinued after 1979, the work of the LSC in the progressive development of the juridical regime of outer space was not interrupted. The five main UN treaties exhausted the basic issues on which States would consent to undertake international legal obligations. During the following period, sets of UN Principles adopted by the UNGA became a suitable form for regulating specific categories of space activities for which the international community was not yet prepared to negotiate legally binding instruments. A new phase began, which witnessed the adoption of "declarations of principles" as the viable solution.

Four sets of Principles were negotiated by the LSC and then approved, through the main Committee, by the GA: the Principles Governing the Use by States of Artificial Earth Satellites for International Direct Television Broadcasting, adopted by UNGA in its resolution 37/92 of 1982; the Principles Relating to Remote Sensing of the Earth from Outer Space, adopted by the UNGA in its resolution 41/65 of 1986 (Remote Sensing Principles); the Principles Relevant to the Use of Nuclear Power Sources in Outer Space, adopted by the UNGA in its resolution 47/68 of 1992 completed by the Safety Framework for Nuclear Power Source Applications in Outer Space, as endorsed by UNGA resolution 64/86 of 2010; and the Declaration on International Cooperation in the Exploration and Use of Outer Space for the Benefit and in the Interest of All States, Taking into Particular Account the Needs of Developing Countries, as adopted by UNGA resolution 51/122 of 1996 (Space benefits Declaration).

With regard to the legal status of these Principles, although being merely recommendations, they have played a role in the consolidation of customary rules of international law. In this perspective, the decisive element comes from the analysis of the practice of States prior to, concomitant with, and following the UN declarations. Therefore, some of them are more firmly established in law, like the freedom of Earth's observation from space, while others seem to be less consolidated, and still in the process of gaining complete legal relevance. Some of them appear frankly outdated.

If we look at the Principles Relating to Remote Sensing of the Earth from Outer Space, they were a successful achievement in which a fair compromise was found between the interests of the sensing States and the needs of the sensed States, including most of the developing countries. At the time of their adoption, the principles did not prohibit activities that had been going on for a long time: on the contrary, they accepted the fact that sensing States were committed to the view that their activity required no consent, including no preliminary consent from sensed States. Therefore, the principles merely codified well-established conduct of States prior to 1986; the UNGA resolution created no new law, but simply gave greater legitimacy to the already existing practices.

Besides, the practice seems to have confirmed the general and main aspects of this legal regime set forth in 1986. A cursory look at the practice of States and international organizations shows a situation in which the core tenets of the UN Principles have maintained their importance, even in an emerging commercialized remote sensing system of service. Indeed, the principles appear relevant to the expansion of those very services and have been consistently reaffirmed. The basic international regime of remote sensing is recognized, promoting the broadest possible use of data (Marchisio 2004; Achilleas 2011).

On the other hand, some of the most prominent issues connected to ongoing developments in the field of remote sensing, mainly societal demands and technological developments, are not fully regulated by the UN declaration on remote sensing. The principles do not provide clear and specific regulations for new issues, such as the focus on global systems, the access to data by the sensing States and the legal protection of data, which is increasingly necessary to promote the costly investments required by remote sensing activities and the expansion of the related market. Nor do they provide an adequate discipline with regard to the production, use and treatment of highly sophisticated and detailed imagery, especially in relation to their potential implications for national security and individual privacy.

The set of Principles Relevant to the Use of Nuclear Power Sources in Outer Space, briefly called the NPS Principles, was but a limited achievement in space legislation. Some innovatory elements were brought into the regulation of this kind of activities, such as the storing of NPS objects in sufficiently high orbits after the operational part of their missions, the safety assessment and notification of re-entry. The NPS Principles, however, apply only to nuclear power sources devoted to the generation of electric power on board space objects for non-propulsive purposes, which have characteristics generally comparable to those of systems used and missions performed at the time of the adoption of the principles.

Therefore, the principles are not applicable to the NPS serving other purposes, including nuclear propulsion for long-distance flights into interplanetary space and to the celestial bodies of our solar system. The expected reopening of these principles no later than 2 years after their adoption has been delayed several times. In 2010, the Safety Framework for Nuclear Power Source Applications in Outer Space, complementary to the NPS, was endorsed by COPUOS and then by UNGA resolution 64/86 of 2010.

The final document of this series, the so-called 1996 Benefit Principles, mostly reflects the existing practice of international space cooperation and does not include new regulatory principles.

1.6. Other UN resolutions

Recently, the COPUOS is mainly engaged in the assessment of the existing legal regimes and oriented toward the formulation of non-binding documents that work upon the rights and obligations as provided by the treaties already in force. After the Third United Nations Conference on the Exploration and Peaceful Uses of Outer Space (UNISPACE III), held in Vienna from July 19–30, 1999, some objectives for further development of legal matters to be initiated through the LSC were agreed upon. In the absence of a conference of States parties (COP) to the OST, the COPUOS and its Subcommittees have often carried out a vicarious role in this field, interpreting the treaties through the adoption of resolutions by the UNGA. While not constituting "authoritative interpretations" of the treaties, these resolutions are nonetheless important tools to assist States in better abiding by their international obligations.

In this perspective, the LSC has moved toward the assessment of several topic of relevance, starting with the concept of the "launching State". The purpose of this work was to clarify all aspects of this concept as contained in the Liability Convention and in the Registration Convention, and as applied by States and international organizations, in the light of new and expected practices in space activities. A resolution on the "Application of the concept of the 'launching State'" was finally approved by the UNGA resolution 59/115 of 2004. The resolution recommends that States consider enacting national legislation on authorization and supervision of space activities by private entities and the conclusion of agreements with respect to joint launches.

Thereafter, the Subcommittee considered the practice of States and international organizations in registering space objects, an issue characterized by new aspects not covered by the Registration Convention, mainly related to the commercial uses of outer space as well as to the privatization of space activities. The assessment of this practice by States revealed strong disparities touching at the information concerning the time of launch, the territory of launch, the basic orbital parameters and the general function of a spacecraft. Moreover, it showed that several space objects were not registered or even registered by more than one State. These points were considered by UNGA resolution 62/101 of 2007, titled "Recommendations on enhancing the practice of States and international intergovernmental organizations in registering space objects". Finally, UNGA resolution 68/74 of 2013 adopted "Recommendations on national legislation relevant to the peaceful exploration and use of outer space".

Soft law seems better able to accommodate the ongoing evolution in the field of technology and landscape. As the experience of the UN specialized institutions shows, a real drive in this sense can come by technical norms. Specialized agencies have truly contributed, and continue to do so, to the evolution of law, by means of regulatory standards and recommended practices. Some of them have binding effectiveness, some others must be implemented by States through domestic acts. The COPUOS is considering its possible role in the elaboration of technical norms on space matters. Thus, it has been involved with the elaboration of "guidelines" related to the sustainability, safety and security of space activities.

In fact, threats in outer space are already a reality: space debris, fragmentations in space, frequencies overlapping, collisions among space objects, intentional and unintentional harmful interferences and deliberate destruction of satellites. No one denies that accidents in outer space must be avoided in order to prevent loss of life and creation of damaging orbital debris. Technical rules on fundamental mitigation and safety measures, such as management, design and operational measures, to limit debris released during normal operations and minimize potential for break-ups during operational phases, have already been adopted. The Space Debris Mitigation Guidelines approved by the COPUOS in 2007 and endorsed by UNGA resolution 62/217 of 2007 have defined the notion of space debris as "all man-made objects, including fragments and elements thereof, in Earth orbit or re-entering the atmosphere, that are non-functional". New initiatives are now considered toward active space debris remediation/removal and development of related technologies for services in orbit and ground-based lasers. A compendium of space debris mitigation standards adopted by States and international organizations has been published as a contribution of Canada, the Czech Republic and Germany to the COPUOS, in reference to the agenda item of its LSC on "General exchange of information and views on legal mechanisms relating to space debris mitigation measures, taking into account the work of the STSC".

In 2010, the STSC began considering as an agenda item the long-term sustainability of outer space activities (LTS). Core thematic areas were the sustainable space utilization supporting sustainable development on Earth; space debris mitigation, safety of space operations, tools to support space situational awareness; space weather: regulatory regimes and guidance for actors in the space arena. This process interacted also with the Group of Governmental Experts on Transparency and Confidence-building Measures in Outer Space Activities (TCBMs), which released in 2013 a report containing several recommendations in matters related to the predictability and security of space activities. In June 2019, the COPUOS reached a further step, adopting the preamble and 21 guidelines on the long-term sustainability of outer space activities. It decided to establish a 5-year working group under the agenda item on LTS of its STSC for continued institutionalized dialogue on issues related to the implementation and review of the guidelines.

Space sustainability is about ensuring that all humanity can continue to use space for peaceful purposes and socioeconomic benefit. It means the use of outer space in a way that maintains its potential to meet the needs and aspirations of present and future generations, ensuring that all humanity continue to use it for peaceful purposes, scientific and technological advancements and socioeconomic benefits (Marchisio 2012).

1.7. References

Achilleas, P. (ed.) (2009). *Droit de l'espace : télécommunications, observation, navigation, défense, exploration*. Larcier, Brussels.

Achilleas, P. (2011). Le cadre juridique international de la collecte et de la distribution des données de télédétection [Online]. Available at: http://www.afri-ct.org/wp-content/uploads/2015/03/68-_Article_Achilleas.pdf [Accessed 21 January 2020].

Couston, M. (2014). *Droit Spatial*. Ellipses, Paris.

Dembling, P.G. and Arons, D.M. (1967). The evolution of the outer space treaty. *Journal of Air Law and Commerce*, 33, 419–456.

Di Pippo, S. (2016). Registration of space objects with the United Nations Secretary-General. *Zeitschrift für Luft- und Weltraumrecht*, 65, 364–374.

von der Dunk, F. (ed.) (2011). *National Space Legislation in Europe: Issues of Authorization of Private Space in the Light of Developments in European Space Cooperation*. Martinus Nijhoff, Leiden.

von der Dunk, F. and Brus, M. (eds) (2006). *The International Space Station: Commercial Utilisation from a European Legal Perspective*. Martinus Nijhoff, Leiden.

von der Dunk, F. and Tronchetti, F. (eds) (2015). *Handbook of Space Law*. Edward Elgar Publishing, Cheltenham.

Hobe, S. (ed.) (2013). *Pioneers of Space Law*. Martinus Nijhoff, Leiden-Boston.

Hobe, S. (2019). *Space Law*. Hart Publishing, Oxford.

Hobe, S., Schmidtt Tedd, B., Schrogl, K.U., Goh, G.M. (eds) (2009–2015). *Cologne Commentary on Space Law: In Three Volumes*. Carl Heymanns Verlag, Cologne.

Jakhu, R. and Freeland, S. (2014). The sources of international space law. In *Proceedings of the International Institute of Space Law 2013*, Jorgenson, C.M. (ed.). Eleven International Publishing, The Hague.

Kerrest, A. (2005). Liability for damage caused by outer space activities. In *Space Law: Current Problems and Perspectives for Future Regulation*, Benkö, M. and Schrogl, K.U. (eds). Eleven International Publishing, The Hague.

Kerrest, A. (2007). Droit de l'espace. Droit des activités spatiales. Quelques définitions et remarques sur une approche pluridisciplinaire [Online]. Available at: https://www.unoosa.org/pdf/pres/lsc2007/symp-02-1F.pdf [Accessed 21 January 2020].

Kerrest, A. (2009). Actualités du droit de l'espace : la responsabilité des États du fait de la destruction de satellites dans l'espace. *Annuaire français de droit international*, LV, 615–626.

Kerrest, A. (2017). The concept of the "launching state" in commercial launch ventures. In *Commercial Uses of Space and Space Tourism*, Wouters, J., De Man, P., Hansen, R. (eds). Edward Elgar Publishing, Cheltenham.

Kerrest, A. and Thro, C. (2017). Liability for damage caused by space activities. In *Routledge Handbook of Space Law*, Jakhu, R. and Dempsey P.S. (eds). Routledge, New York.

Kiss, A.C. (1968). L'accord sur le retour et le sauvetage des astronautes et la restitution des objets lancés dans l'espace extra-atmosphérique. *Annuaire français de droit international*, 14, 736–746.

Kopal, V. (1969). The agreement on rescue of astronauts and return of space objects. In *New Frontiers in Space Law*, McWhinney, E., Martin, A., Bradley, A.W. (eds). Sijthoff, Leiden.

Kopal, V. (2011). Origins of space law and the role of the United Nations. In *Outer Space in Society, Politics and Law*, Brunner, C. and Soucek, A. (eds). Springer Verlag, Vienna.

Kopal, V. and Diederiks-Verschoor, I.H.P. (2008). *An Introduction to Space Law*, 3rd edition. Kluwer Law International, The Hague.

Kyriakopoulos, G.D. and Manoli, M. (eds) (2019). *The Space Treaties at Crossroads: Considerations de Lege Ferenda*. Springer International Publishing, Cham.

Lachs, M. (2010). *The Law of Outer Space: An Experience in Contemporary Law-Making, Reissued on the Occasion of the 50th Anniversary of the International Institute of Space Law*. Martinus Nijhoff Publishers, The Hague.

Lafferranderie, G. and Crowther, D. (eds) (1997). *Outlook on Space Law over the Next 30 Years*. Kluwer Law International, The Hague/London/Boston.

Lyall, F. and Larsen, P.B. (2018). *Space Law: A Treatise*, 2nd edition. Routledge, London.

Marboe, I. (ed.) (2012). *Soft Law in Outer Space. The Function of Non-binding Norms in International Space Law*. Böhlau Verlag, Vienna/Cologne/Graz.

Marchisio, S. (1986). Accords de financement multibilatéral : règles et pratiques des organisations internationales. In *Le droit international à l'heure de sa codification/Études en l'honneur*, vol. 2, Ago, R. (ed.). Giuffré, Milan.

Marchisio, S. (2004). The 1986 United Nations principles on remote sensing: A critical assessment. In *Scritti in onore*, Arangio-Ruiz, G. (ed.). Editoriale Scientifica, Naples.

Marchisio, S. (2005). The evolutionary stages of the legal subcommittee of the United Nations Committee on the Peaceful Uses of Outer Space (COPUOS). *Journal of Space Law*, 31(1), 219–242.

Marchisio, S. (2009). Le régime juridique de la télédétection. In *Droit de l'espace : télécommunications, observation, navigation, défense, exploration*, Achilleas, P. (ed.). Larcier, Brussels.

Marchisio, S. (2011). Reviewing the astronauts agreement: The role of COPUOS. In *Astronauts and Rescue Agreement: Lessons Learned*, Marchisio, S. and Lafferranderie, G. (eds). ESA, Paris.

Marchisio, S. (2012). The legal dimension of the sustainability of space activities. In *Current Problems of Modern International Law. Studies in Memory of Professor Igor Pavlovich Blishchenko*, Abashidze, A., Zhukov, G.P., Solntsev, A.M. (eds). IPK-RUDN, Moscow.

Marchisio, S. (2018). Il Trattato sullo spazio: passato, presente e futuro. *Rivista di diritto internazionale*, 1, 205–213.

Marchisio, S. (2022). *The Law of Outer Space Activities*. Edizioni Nuova Cultura, Rome.

McDougal, M.S., Lasswell, H.D., Vlasic, I.A. (eds) (1963). *Law and Public Order in Space*. Yale University Press, New Haven.

Purdy, R. and Leung, D. (eds) (2013). *Evidence from Earth Observation Satellites: Emerging Legal Issues*. Martinus Nijhoff, Leiden.

Ravillon, L. (2004). *Droit des activités spatiales. Adaptation aux phénomènes de commercialisation et de privatisations*. Litec Lexis Nexis, Paris.

Tronchetti, F. (2009). *The Exploitation of Natural Resources of the Moon and other Celestial Bodies: A Proposal for a Legal Regime*. Martinus Nijhoff, Leiden.

Zukhov, G. and Kolosov, Y. (2014). *International Space Law*, 2nd edition. Translated by B. Belitzky. Statut Publishing House, Moscow.

2

National Space Law

Frans VON DER DUNK[1,2]
[1] *Nebraska College of Law, Lincoln, USA*
[2] *Lazarski University, Warsaw, Poland*

2.1. National space law in the context of international space law

This chapter addresses national space law. While often slightly different terms such as *national space legislation* or *national space regulation* are used, the term *national space law* initially should be interpreted in the broadest possible sense. *National law* would thus refer to the totality of all relevant legal documents at the national level, whether entitled *law*, *statute*, *act*, *decree*, *directive* or *regulation*, plus customary law, implementation by way of government authorization and jurisprudence, as relevant further explained through the writings of recognized experts[1].

Then, as *space law* encompasses "every legal or regulatory regime having a significant impact, even if implicitly or indirectly, on at least one type of space activity or major space application" (von der Dunk 2015), national space law should refer to all such regimes at the domestic level (Gerhard 2008; Spencer 2010; Marboe 2011, 2015). Space activity, moreover, in this context should be defined as every activity "at least partly taking place in outer space" or "predominantly and intentionally *directed* at outer space" (von der Dunk 1998). Following this same

1 Note that this essentially reflects a domestic version of the provisions of Article 38 of the Statute of the International Court of Justice, entered into force on October 24, 1945 (ICJ Statute). This clause is widely considered to reflect the major sources of public international law more generally speaking.

logic, space applications refers to activities resulting, to a major extent, from space activities.

However, in the context of this encyclopedia, in this chapter national space law is addressed in particular in its relation to international space law, in turn presenting a category of public international law more generally (Gál 1969; Zhukov and Kolosov 1984; Malanczuk 1994; Cheng 1997; Matte 2000; Lachs 2010). This means that national space law in particular will be addressed in its capacity as part of *customary* international law, as a form of State practice combined with *opinio iuris*, taking the form here of national interpretation and implementation of international *treaty* law (ICJ Statute, Art. 38, 1, a, b; Wallace 1997, pp. 7–19; Cassese 2001, pp. 119–126; Lepard 2017, pp. 1–12).

The current chapter therefore focuses in a more limited fashion on that part of national space law that directly follows from, and in turn directly impacts, international space law. It provides for interpretation and implementation of the latter on a domestic plane notably vis-à-vis private entities, who do not themselves constitute subjects of public international space law, yet wherever relevant would have to be subjected to it.

2.2. From international to national space law

The starting point for analysis is the State-oriented character of international space law: the main United Nations (UN) treaties almost exclusively focused on activities of States and their rights and obligations (Hobe 2009, pp. 1–17; von der Dunk 2015b, pp. 44–49). The Treaty on Principles Governing the Activities of States in the Exploration and Use of Outer Space, including the Moon and Other Celestial Bodies of 16 December 1966 (OST)[2], mentioned intergovernmental (therefore still public) organizations only twice and in a clearly secondary capacity only (OST, Arts. VI and XIII) and private actors only once – in an even more subsidiary capacity and by using the abstract term "non-governmental entities" (OST, Art. VI). Likewise, the three generally ratified follow-on treaties drafted in the bosom of the UN, the Agreement on the Rescue of Astronauts, the Return of Astronauts and the Return of Objects Launched into Outer Space of 19 December 1967 (ARRA), the Convention on International Liability for Damage Caused by Space Objects of 29 November 1971 (Liability Convention)[3] and the Convention on

2 See, in general, Hobe et al. 2009; von der Dunk 2015b, pp. 43–78; Dembling 1967, pp. 419–456; Cheng 1997, pp. 215–264; Lyall 2009, pp. 53–80.

3 See, in general, Hobe et al. 2009, pp. 83–226; von der Dunk 2015b, pp. 82–94; Cheng 1997, pp. 286–356; Lyall 2009, pp. 105–114.

Registration of Objects Launched into Outer Space of 12 November 1974 (Registration Convention)[4] were very much focused on State activities in outer space and the concurrent State responsibilities and liabilities.

At the same time, perhaps the largest single paradigm change in the Space Age concerns precisely the ever increasing involvement of the private sector due to the increasing commercialization and practicalization of space activities and their more down-to-earth applications (Böckstiegel 2002; von der Dunk 2008a, pp. 498–515; Marboe 2011, pp. 29–71). This meant that ensuring the proper integration of non-State actors and their activities in the State-oriented regime of international space law has become ever more important. Exactly in this area therefore national space law plays its key role in domestic implementation of international space law by regulating certain categories of private space activities.

Thus, following increasing awareness of the importance of national law in the interpretation and implementation of international space law, since roughly two decades fundamental attention started to be paid also at the international level to national space law[5]. The approach to try and bring some uniformity into the growing body of national space law in this context so far culminated in 2013 with the adoption of the UNGA resolution 68/74 of 2013 entitled "Recommendations on national legislation relevant to the peaceful exploration and use of outer space" (National Space Law Resolution)[6]. They interestingly provide:

> The State, taking into account its obligations as a launching State and as a State responsible for national activities in outer space under the United Nations treaties on outer space, should ascertain national

4 See, in general, Hobe et al. 2009, pp. 227–324; von der Dunk 2015b, pp. 94–99; Lyall 2009, pp. 84–96; Cocca 1979, pp. 173–193.

5 Notably, a German academic research project resulted in the formulation, in 2002, of five "building blocks" which should form the baseline for any national space law in this context; and along similar lines, the Space Law Committee of the International Law Association (ILA), in 2012 adopted a "Model Law on National Space Legislation". See Schrogl 2002 (pp. 552–558) and ILA (2012). Resolution No. 6/2012 adopting the "Sofia Guidelines for a Model Law on National Space Legislation" [Online]. Available at: www.ila-hq.org/en/committees/index.cfm/cid/29 [Accessed 22 April 2022].

6 See, also, COPUOS (2012). Report of the Working Group on National Legislation Relevant to the Peaceful Exploration and Use of Outer Space on the work conducted under its multi-year workplan, A/AC.105/C.2/101. Note that UNGA resolution 59/115 of 2005 (Launching State Resolution) already urged Member States to "consider enacting and implementing national laws authorizing and providing for continuing supervision of the activities in outer space of non-governmental entities under their jurisdiction" (Launching State Resolution, paragraph 1).

> jurisdiction over space activities carried out from territory under its jurisdiction and/or control; likewise, it should issue authorizations for and ensure supervision over space activities carried out elsewhere by its citizens and/or legal persons established, registered or seated in territory under its jurisdiction and/or control. (National Space Law Resolution)

By that token, international space law itself already provides for a few key concepts fundamentally structuring how the private sector should be tied into the international space law regime by way of national space law. Most importantly, this concerns the international responsibility of States for private space activities and the attendant obligation for States to authorize and continuously supervise private sector space activities (OST, Art. VI) (Gerhard 2009, pp. 103–125; von der Dunk 2015b, pp. 45–55; Stubbe 2017, pp. 63–104) and to thereby make sure that such activities comply with international space law[7]. The concept, however, is not without its uncertainties and peculiarities, leaving States somewhat in the dark in a number of respects as to how exactly to draft and apply their national laws and regulations.

The main problem arises with respect to the proper interpretation of "*national* activities in outer space" (OST, Art. VI)[8], by means of which Article VI of the OST scopes the international responsibility, and the concurring definition of "the appropriate State" (OST, Art. VI) which then has to authorize and continuously supervise such activities. While three general approaches were discerned in academic discourse, it is of course the States which by way of their legal interpretations – read by way of their national space legislation – ultimately determine what the correct international (customary) legal interpretation would be. A further key question then regards *how* and *to what extent* the relevant State has actually translated its obligations under international space law in terms of substance to the private sector to cover their international responsibilities as interpreted.

Further to the National Space Law Resolution, this fundamental structuring concept of State responsibility informs any national space law which undertakes a comprehensive effort to address the involvement of the private sector in space activities. Ultimately, of course, the concept serves to translate the substantive rules, rights and obligations of international space law into a domestic setting, and to

7 Note that Article VI of the OST effectively requires responsibility "for assuring that national activities are carried out in conformity with the provisions set forth in the present Treaty", which is generally, following also Article III of the OST, considered to call by extension for compliance with *all* applicable rules, rights and obligations of international (space) law (von der Dunk 2015b, p. 51).

8 Emphasis added.

ensure that also national non-State entities will be subjected to those rules, rights and obligations. The following will provide a general overview of how national States dealt with that translation.

2.3. National space law and licensing regimes addressing private sector space activities

Currently more than two dozen States have established national laws, often further elaborated by second-tier regulation, addressing private sector space activities by way of a licensing regime[9], by which the State concerned enables certain private sector entities to undertake certain space activities and/or offer certain space applications subject to certain conditions protecting a range of public interests.

In this respect, the National Space Law Resolution further exhorts national regimes to be basically comprehensive in scope, and to

> include, as appropriate, the launch of objects into and their return from outer space, the operation of a launch or re-entry site and the operation and control of space objects in orbit; other issues for consideration may include the design and manufacture of spacecraft, the application of space science and technology, and exploration activities and research. (National Space Law Resolution)

By contrast, in – so far – five cases States have limited themselves *ratione materiae* to handling private involvement in launch activities only: Norway[10], Australia[11], Brazil[12], South Korea[13] and New Zealand[14]. Given the close link

9 The term "licensing" is used here in a general sense; many States and their laws and regulations refer to "authorizations", "permissions", "approvals" or other terms, respectively, their translation into the national language, but they would all boil down to the same: the formalized and institutionalized consent of a relevant sovereign State given to private sector entities to conduct certain space activities subject to certain conditions.

10 Act on launching objects from Norwegian territory into outer space of 1969 (Norwegian Act on Launching). For a compilation of national space laws and regulations, see UNOOSA (2023).

11 An act about space activities, and for related purposes of 1998 (Australian Space Activities Act).

12 Administrative Edict No. 27 of 2001(Brazilian Administrative Edict).

13 Space Development Promotion Act of 2005 (Korean Space Development Promotion Act).

14 Outer Space and High-altitude Activities Act of 2017 (New Zealand Space Act).

between launching and liability, those States use their licensing regimes primarily to address possible consequences of their obligations following from Article VII of the OST and the Liability Convention, not so much of the general "authorization and supervision" obligation of Article VI of the OST. Hence, their value as State practice-*cum-opinio iuris* in terms of the proper scope of the latter is fairly limited; the best that could be said is that all of them include relevant activities conducted from national territory.

From the present perspective, the national space laws which *ratione materiae* address all space activities at least in principle most prominently represent relevant State practice and attendant *opinio iuris* on how the key clause of Article VI of the OST on "national activities in outer space" should be interpreted.

This concerns first of all the United States, which represents a unique case in having an array of national laws and regulations dealing with various categories of space activities and the role of the private sector in them, even if more recently efforts have been made at codifying and streamlining this rather dispersed legislation[15].

The other comprehensive national regimes by contrast address private space sector participation by one overarching national space law providing for a licensing system, whether or not further elaborated by secondary legislation and regulation. This concerns, chronologically speaking, Sweden[16], the United Kingdom[17], South Africa[18], Russia[19], Ukraine[20], Belgium[21], the Netherlands[22], France[23], Nigeria[24],

15 Most importantly, private satellite communication activities were formally addressed since 1970, when an FCC report (Communications Satellite Facilities (1970). *First Report and Order*, 22 FCC 2d 86, Appendix C, p. 1) confirmed that the Communications Act of 1934 (47 U.S.C. 151 (1988); 48 Stat. 1064) was applicable also to *satellite* communications; since 1984, the private launch service sector has been regulated by a Commercial Space Launch Act (following several major amendments now codified as 51 U.S.C. 509); in the same year regulation of the private satellite remote sensing sector kicked off with the first Land Remote Sensing Act (following major overhauls now codified as 51 U.S.C. 601); major later legislation relevant here consisted of the Commercial Space Act of 1998 and the Commercial Space Launch Competitiveness Act of 2015.

16 Act on Space Activities of 1982 (Swedish Act on Space Activities).

17 Outer Space Act of 1986 (UK Outer Space Act).

18 Space Affairs Act of 1993 (South African Space Affairs Act).

19 Law of the Russian Federation on Space Activities of 1993 (Russian Law on Space Activities).

20 Law of Ukraine on Space Activities of 1996 (Ukrainian Law on Space Activities).

Austria[25], Kazakhstan[26], Indonesia[27], Turkmenistan[28], and then also Denmark[29], Japan[30], Luxembourg[31], Greece[32], Finland[33], Portugal[34] and the United Arab Emirates[35].

Finally, as to domestic space legislation comprehensively addressing all categories of space activities, a – so far unique – case is represented by Hong Kong. Hong Kong is, of course, not a State as such but merely a special region to which, upon its transfer from the UK to China in 1997, a special ordinance was applied transposing the main substance of the hitherto applicable UK Outer Space Act to the

21 Law on the Activities of Launching, Flight Operations or Guidance of Space Objects of 2005 (Belgian Space Law).

22 Law Incorporating Rules Concerning Space Activities and the Establishment of a Registry of Space Objects of 2007 (Dutch Space Law).

23 Law on Space Operations of 2008 (French Law on Space Operations).

24 National Space Research and Development Agency Act of 2010 (Nigerian NASRDA Act).

25 Austrian Federal Law on the Authorization of Space Activities and the Establishment of a National Space Registry of 2011 (Austrian Outer Space Act).

26 Law of the Republic of Kazakhstan on Space Activities of 2012 (Kazakh Law on Space Activities).

27 Law of the Republic of Indonesia on Space Activities of 2013 (Indonesian Law on Space Activities).

28 Law of Turkmenistan about Space Activities of 2015 (Turkmen Law on Space Activities). See Malysheva 2018, p. 543.

29 Outer Space Act of 2016 (Danish Outer Space Act).

30 Act on Launching of Spacecraft, etc., and Control of Spacecraft of 2016 (Japanese Space Activities Act).

31 Following the Law on the exploration and utilization of space resources of 2017 (Luxembourgish Law on Space Resources) addressing space resource exploitation only, a general space law was adopted in 2020: the Law on space activities of 2020 (Luxembourgish Law on Space Activities) amending 1. the amended Law of 9 July 1937 on insurance tax known as "Versicherungsgesetz"; 2. the amended Law of 4 December 1967 on income tax.

32 Law # 4508; Licensing of space activities – Registration in the National Register of Space Objects – Establishment of Hellenic Space Agency and other provisions of 2017 (Greek National Space Law). See Hellenic Parliament (2017).

33 Act on space activities of 2018 (Finnish Act on Space Activities).

34 Decree-Law No. 16/2019 of 2019 (Portuguese Law on Space Activities).

35 Federal Law No. 12 of 2019 on the Regulation of the Space Sector of 2919 (UAE Law on the Space Sector). Available at: https://www.moj.gov.ae/assets/2020/Federal%20Law%20No%2012%20of%202019%20on%20THE%20REGULATION%20OF%20THE%20SPACE%20SECTOR.pdf.aspx [Accessed 22 April 2022].

new situation, where now China is the State responsible and liable for space activities undertaken by or from Hong Kong[36].

This chapter will briefly survey those comprehensive national legal regimes on private sector licensing from the perspective of international space law, regarding how these States apply their obligations of authorization and continuing supervision under Article VI of the OST and, more generally, take care of their international responsibility for national private sector space activities and ensuring compliance by that private sector with the major substantive rules, rights and obligations of international space law.

2.4. National space law and international responsibility, authorization and continuing supervision

2.4.1. *Introductory remarks*

Essentially from the moment on that the OST was promulgated, three general approaches had been proposed in literature (von der Dunk 2015b, pp. 53–54) on how to sensibly interpret the key clause determining the scope of Article VI of the OST, the one referring to "national activities in outer space" for which States are internationally responsible and of which they are consequently required to ensure authorization and continuing supervision[37]. In other words, on what the scope *ratione personae* of the licensing regime established pursuant to Article VI is: which categories of legal persons are subjected to relevant obligations to take out a license?

The ultimate proof of what this key clause of the OST means is, of course, not so much the interpretation by individual authors – noting that the aforementioned clause of the ICJ Statute, listing what is generally considered to constitute the

36 Outer Space Ordinance, An Ordinance to confer licensing and other powers on the Chief Executive to secure compliance with the international obligations of the People's Republic of China with respect to the launching and operation of space objects and the carrying on of other activities in outer space of 1997 (Hong Kong Outer Space Ordinance), as amended 1999, Chapter 523; *National Space Legislation of the World*, Vol. II (2002), at p. 403; (2002) *Zeitschrift für Luft- und Weltraumrecht*, 51, 50. See, in general, on Hong Kong's space legislation, Zhao 2015, pp. 255–260; Reif 2002, pp. 47–56; Marboe 2015, p. 178.

37 Strictly speaking, while the provision referring to "national activities in outer space" would allow more States to qualify as the responsible State, the reference to the "appropriate State" which has to authorize and continuously supervise relevant non-governmental activities is in the singular; thus, at least the appropriate State would be one of the responsible States, which in other words among themselves should basically agree which one would take that special accountability to authorize and supervise (von der Dunk 1998, pp. 18–21).

sources of public international law, only refers to "the teachings of the most highly qualified publicists of the various nations (...) as *subsidiary* means for the determination of rules of law" (ICJ Statue, Art. 38, 1, d)[38] – but what States have actually put into (State) practice in a manner evidencing their *opinio iuris*, read how they addressed this issue in their national space legislation. In the following overview, the order in which these national laws were introduced above (see *supra*) will be generally maintained.

2.4.2. *The complex US national space law regime*

First, the Commercial Space Launch Act of 1984 applies a licensing requirement to both launch activities conducted by US nationals (including companies) and to such activities conducted from US territory. In addition, however, launch activities conducted by foreign entities *controlled* by US nationals (including companies) would be covered by the Act (Commercial Space Launch Act, Sec. 50904(a), 51 U.S.C.)[39]

In terms of substance, such US interests as "the public health and safety, safety of property, and national security and foreign policy interests of the United States" (Commercial Space Launch Act, Sec. 50905(a), 51 U.S.C.) have to be protected and served by any operations licensed under the Commercial Space Launch Act, the phrase "foreign policy interests" including US interests in complying with international space law and ensuring compliance therewith also by its private licensees. Some details, in particular with regard to technical and operational safety and national security issues, are provided by implementing regulations or other statutes, but beyond those the extent to which international space law is made directly applicable to private licensees is either left to specific licenses or to general applicability of international law, including space law, in the US legal order[40].

Second, the US Communications Act, formally declared applicable to satellite communications in 1970, applies the licensing obligation to any "person [that wants to] use or operate any apparatus for the transmission of energy or communications

38 Emphasis added.

39 See, further, Commercial Space Act, Sec. 50902(1)(C), 51 U.S.C. *j°* Sec. 50904(a).

40 See, on safety, esp. 14 C.F.R. Chapter III. See, on security, esp. Arms Export Control Act of 1976 (22 U.S.C. 2751) and Export Administration Act of 1979 (50 U.S.C. 2401); the former has further been elaborated by the United States Munitions List (USML; 22 C.F.R. 121) and the International Traffic in Arms Regulations (ITARs), the latter by the Commerce Control List (CCL; 15 C.F.R. 774) and the Export Administration Regulations (EARs). See, further, US Constitution, Art. VI(2).

or signals by radio (...) from (...) the United States" (Communications Act, Sec. 301, 47 U.S.C.), applying the territorial criterion only – however, not to the launch, but to the control center of the satellite operation. In any event, as the Act had been adopted in 1934, it would be difficult to argue that this interpretation provides any form of US State practice or *opinio iuris* relevant for the interpretation of Article VI of the OST drafted four decades later.

Nevertheless, the Communications Act, while not including any clause dictating specific relevant license conditions on compliance with international law, does include a clause allowing the Federal Communications Commission (FCC), the US federal agency bestowed with the licensing authority under the Act, "to suspend the license of any operator" in case it "has violated, or caused, aided, or abetted the violation of, any provision of any Act, treaty, or convention binding on the United States" (Communications Act, Sec. 303(m)(1), 47 U.S.C.). In compliance furthermore with the aforementioned general applicability of international (space) law in the US legal order, one can safely assume that no license would be granted and no activity under a license would be allowed which would be in clear violation of international space law even under this Act dating back to 1934[41].

Third, the US Land Remote Sensing Policy Act, the most recent piece of US federal legislation focusing on satellite remote sensing, applies its licensing regime to any "person that is subject to the jurisdiction or control of the United States" (Land Remote Sensing Policy Act, Sec. 60122(a), 51 U.S.C.). This refers to persons (natural as well as legal) with US nationality as well as persons operating from US territory, or even persons operating satellite remote sensing systems with substantial connections to, or deriving substantial benefits from, the United States. In short, it is a rather comprehensive approach.

In terms of substance, this Act specifically mentions the requirement for licensees-to-be "to observe the international obligations of the United States in accordance with section 60146 of this title" (Land Remote Sensing Policy Act, Sec. 60122(b)(1)), which latter section however refers to the proper use of radio frequencies only. Also, in deference to the Principles Relating to Remote Sensing of the Earth from Outer Space, UNGA resolution 41/65 of 1986 (Remote Sensing Principles) and its handling of sensing data vis-à-vis the sensed State, the Act provides for an obligation to "make available to the government of any State (including the United States) unenhanced data collected by the system concerning the territory under the jurisdiction of such government as soon as such data are

41 In addition, when it comes to allocation of electromagnetic spectrum, Sec. 303(y)(1) explicitly refers to consistency "with international agreements to which the United States is a party".

available and on reasonable terms and conditions" (Land Remote Sensing Policy Act, Sec. 60122(b)(2), 51 U.S.C.)[42].

Fourth, the US Commercial Space Launch Competitiveness Act includes a chapter presaging another licensing regime – for space resource exploration and exploitation activities. In terms of its scope *ratione personae*, this regime principally addresses "United States citizens" (Commercial Space Launch Competitiveness Act, Sec. 51302(a)(1), 51 U.S.C.), whereby the licensing regime to be developed for these activities will also primarily address US nationals, in line with the first approach[43].

In terms of substance, the only thing which can currently be said about such a future licensing regime is that should be "in accordance with applicable law, including the international obligations of the United States" (Commercial Space Launch Competitiveness Act, Sec. 51303, 51 U.S.C.), effectively a mere specific manifestation of the general requirement of US law to be compliant with its international obligations as per the aforementioned fundamental recognition of international (space) law as part of the US legal order.

2.4.3. *National space law regimes under a single comprehensive national space law*

2.4.3.1. *Sweden*

Sweden requires a license for any space activity by virtue of the Swedish Act on Space Activities to "be carried on from Swedish territory [or] by any (…) Swedish natural or juridical person (…) anywhere else" (Swedish Act on Space Activities, Sec. 2). This is a straightforward reflection of both the territorial and the nationality-based application of national jurisdiction as proposed by the National Space Law Resolution.

The Swedish Act on Space Activities further states in very general terms that "[a] licence may be restricted in the way deemed appropriate with regard to the circumstances. It may also be subject to required conditions with regard to control of the activity or for other reasons" (Swedish Act on Space Activities, Sec. 3). As the

42 Note that Principle XII of the Remote Sensing Principles provides: "as soon as the primary data and the processed data concerning the territory under its jurisdiction are produced, the sensed State shall have access to them on a non-discriminatory basis and on reasonable cost terms".

43 Referring back for the definition of the term to Commercial Space Launch Act, Sec. 50902. See, also, Sec. 51302(b)(1), 51 U.S.C.

Act was principally established to ensure Swedish compliance with international space law, it may indeed be assumed that relevant conditions will be included to that end if necessary.

2.4.3.2. *United Kingdom*

Originally, the UK took a different approach than Sweden, as the 1986 Outer Space Act applied its licensing regime only to persons with UK nationality regardless of where they conducted their space activities (UK Outer Space Act, Secs. 1 *j°* 2)[44] However, when more recently serious plans were developed to start launching in particular suborbital flights from the State using foreign technology and foreign operators, in 2018 a Space Industry Act was adopted to extend the scope of UK national space legislation accordingly. This Act indeed was made to apply to space activities "carried out in the United Kingdom". In order to avoid complicating overlaps, the scope of the UK Outer Space Act was now limited to activities by UK nationals "not (…) carried on in the United Kingdom" (Sec. 1, 1 and Sec. 1, 3, b).

Both Acts, moreover, explicitly include among the licensing obligations the need to take any international obligations of the United Kingdom into account, although not in exactly identical terms: the UK Outer Space Act *requires consistency* with such obligations of any licensed activities, whereas the UK Space Industry Act merely requires the regulator to *take* those obligations more generally *into account* (UK Outer Space Act, Sec. 4, 2, b and Sec. 2, 2, g).

2.4.3.3. *South Africa*

The South African Space Affairs Act applied its licensing regime to activities both if conducted from national territory and if conducted by national citizens, although interestingly this applies only to launch activities, including the operation of a launch facility. Other space activities, that is all other "activities directly contributing to (…) the operation of [space]craft in outer space" (South African Space Affairs Act, Sec. 11, 1, e), require a license under the Act to the extent of "participation by any juristic person incorporated or registered in the Republic, in space activities – (i) entailing obligations to the State in terms of international conventions, treaties or agreements entered into or ratified by the Government of the Republic; or (ii) which may affect national interests" (Sec. 1, 19th bullet), which leaves it to the proper international interpretation of the key clause of Article VI of the OST whether South Africa considers itself to be potentially held responsible for (all) space activities of legal persons with its nationality other than launching activities. However, the government is mandated to extend the license obligations to

44 Section 2 defines the categories of natural and legal persons comprised by these clauses in greater detail.

"any other space or space-related activities" that could also cause all non-launching activities on national territory to become licensed after all (Sec. 11, 1, d).

In terms of substance, South Africa takes its responsibility seriously and imposes conditions "taking into account (...) the international obligations and responsibilities of the Republic" (South African Space Affairs Act, Sec. 11, 2, c), whereas in the context of international peace and security the South African Space Affairs Act even provides for further references and definitions[45].

2.4.3.4. *The Russian Federation*

The Russian Law on Space Activities applies its licensing regime both to activities conducted by Russian "organizations and citizens", without any territorial limitation in this respect, and to those conducted by "foreign organizations and citizens under the jurisdiction of the Russian Federation", the latter not only referring to activities undertaken on Russian territory, but also to those involving Russia-registered space objects pursuant to Article VIII of the OST (Russian Law on Space Activities, Art. 9)[46].

In terms of licensing requirements preserving compliance with international space law, the Russian Law on Space Activities refuses the granting of licenses to undertake "space activities (...) prohibited by international treaties of the Russian Federation", requires "promotion of international cooperation" in line with Article III of the OST, and includes several prohibitions addressing military activities in outer space unlawful under Article IV of the OST (Russian Law on Space Activities, Art. 4, 2)[47].

2.4.3.5. *Ukraine*

Ukraine, the second-largest new State after Russia to arise from the ashes of the Soviet Union, has tended to follow its larger neighbor in interpreting and implementing the space treaties vis-à-vis the space sector. As to the scope *ratione personae* of the Ukrainian Law on Space Activities, therefore, everyone undertaking space activities from Ukrainian territory or otherwise operating under Ukrainian jurisdiction requires a license. The latter concept refers not only to everyone with Ukrainian nationality, natural and legal persons alike, but also, like in the Russian

45 See, further, South African Space Affairs Act, Secs. 1, 6th & 27th bullets, *j°* 2(1).

46 See, further, Russian Law on Space Activities, Art. 17(2), referring to Russian registration of space objects in Art. 17(1).

47 See, also, Russian Law on Space Activities, Art. 26, further confirming the applicability of international obligations in the Russian legal order in this context.

case, everyone operating a satellite registered by Ukraine (Ukrainian Law on Space Activities, Arts. 10 and 13).

The licensing regime established by the Ukrainian Law on Space Activities does not provide any further substance as to the translation of international space law to the private space sector. However, Article 9 already provides the first list of space activities generally prohibited by the Law, which largely echo the provisions of the Russian Law on Space Activities as far as the military uses of outer space are concerned, and adds among others unsafe and environmentally unfriendly activities, as well as "other actions connected with space activity which are not admitted by international law" (Ukrainian Law on Space Activities, Art. 10)[48].

2.4.3.6. *Belgium*

Belgium is the first State which fundamentally diverges from general application of the Belgian Space Law to nationals and territory, as it limited its scope *ratione personae* in first instance to activities conducted by anyone operating from Belgian territory only. This baseline approach however is then extended by a provision allowing for extension to such activities conducted elsewhere by Belgian nationals if called for by an international agreement to which Belgium is a party (Belgian Space Law, Art. 2, 1)[49].

While a general authority to impose "conditions for granting authorizations with a view to ensuring the safety of people and property, protecting the environment, ensuring the optimal use of air space and outer space, protecting the strategic, economic and financial interests of the Belgian State" (Belgian Space Law, Art. 5, 1) was provided for by the Belgian Space Law, authorized activities specifically "must be carried out in accordance with international law and, in particular, with the principles laid down in the OST and the other treaties and agreements to which Belgium is a party" (Art. 4, 3), providing for rather comprehensive requirement of compliance with international (space) law.

48 As Article 10 merely provides for the generic establishment of such a licensing regime, further details were worked out by the Law of Ukraine on Licensing for Certain Types of Economic Activities of 2000.

49 This clause also seems to allow for accommodation of an interpretation of "national activities" as per Article VI of the OST, as also encompassing space activities by Belgian nationals, should such an interpretation come to be accepted at the international level; Belgium, however, apparently proceeds from the assumption that currently such interpretation would not be warranted.

2.4.3.7. *Netherlands*

The Netherlands followed largely the same approach as its Southern neighbor: first applying the licensing obligation to all "space activities that are performed in or from within the Netherlands or else on or from a Dutch ship or Dutch aircraft", using Dutch territory and quasi-territory as the basis for jurisdiction, but then allowing for the possibility of implementing regulations to extend the scope of the Dutch Space Law to Dutch nationals undertaking space activities "on or from the territory of a State that is not party to the Outer Space Treaty or on or from a ship or aircraft that falls under the jurisdiction of a State that is not party to the Outer Space Treaty" (Dutch Space Law, Sec. 2, 2, a).

The licensing regime under the Dutch Space Law allows for regulations and restrictions as necessary for "a. the safety of persons and goods; b. protection of the environment in outer space; c. financial security; d. protection of public order; e. security of the State; [and] f. fulfilment of the international obligations of the State" (Dutch Space Law, Sec. 3, 3). Once more, this represents a comprehensive list of interests to be protected by national space law, including explicitly compliance with international (space) law.

2.4.3.8. *France*

The French Law on Space Operations opted for an incomplete version of the approach of the National Space Law Resolution much like South Africa: the licensing regime makes a distinction between launching and return activities, on the one hand, and all other space activities, on the other hand. For the former, the authorization obligation extends to anyone undertaking such activities from French territory or facilities (thus including quasi-territorial jurisdiction) *and* any French operator undertaking such activities elsewhere; for the latter, only French operators require an authorization (French Law on Space Operations, Art. 2(1) & (2), respectively, (3)).

Also the French Law on Space Operations provides for a comprehensive but generalized list of requirements which an authorization should comply with, such as "moral, financial and professional guarantees", compliance "with the technical regulations set forth, in particular for the safety of persons and property, the protection of public health and the environment", including limiting risks pertaining to space debris, and compliance with "national defence interests or the respect by France of its international commitments" (French Law on Space Operations, Art. 4)[50].

50 See, also, French Law on Space Operations, Art. 5.

2.4.3.9. *Nigeria*

The Nigerian NASRDA Act had as its major target the establishment of the National Space Research and Development Agency (NASRDA); however, the Agency at the same time was specifically given the power to “grant license to any person or body corporate for activities stated in section 6(k) of this Act” (Nigerian NASRDA Act, Sec. 9, 1), which though in fact (mainly) targeting satellite remote sensing operations as per Section 6(k), might well be used for instance to address positioning and navigation data generated by satellites as well, so that it is included here in the list of comprehensive licensing schemes. While the Act itself only applied personal jurisdiction, the Regulations on the Licensing and Supervision of Space Activities currently being drafted extend that scope by requiring licenses from all “[c]orporations registered in Nigeria with ownership of space object(s)” (Secs. 1, a) – that is, private operators with the legal nationality of Nigeria – *and* from all “[o]perators (…) of space object(s) and launch vehicle(s) within Nigerian territory” (Secs. 1, b) – that is, private operators acting from Nigerian territory, as well from Nigerian ships or aircraft. In short, Nigeria is currently in the process of moving over to the approach of the National Space Law Resolution as well[51].

In terms of substance, the license conditions under the Nigerian NASRDA Act include lack of jeopardy of public health, safety of persons and property; consistency with the international obligations and national security of Nigeria; as well as requirements to prevent contamination of outer space or interference with the space activities of others, and to dispose of a payload after termination of operations (Nigerian NASRDA Act, Sec. 9, 2 and Sec 9, 4)[52].

2.4.3.10. *Austria*

The Austrian Outer Space Act applied the obligation to authorize on a territorial, quasi-territorial and personal basis, to “space activities carried out 1. on Austrian territory, 2. on board of vessels or airplanes, registered in Austria or 3. by a natural person with Austrian citizenship or legal persons seated in Austria” (Austrian Outer Space Act, Sec. 1, 1). With specific reference to Article VIII of the OST and the Registration Convention furthermore, “[d]uring the presence in outer space and on celestial bodies, a registrable space object as well as its personnel remain under the jurisdiction and control of Austria” (Sec. 9, 4).

---

51 See, also, Nigerian NASRDA Act, Secs. 2(1) respectively (2), Draft Regulations on the Licensing and Supervision of Space Activities of 2015 (Nigerian Draft Regulations).

52 See, also, Nigerian NASRDA Act, Sec. 9, 3.

Obtaining an authorization under the Austrian Outer Space Act requires operational "reliability, capability and expertise" (Austrian Outer Space Act, Sec. 4, 1, 1), lack of "any immediate threat to the public order, to the safety of persons and property and to public health" (Sec. 4, 1, 2), lack of threats to "national security, Austria's obligations under international law or Austrian foreign policy interests" (Sec. 4, 1, 3), the need to address "the mitigation of space debris" (Sec. 4, 1, 4) as further detailed in a separate clause, absence of "harmful contamination of outer space or celestial bodies or adverse changes in the environment" (Sec. 4, 1, 5) and compliance with "the requirements of the ITU concerning orbital positions and frequency assignments" (Sec. 4, 1, 6).

2.4.3.11. *Kazakhstan*

As for the Kazakh Law on Space Activities, the statutory scope of the licensing obligation *ratione personae* is yet to be determined by "legislation of the Republic of Kazakhstan on licensing". In view of the registration requirements, however, it may be assumed that both personal and territorial jurisdiction are applied (Kazakh Law on Space Activities, Art. 3)[53].

Similarly, further law is to specify the conditions for obtaining a license, while in view of more general provisions of the Kazakh Law on Space Activities it may be assumed that those would include "compliance with national interests, provisions of defense and national security" (Kazakh Law on Space Activities, Art. 3, 1), "compliance with ecological requirements, requirements in the field of technical regulation and provision of sanitary and epidemiological wellbeing of the population" (Art. 3, 5), "compliance with the international law norms in the field of space activities" (Art. 3, 6) and "efficient and rational use of outer space" (Art. 3, 7)[54].

2.4.3.12. *Indonesia*

The Indonesian Law on Space Activities applies *ratione personae* to "all of the Space Activities which are carried out in and/or from the territorial sovereignty and the jurisdiction of the Republic Indonesia" (Indonesian Law on Space Activities, Art. 5, a), covering territory and quasi-territory, as well as to "Indonesian citizens or Indonesian legal entities" (Art. 5, c), taking care of the nationality criterion. Indonesia thus takes a very broad approach to interpretation of the concept of "national activities" pursuant to Article VI of the OST.

53 See, also, Kazakh Law on Space Activities, Art. 11, 1.

54 See, further, Kazakh Law on Space Activities, Art. 13.

The conditions which the Indonesian Law on Space Activities then imposes on licensed space activities will include prohibitions relating to weapons of mass-destruction closely following Article IV of the OST, to threats to security, safety and the public interest as well as to the environment. More generally, Article 3 of the Law echoes Article I of the OST concerning the general legal status of outer space and activities therein, whereas Article 7(2) requires space activities conducted under the Law to take into account inter alia international law (Indonesian Law on Space Activities, Art. 8).

2.4.3.13. *Turkmenistan*

The Turkmen Law on Space Activities addresses, *ratione materiae*, all "activities aimed at the exploration and use of outer space" (Turkmen Law on Space Activities, Art. 1, 1), and *ratione personae* will apply to "legal entities and/or individuals of Turkmenistan, including foreign ones (…) that carry out space activities from the territory of Turkmenistan" (Art. 1, 3), in conformity with the baseline scope proposed by the National Space Law Resolution.

While Article 17 of the Turkmen Law on Space Activities refers to future legislation for the details regarding the licensing of such activities, a list is provided of the basic principles that are to be reflected in the licensing conditions, referring to such standard ones as compliance with international law, international peace and security, national security and safety and environmental considerations (Turkmen Law on Space Activities, Art. 4).

2.4.3.14. *Denmark*

The Danish Outer Space Act "applies to space activities carried out within the Danish State" (Danish Outer Space Act, Sec. 2, 1) as well as outside Denmark if undertaken "on Danish craft or facilities; or (…) by Danish operators" (Sec. 2, 2). Thus, also Denmark interprets Article VI of the OST's concept of "national activities in outer space" as encompassing activities conducted within its territorial, quasi-territorial and personal jurisdiction.

The Danish Outer Space Act also provides a generic list of conditions, including technical and financial capabilities to undertake relevant operations safely and in compliance with further standards and guidelines, the taking of appropriate measures with respect to space debris mitigation and environmental safety, and compliance with the international telecommunications legal framework developed under the *aegis* of the ITU. Finally, the approved activities should "not conflict with national security interests, Denmark's international obligations or foreign-policy interests" (Danish Outer Space Act, Sec. 6, 1).

2.4.3.15. *Japan*

The Japanese Space Activities Act *ratione materiae* addresses, already as per its title, both the launching of spacecraft and the control thereof in outer space. *Ratione personae*, permissions for launching licenses for operating spacecraft are required for anyone launching from "a launch site located in Japan or onboard a ship or aircraft with Japanese nationality" (Japanese Space Activities Act, Art. 4, 1), or operating the spacecraft from "a spacecraft control facility located in Japan" (Art. 20, 1).

As to the conditions for permissions and licenses, the Japanese Space Activities Act generally refers to the need "to ensure the accurate and smooth implementation of conventions concerning the development and use of outer space, ensure public safety and protect people affected by relevant damage by establishing a system for permission and license related to the launching of spacecraft, etc. and the control of spacecraft in Japan" (Japanese Space Activities Act, Art. 1); the Act itself then very much focuses on safety details.

2.4.3.16. *Luxembourg*

The Luxembourgish Law on Space Activities addresses all space activities excluding those related to space resource utilization (as being subject instead to the Luxembourgish Law on Space Resources) and requires an authorization from anyone undertaking them from Luxembourg or being of Luxembourgish nationality (Luxembourgish Law on Space Activities, Art. 1)[55].

For an authorization, the Luxembourgish Law on Space Activities requires compliance with the international obligations of the State and conditions to be spelled out, which largely focus on the safety of operations (Luxembourgish Law on Space Activities, Arts. 3, 6-8); the Luxembourgish Law on Space Resources generally has a similar focus in terms of conditions (Art. 7).

2.4.3.17. *Greece*

The Greek National Space Law essentially applies its licensing obligations to all space activities, whether undertaken from Greek territory, from facilities under Greek jurisdiction or by Greek citizens, natural or legal, as well as in any (other) scenario potentially giving rise to Greek liability under the space treaties. In this sense, the Law represents one of the most accurately comprehensive examples of national space legislation (Greek National Space Law, Art. 1, 2 and Art. 2, 6).

55 The scope *ratione personae* of the Luxembourgish Law on Space Resources by contrast is limited to Luxembourgish citizens only (Luxembourgish Law on Space Resources, Art. 2).

As to the conditions for a license, those prominently include safety- and security-related requirements, compliance with the obligations of the State under international law in general and the ITU regime in particular, as well as specific requirements related to space debris and the environment (Greek National Space Law, Art. 4.1).

2.4.3.18. *Finland*

The Finnish Act on Space Activities uses almost the exact same language as the Danish Outer Space Act and addresses "space activities carried on within the territory of the State of Finland" and "space activities outside the territory of the State of Finland if they are carried on (1) on board a vessel or aircraft registered in Finland or (2) by a Finnish citizen or a legal person incorporated in Finland" (Finnish Act on Space Activities, Sec. 1).

In terms of requirements the Finnish Act on Space Activities is largely identical in its comprehensiveness to the Danish Outer Space Act, but it is considerably more detailed on a number of issues when addressing the standard list of capability and reliability, safety- and environment-related concerns, and the broader national security, international law (including the ITU regime) and foreign policy implications of licensed space activities (Finnish Act on Space Activities, Sec. 5)[56]. Thus, the Act represents one of the most detailed examples of implementing general obligations following from the space treaties in the context of domestic licensing of private actors.

2.4.3.19. *Portugal*

The Portuguese Law on Space Activities is quite straightforward in determining both its scope *ratione materiae* and its scope *ratione personae*: it "applies to space activities" undertaken from "national territory, (…) on board Portuguese vessels and aircraft or in facilities under Portuguese jurisdiction or sovereignty" or undertaken "outside the national territory by Portuguese operators" (Portuguese Law on Space Activities, Art. 2, 1).

Further to Article 1, Article 7 provides for the licensing requirements which address concerns regarding "technical, economic and financial capacity" (Portuguese Law on Space Activities, Art. 7, 1, a), safety and space debris, compliance with international law, national security and other "strategic interests" (Art. 7, 1, e) of Portugal.

56 See, also, e.g. Secs. 9, 10.

2.4.3.20. *United Arab Emirates*

The UAE Law on the Space Sector is very comprehensive both in its scope *ratione materiae*, encompassing all thinkable space activities and even space-related activities as per Article 4, and in its scope *ratione personae*, as it applies to all such activities "carried out (…) (1) In the State's Territory or the State's facilities outside the State's Territory. (2) From vessels or aircraft registered with the State or Space Objects registered by the State. (3) By persons who hold the nationality of the State, or companies that have a main headquarters in the State" (UAE Law on the Space Sector, Art. 3).

As the UAE Law on the Space Sector aims "to create an appropriate regulatory environment to achieve the objectives of the State's national space policy", focusing on "the necessary safety, security and environmental protection measures to enhance the long-term stability and sustainability" and "the commitment of the State to implement the provisions of international conventions and treaties related to Outer Space and to which the State is a party" (UAE Law on the Space Sector, Art. 2), it also covers the standard set of concerns for States in domestically legislating the activities of the private space sector[57].

2.4.3.21. *Hong Kong*

As discussed, strictly speaking Hong Kong is not a national State and the Hong Kong Outer Space Ordinance therefore not a *national* space law. However, since it basically fulfills the same function for the Chinese Special Administrative Region Hong Kong, it will be included in the present survey. While drafted as an adaptation of the UK Outer Space Act to Hong Kong's transition from the United Kingdom back to China, it did not copy the former's approach to addressing only activities subject to personal jurisdiction as it applies to relevant "activities whether carried on in Hong Kong or elsewhere" (Hong Kong Outer Space Ordinance, Sec. 3), the first part of the phrase referring to Hong Kong territory, the last part only possibly referencing Hong Kong citizens and companies.

In terms of the substance of international space law, rather similar to the UK Outer Space Act, under the Hong Kong Outer Space Ordinance the Chief Executive, now in charge on behalf of the Chinese government, may grant a license "if he thinks fit" (Hong Kong Outer Space Ordinance, Sec. 5(1)). The Ordinance itself spells out in considerable detail the general requirements to be complied with, focusing on public health, safety of persons and property, the obligations of China

57 See, also, UAE Law on the Space Sector, Arts. 14 *ff.*

under international law, national security, the contamination of outer space and interference with other activities in the peaceful use and exploration of outer space[58].

2.4.4. *Concluding remarks*

Summarizing the above analysis of national space legislation with a view to the proper interpretation and implementation of Article VI of the OST, at least as far as the set of comprehensive national space laws under discussion is concerned, no State applies its licensing regime to activities of nationals *only*. The United Kingdom, and Nigeria (and Luxembourg as long as only dealing with space resource utilization), were alone in at least originally doing so, but fundamentally changed their approach in this respect more recently.

Another approach, of mainly targeting those private space activities that already give rise to a State's liability, is essentially followed by a very limited set of national space laws. This applies to two national space laws only partially: those of Belgium and the Netherlands, not entirely limiting themselves moreover to territorial or quasi-territorial application only. The true stand-out case is Japan, which (still) exclusively follows a territorial approach even if not only addressing launch activities, where such focus on territorial jurisdiction would be obvious. (As indicated, the five national regimes addressing launching only, of Norway, Australia, Brazil, South Korea and New Zealand, also all take care of such activities as conducted from national territory.)

Consequently, the approach suggested by the National Space Law Resolution, of applying the licensing regime to those non-governmental activities already subject to a State's jurisdiction anyway, has by far the largest score of adherent States. Except for Belgium, the Netherlands and Japan, all States with a comprehensive licensing scheme now apply both personal and territorial jurisdiction, even as a number of lower-level differences – for instance, in adding quasi-territorial jurisdiction (with Russia, Ukraine and Austria most explicitly applying that to registered space objects in addition to ships and aircraft) or excluding certain types of activities – can still be seen. Except for the United States (largely for historical reasons), South Africa and France, they do so across the board, regardless of what particular category of space activities is concerned.

58 See, also, Hong Kong Outer Space Ordinance, Secs. 5(2), 6(2)(e) & (g), 7(2). The last clause essentially mandates "the disposal in outer space of the payload of any space object to which the licence relates on the termination of activities under the licence".

In general, there can therefore be little doubt any longer that the most legitimate interpretation of "national activities in outer space", as evidenced by State practice and attendant *opinio iuris* expressed through national space legislation, is that it encompasses all private sector space activities conducted from within the territorial, quasi-territorial (including at least ships, aircraft and mobile platforms, but likely also registered space objects themselves) and personal jurisdiction of the State at issue.

When it comes to the substance of international space law, all States concerned follow the same approach as well: the highest level legislation merely refers to general categories of public interests, more or less all of them referring comprehensively to the international legal obligations of the State concerned, to the (national and international) security and safety issues involved, to the ITU frequency and orbital coordination regime, and to other general international/foreign policy interests and public benefits such as science. A good number of them are slightly more explicit on such issues as the weaponization and militarization of outer space or the issue of space debris and the environment in general. Beyond that, all leave any specific details to lower level regulation or even the individual licenses, giving the State authorities the requisite flexibility in imposing actual precise licensing conditions upon applicants.

In sum, national space law indeed – at least at a general, abstract level – takes comprehensive care throughout of ensuring that relevant private entities active in outer space are directly and unequivocally bound to international space law through respective licensing regimes, albeit that much is left to the lower level implementation processes and procedures, allowing for the necessary discretion and flexibility in view of the large variety of possible private interests and activities in outer space.

2.5. Final remarks

The key role of national space law in ensuring the proper integration of private space activities into public international space law, in particular the fundamental UN treaties of the 1960s and 1970s, is by now incontrovertible. In the framework of COPUOS, which operates on the basis of almost absolute respect for State sovereignty and the resulting discretion of States to implement or not implement international obligations as they see fit, it is by now axiomatic that, unless States have no private space activities at all taking place under their *aegis* and/or prohibited them fundamentally, they should ensure such a proper integration by way of national space laws. The States, which do not yet have such national space laws are gently urged to establish them, are handed "model laws" and "model building blocks" to work with and can always access a growing depository of existing national space law to see how other States have tackled the issue.

At the same time, the States, which do have national space legislation in place addressing private participation in space activities, have been confronted with the fact that the major treaties requiring domestic interpretation and implementation do not provide much details in terms of a standard model. As these treaties were drafted in the days when not even in the United States, that champion of private enterprise, were private space activities substantively foreseen, a key notion such as "international responsibility for national activities in outer space" remained, as to its detailed interpretation, application and implementation, essentially undefined.

The present chapter has undertaken an effort to analyze the national space laws established so far, at least those which attempt to cover the field *ratione materiae* completely, in particular from this vantage point: to what extent has the State practice and *opinio iuris* which these laws embody, been consistent enough to arrive at further definitions of an international customary nature of these concepts? Any such consistency and customary international law value would certainly make it easier for States not yet in the possession of a national space law to interpret and implement their international obligations in the context of private sector space activities – but would at the same time give them less discretion to interpret and implement them, giving rise to a more coherent global environment for private space activities.

The – limited – analysis of this chapter results in a mixed picture. Still, the clause of Article VI of the OST on "national activities in outer space" can be seen to be almost uniformly interpreted as including activities conducted by nationals, from national territory or (in most cases at least) from national quasi-territory, whereas the substantive approach to the transferal of international space law obligations to private sector entities also knows a large measure of commonality.

It is, however, finally also incontrovertible that the study of national space legislation, in particular as to its handling of the international responsibilities of the States concerned, has become an indispensable element of the study of international space law: substantial interpretation and implementation thereof requires analyses of substantial domestic space law, and its rules, rights and obligations.

2.6. References

Abhijeet, K. (2018). The Indonesian space act. In *Proceedings of the International Institute of Space Law 2017*, Blount, P.J., Masson-Zwaan, T.L., Moro-Aguilar, R., Schrogl, K.-U., (eds). Eleven International Publishing, The Hague.

Achilleas, P. (2010). Regulation of space activities in France. In *National Regulation of Space Activities*, Jakhu, R.S. (ed.). Springer, Dordrecht.

Bittencourt Neto, O. (2009). Private launch activities on Brazilian territory: Current legal framework. *Zeitschrift für Luft- und Weltraumrecht*, 58, 429–449.

Bjornerud, M. (2004). Baikonur continues: The new lease agreement between Russia and Kazakhstan. *Journal of Space Law*, 30(1), 13–36.

Cassese, A. (2001). *International Law*. Oxford University Press, Oxford.

Cheng, B. (1997). *Studies in International Space Law*. Clarendon Press, Oxford.

Cocca, A.A. (1979). Registration of space objects. In *Manual on Space Law*, Jasentuliyana, N. and Lee, R.S.K. (eds). Kluwer Academic Publishers, Dordrecht.

Dembling, P.G. and Arons, D.M. (1967). The evolution of the outer space treaty. *Journal of Air Law and Commerce*, 33, 419–456.

von der Dunk, F.G. (1998). *Private Enterprise and Public Interest in the European "Spacescape"*. International Institute of Air and Space Law, Leiden.

von der Dunk, F.G. (2001). Launching from "down under": The new Australian space activities act of 1998. In *Proceedings of the Forty-Third Colloquium on the Law of Outer Space*. American Institute of Aeronautics and Astronautics, Reston.

von der Dunk, F.G. (2008a). As space law comes to Nebraska, space comes down to earth. *Nebraska Law Review*, 87, 498–515.

von der Dunk, F.G. (2008b). Implementing the United Nations outer space treaties – The case of the Netherlands. In *Nationales Weltraumrecht/National Space Law*, Brünner, C. and Walter, E. (eds). Böhlau Verlag, Vienna.

von der Dunk, F.G. (2013a). Another addition to national space legislation: The Austrian outer space act, adopted 6 December 2011. In *Proceedings of the International Institute of Space Law 2012*, Jorgenson, C.M. (ed.). Eleven International Publishing, The Hague.

von der Dunk, F.G. (2013b). Towards "flags of convenience" in space? In *Proceedings of the International Institute of Space Law 2012*, Jorgenson, C.M. (ed.). Eleven International Publishing, The Hague.

von der Dunk, F.G. (2015a). Preface. In *Handbook of Space Law*, von der Dunk, F.G. (ed.). Edward Elgar Publishing, Cheltenham.

von der Dunk, F.G. (2015b). International space law. In *Handbook of Space Law*, von der Dunk, F.G. (ed.). Edward Elgar Publishing, Cheltenham.

von der Dunk, F.G. (2018). Kiwi's in space. In *Proceedings of the International Institute of Space Law 2017*, Blount, P.J., Masson-Zwaan, T.L., Moro-Aguilar, R., Schrogl, K.-U. (eds). Eleven International Publishing, The Hague.

von der Dunk, F.G. and Negoda, S.A. (2002). Ukrainian national space law from an international perspective. *Space Policy*, 181, 15–23.

von der Dunk, F.G. and Nikolaisen, A. (2002). Vikings first in national space law: Other Europeans to follow – The continuing story of national implementation of international responsibility and liability. In *Proceedings of the Forty-Fourth Colloquium on the Law of Outer Space*. American Institute of Aeronautics and Astronautics, Reston.

Gál, G. (1969). *Space Law*. Oceana Publications, Dobbs Ferry.

Gerhard, M. (2002). The law of Ukraine on space activities. *Zeitschrift fur Luft- und Weltraumrecht*, 51(1), 57–59.

Gerhard, M. (2008). The state of the art and recent development of national space legislation. In *Nationales Weltraumrecht/National Space Law*, Brünner, C. and Walter, E. (eds). Böhlau Verlag, Vienna.

Gerhard, M. (2009). Article VI. In *Cologne Commentary on Space Law: In Three Volumes*, Vol. I, Hobe, S., Schmidt-Tedd, B., Schrogl, K.-U. (eds). Carl Heymanns Verlag, Cologne.

Gerhard, M. and Schrogl, K.-U. (2002). Report of the "Project 2001" working group on national space legislation. In *"Project 2001" – Legal Framework for the Commercial Use of Outer Space*, Böckstiegel, K.H. (ed.). Carl Heymanns Verlag, Cologne.

Gillon, T. (2008). Regulating remote sensing space systems in Canada – New legislation for a new era. *Journal of Space Law*, 34(1), 19–32.

Hedman, N. (2008). Swedish legislation on space activities. In *Nationales Weltraumrecht/ National Space Law*, Brünner, C. and Walter, E. (eds). Böhlau Verlag, Vienna.

Hellenic Parliament (2017). 20/12/2017 [Online]. Available at: https://www.hellenicparliament.gr/en/Nomothetiko-Ergo/Psifisthenta-Nomoschedia [Accessed 22 April 2022].

Hobe, S. (2009). Historical background. In *Cologne Commentary on Space Law: In Three Volumes*, Vol. I, Hobe, S., Schmidt-Tedd, B., Schrogl, K.-U. (eds). Carl Heymanns Verlag, Cologne.

Hobe, S. and Neumann, J. (2010). Regulation of space activities in Germany. In *National Regulation of Space Activities*, Jakhu, R.S. (ed.). Springer, Dordrecht.

Hosková, M. (2000). The 1994 Baikonour agreements in operation. In *Proceedings of the Forty-Second Colloquium on the Law of Outer Space*. American Institute of Aeronautics and Astronautics, Reston.

Hulsroj, P. and Pecujlic, A.N. (2016). New in the nest: The Danish space act. *Air & Space Law*, 41(6), 503–511.

Jakhu, R.S. (2010). Regulation of space activities in Canada. In *National Regulation of Space Activities*, Jakhu, R.S. (ed.). Springer, Dordrecht.

Kerrest de Rozavel, A. (1997). The launch of spacecraft from the sea. In *Outlook on Space Law Over the Next 30 Years*, Lafferranderie, G. and Crowther, D. (eds). Kluwer Law International, The Hague.

Kerrest de Rozavel, A. (1998). Remarks on the responsibility and liability for damages caused by private activity in outer space. In *Proceedings of the Fortieth Colloquium on the Law of Outer Space*. American Institute of Aeronautics and Astronautics, Reston.

Kerrest de Rozavel, A. and Smith, L.J. (2009). Article VII. In *Cologne Commentary on Space Law: In Three Volumes*, Vol. I, Hobe, S., Schmidt-Tedd, B., Schrogl, K.-U. (eds). Carl Heymanns Verlag, Cologne.

Kerrest de Rozavel, A. and von der Dunk, F.G. (2011). Liability and insurance in the context of national authorisation. In *National Space Legislation in Europe*, von der Dunk, F.G. (ed.). Brill, Nijhoff.

Lachs, M. (2010). *The Law of Outer Space*. Martinus Nijhoff Publishers, Leiden.

Lee, Y. (2007). A review of the space development promotion act of the Republic of Korea. *Journal of Space Law*, 33, 123–156.

Leister, V. and Monserrat, J. (2003). Brazil-Ukraine partnership for the use of the Alcantara launch center. In *Proceedings of the Forty-Fifth Colloquium on the Law of Outer Space*. American Institute of Aeronautics and Astronautics, Reston.

Lepard, B.D. (2017). Introduction – Why does customary international law need reexamining? In *Reexamining Customary International Law*, Lepard, B.D. (ed.). Cambridge University Press.

Limpitlaw, J. (2010). Regulation of space activities in South Africa. In *National Regulation of Space Activities*, Jakhu, R.S. (ed.). Springer, Dordrecht.

Lyall, F. and Larsen, P.B. (2009). *Space Law – A Treatise*. Routledge, London.

Malanczuk, P. (1994). Space law as a branch of international law. *Netherlands International Law Yearbook*, 25, 143–180.

Malkov, S.P. and Doldirina, C. (2010). Regulation of space activities in the Russian Federation. In *National Regulation of Space Activities*, Jakhu, R.S. (ed.). Springer, Dordrecht.

Malysheva, N.R. (2010). Regulation of space activities in Ukraine. In *National Regulation of Space Activities*, Jakhu, R.S. (ed.). Springer, Dordrecht.

Malysheva, N.R. (2018). *Space Law and Policy in the Post-Soviet States*. Eleven International Publishing, The Hague.

Mann, B.W. (2008). First license issued under Canada's remote sensing satellite legislation. *Journal of Space Law*, 34(1), 67–88.

Marboe, I. (2011). National space legislation. In *Outer Space in Society, Politics and Law*, Brünner, C. and Soucek, A. (eds). Springer, Vienna.

Marboe, I. (2013). Culmination of efforts in the area of national space legislation in 2012. In *Proceedings of the International Institute of Space Law 2012*, Jorgenson, C.M. (ed.). Eleven International Publishing, The Hague.

Marboe, I. (2015). National space law. In *Handbook of Space Law*, von der Dunk, F.G. (ed.). Edward Elgar Publishing, Cheltenham.

Marboe, I. and Hafner, F. (2011). Brief overview over national authorization mechanisms in implementation of the UN international space treaties. In *National Space Legislation in Europe*, von der Dunk, F.G. (ed.). Brill, Nijhoff.

Masson-Zwaan, T.L. (2016). Registration of small satellites and the case of the Netherlands. In *Small Satellites – Regulatory Challenges and Chances*, Marboe, I. (ed.). Brill, Nijhoff.

Matte, N.M. (2000). Space law. In *Encyclopedia of Public International Law*, Vol. 4, Bernhardt, R. (ed.). North Holland Publishing, Amsterdam.

Mayence, J.F. (2011). Granting access to outer space: Rights and responsibilities for states and their citizens. In *National Space Legislation in Europe*, von der Dunk, F.G. (ed.). Brill, Nijhoff.

Michielsen, P. (2016). The Belgian space act: An innovative legal safeguard to boost the space industry. *Air & Space Law*, 41(2), 89–117.

Monserrat, J. (2003). Brazilian launch licensing and authorizing regimes. In *Proceedings of the United Nations/International Institute of Air and Space Law Workshop: Capacity Building in Space Law*, United Nations.

Monserrat, J. (2010). Regulation of space activities in Brazil. In *National Regulation of Space Activities*, Jakhu, R.S. (ed.). Springer, Dordrecht.

Mosteshar, S. (2010). Regulation of space activities in the United Kingdom. In *National Regulation of Space Activities*, Jakhu, R.S. (ed.). Springer, Dordrecht.

Omarova, G. (2006). Perspectives for national space legislation in Kazakhstan. In *"Project 2001 Plus" – Global and European Challenges for Air and Space Law at the Edge of the 21st Century*, Hobe, S. (ed.). Carl Heymanns Verlag, Cologne.

Rhee, S.M. (2009). Current status and recent developments in Korea's national space laws. *Journal of Space Law*, 35(2), 523–538.

Reif, S. (2002). Space law in the People's Republic of China – Hong Kong Special Administrative Region (HKSAR) government: Outer space ordinance (last amended 55 of 1999 s. 3). *Zeitschrift für Luft- und Weltraumrecht*, 51, 47–56.

Schmidt-Tedd, B. and Stelmakh, O. (2012). Einführung in das Gesetz der Republik Kasachstan über Weltraumaktivitäten. *Zeitschrift für Luft- und Weltraumrecht*, 61, 426–450.

Smith, L.J. and Kerrest de Rozavel, A. (2013). The 1972 convention on international liability for damage caused by space objects. In *Cologne Commentary on Space Law: In Three Volumes*, Vol. II., Hobe, S., Schmidt-Tedd, B., Schrogl, K.-U. (eds). Carl Heymanns Verlag, Cologne.

Spencer, R.L. (2010). International space law: A basis for national regulation. In *National Regulation of Space Activities*, Jakhu, R.S. (ed.). Springer, Dordrecht.

Stubbe, P. (2017). *State Accountability for Space Debris*. Brill, Nijhoff.

Supancana, I.B.R. (2006). The development of space law in Indonesia. *Indonesian Law Journal*, 1(1), 40–45.

Tapio, J. (2018). The Finnish space act: En route to promoting sustainable private activities in outer space. *Air & Space Law*, 43, 387–409.

UNOOSA (2023). National space law [Online]. Available at: http://www.unoosa.org/oosa/en/ourwork/spacelaw/nationalspacelaw/index.html [Accessed 22 April 2022].

de Villiers Lessing, I. (1994). South Africa: Recent development in space law. *Telecommunications & Space Journal*, 1, 139–142.

Vorwig, P.A. (2010). Regulation of private launch services in the United States. In *National Regulation of Space Activities*, Jakhu, R.S. (ed.). Springer, Dordrecht.

Wallace, R.M.M. (1997). *International Law*. Sweet and Maxwell, London.

Zhao, Y. (2015). *National Space Law in China*. Brill, Nijhoff.

Zhukov, G. and Kolosov, Y. (1984). *International Space Law*. Cambridge University Press.

de Zwart, M. and Lisk, J. (2018). Development of the New Zealand and Australian space industries. In *Proceedings of the International Institute of Space Law 2017*, Blount, P.J., Masson-Zwaan, T.L., Moro-Aguilar, R., Schrogl, K.-U. (eds). Eleven International Publishing, The Hague.

Part 2

The Main Principles of Space Law

3

Freedom of Exploration and Use of Outer Space

Jenni TAPIO[1,2]

[1] *Ministry of Economic Affairs and Employment, Helsinki, Finland*

[2] *University of Helsinki, Finland**

3.1. Introduction

This chapter discusses one of the founding principles of international space law – the freedom of exploration and use of outer space for the benefit and in the interest of all countries[1]. This principle, set out in Article I of the Treaty on Principles Governing the Activities of States in the Exploration and Use of Outer Space, including the Moon and Other Celestial Bodies of 19 December 1966 (OST)

* This chapter was written by the author in autumn 2019–spring 2020. Since then there have been important developments in space governance, these include *inter alia* the increasing number of national space laws, the evolution of non-legally binding instruments such as the LTS Guidelines and the development on the discussion on legal aspects of space resources, especially in the context of COPUOS. Some reflections on some of these issues have been updated prior to publication, as has been feasible. The views expressed in this chapter are solely those of the author and do not necessarily reflect any of the views of any of the affiliated entities.

1 The common interest of all States in the exploration and use of outer space and partial demilitarization of outer space have been said to be "the underlying motives for the conclusion of the Outer Space Treaty" (Hobe and Hedman 2009, p. 24). The common interest as a determination for the exploration and use of outer space "for peaceful purposes" is set out in the Preamble of the OST, and the substantive principles, including the freedom of exploration and use "derive from it" (Jenks 1965, p. 193).

Space Law,
coordinated by Thomas LECLERC. © ISTE Ltd 2023.

is fundamental to international space law, reflecting the principle of freedom that already existed under customary international law[2]. To ensure that outer space remains free for exploration and use, and that all actors uphold the primary principles, it was considered paramount to take measures to safeguard responsible behavior in outer space. Consequently, the principle of international responsibility of States for the national space activities carried out by governmental and non-governmental actors was included in Article VI of the OST. These important principles, independently and concomitantly, encapsulate many of the core questions of contemporary space law: what is outer space, *who can use it, and on what terms*?

3.2. The basic elements of the freedom of exploration and use

In addition to demonstrating an impressive technological feat and an extension of what could be made possible by human effort in its quest to explore, the launch of Sputnik I on 4 October 1957 by the Soviet Union (USSR) also prompted the need for a collection of guiding principles relevant for human activities in outer space[3]. Issues such as diverse interests and the differences in spaceflight capabilities between States coupled with the strategic importance of outer space meant that many of the debates relating to exploration and use of outer space were, and still are, profoundly rooted in international (geo)politics and the relationship between States. This is also reflected in the wording of Article I of the OST, expressly calling for equality between countries and the sharing of benefits, especially between those States that have the means and resources to implement the freedom of exploration and use, and those still developing their capabilities. In today's reality, there have been added complexities to the understanding and functioning of key principles of space law, such as that of Article I of the OST, as a result of the notable role of new actors and activities, including inter alia commercial actors, public–private ventures and new States becoming increasingly involved in space activities.

Under the umbrella of space freedom, there are two intrinsically intertwined, yet somewhat distinct legal concepts: (i) the freedom of outer space, that is, the (legal)

2 "[…] the principle of the freedom of outer space appears at present as universally accepted norm of common international law with a binding force" (Markov 1963, p. 140).

3 The first fundamental principles of space law were collected in UNGA resolution 1721 (XVI) of 1961, stating inter alia that international law, including the UN Charter applies to outer space; the outer space and the celestial bodies are free for exploration and use by all States in accordance with international law; and the outer space and the celestial bodies are not subject to national appropriation. The principle of freedom of exploration and use was then subsequently reflected in Declaration of Legal Principles Governing the Activities of States in the Exploration and Use of Outer Space, UNGA Resolution 1962 (XVIII) of 1963 (Legal Principles Declaration).

status of outer space as an area beyond sovereignty, and (ii) the freedom of exploration and use – the freedom to conduct space activities *in that area*. Arguably, the space freedom principle is a combination of these both. The former refers to the issues arising from the character of outer space as an international space free from claims of sovereignty by States[4]. The legal questions relating to outer space as a domain are most notably the questions concerning the definition and delimitation of outer space, as well as the prevention of claims of sovereignty (non-appropriation), and jurisdiction. The most pressing legal questions relating to the activities – such as exploration and use (Gorove 1977, pp. 52–53) – of States and their nationals in utilizing outer space concern the scope and content of the activities permitted or prohibited[5] under international law, namely *what is the definition of exploration and use?* In this context, it is noteworthy that freedoms come with conditions and responsibilities, both by one State to another and arguably, even toward outer space as a domain itself[6]. Given the inherently dangerous and hazardous nature of space activities, an important constraint to the freedom of exploration and use is the direct responsibility and liability of States, which extends beyond their own governmental activities to the space activities of their private entities (OST, Arts. VI and VII). To ensure that outer space remains as a domain in which freedoms can be meaningfully exercised, some control over the activities must be maintained to safeguard responsible exploration and use, as well as the maintenance of a sustainable and safe space environment.

The ever-increasing number of private activities, new actors and novel uses of technology in outer space underline the need for continuous discussion and further understanding of the important principles underpinning space activities[7]. However, many of the questions still being discussed today relating to the exploration and use or the legal status of outer space and its consequences, both in academic

4 Accordingly, the outer space has described with notions such as *res extra commercium, res communis omnium* or *res communis humanitatis*. For good summary of the discussion and relevant literature on outer space as an international space, see Frans von der Dunk (2015, pp. 55–60). The high seas and the polar regions (especially Antarctica) are mostly commonly referred to as the other international spaces (Kish 1973). Airspace also often included (Brünner and Soucek 2011, pp. 246–292).

5 Especially activities in relation to Articles II and IV of the OST.

6 The most fundamental substantive principles are included especially in Articles I-XII of the OST.

7 The global space economy was valued at US$ 360 billion in 2019, which represented a 3% growth since the previous year. Much of this value was concentrated in satellite services. It is notable that the bulk of this value comes from private industry, either acting independently or in conjunction or with the support of government. Available at: https://brycetech.com/reports/report-documents/SSIR-2019-2-pager.pdf [Accessed 10 January 2020].

commentary, in government and inter-governmental settings, are not new[8]. Solutions to the issues arising alongside the development of technology and increased human capabilities have been sought based on the collection of principles set forth in the OST: some with better success than others. The need to further define the general principles is not to be seen as a failure of the OST. The fundamental principles have evolved, and have been elaborated on, in subsequent international legally binding and non-legally binding instruments. The successive four space treaties after OST represent a collection of more specific rules dealing with inter alia the rescue and return of space objects and astronauts (Rescue and Return Agreement), liability for damage caused by space objects (Liability Convention), registration of space objects (Registration Convention) and the activities on celestial bodies (Moon Agreement). The number of ratifications of the five United Nations (UN) space treaties is not consistent[9], but they are nevertheless often cited as "the cornerstones of international space law and form the primary international legal source for the governance of outer space activities"[10]. As is the case for the other provisions of the OST, the principle of freedom of exploration and use in Article I of the OST, is a legally binding obligation on State Parties under international law[11]. The principles are written in a generic and unspecific way, necessitating interpretation of the obligations contained therein. The Vienna Convention on the Law of Treaties of May 23, 1969 (VCLT)[12] provides for the agreed technical legal means for treaty interpretation[13]. The means provided can be applied to the Outer Space Treaty to determine the exact meaning of the legal principles, such as the scope and reach of the freedom of exploration and use. However, importantly, in interpreting the

8 This can be seen, for example, in the agenda of the Legal Subcommittee (LSC) of COPUOS in its 957th meeting in 2019, discussing inter alia issues regarding "the definition and delimitation of outer space", "legal mechanisms relating to space debris mitigation and remediation measures", "legal aspects of space traffic management", and "potential legal models for activities in exploration, exploitation and utilization of space resources" (see COPUOS 2019b). Report of the Legal Subcommittee on its 58th session, held in Vienna from 1 to 12 April 2019, A/AC.105/1203.

9 For the status of State ratifications of the five UN space treaties, see UN Office for Outer Space Affairs (n.d.). The list of acceptance of the UN space treaties is updated on this page every year, the most recent list is found behind the link provided in the list of references.

10 This or similar phrasing is often used by delegations in their plenary statements in COPUOS (while some delegations refer in this context to four treaties); The Moon Agreement has not been widely ratified, and its impact consequently may be reduced in comparison to the others.

11 It has been concluded that at least certain elements of the OST have become customary international law, this includes the freedom of exploration and use (Lyall and Larsen 2016, pp. 58, 59, 78).

12 The primary tools of treaty interpretation are contained in Articles 31 and 32 of the VCLT.

13 For a good overview of the plethora of issues relating to the application of the VCLT to the UN space treaties, see Jakhu and Freeland (2013, pp. 375–391).

principles, it must be borne in mind that "it should be inappropriate to 'read into' that provision certain rules so as to reflect what *should* be" if such rules cannot be inferred from the provision through the application of the aforementioned rules of interpretation (Jakhu and Freeland 2013).

After the adoption, and subsequent limited acceptance, of the Moon Agreement in 1979, law making in the form of treaties at the UN appears, for various reason, to have ceased. Consequently, several non-legally binding, soft law instruments have been produced by UN Committee on the Peaceful Uses of Outer Space (COPUOS) and adopted by the UN General Assembly, which have in turn developed the understanding of the key space law principles[14]. The character of these instruments is heterogeneous and differ in various aspects (Tapio and Soucek 2019, pp. 565–582)[15]. Arguably, even all of the instruments are not legally binding, some are more significant than others and have generally been more widely accepted, for example, the Remote Sensing Principles[16], which are referred to in a number of national laws. Another prominent example of the Principle Resolutions[17] is the Space Benefits Declaration, directly connected to the freedom of exploration and use of outer space.

---

14 For detailed discussion of soft law in the context of space activities, see, for example, Marboe (2012); see especially the topic of non-legally binding instruments in space governance in works contributed to by the author, Tapio, J. and Soucek, A., *Standardization as an Instrument of Cooperation: A Silver Lining for Harvesting Common Benefits on the Way Back to the Moon?*, Proceedings of the International Institute for Space Law 2021 (Eleven, 2022), pp. 79–91; Palmroth, M., Tapio, J., Soucek, A. et al.: *Toward Sustainable Use of Space: Economic, Technological, and Legal Perspectives.* Space Policy (2021), vol. 57; Soucek, A. & Tapio, J.: Does the End Justify the Means? *A Legal Study on the Role and Consequences of Normative Pluralism in International Space Governance*, Proceedings of the International Institute of Space Law, vol. 2020, no. 6, 399–418; Tapio, J. & Soucek, A.: *National Implementation of Non-Legally Binding Instruments: Managing Uncertainty in Space Law?*, (2019) Air & Space Law, 44(6), 565–582; Soucek, A. & Tapio, J.: *Normative References to Non-Legally Binding Instruments in National Space Laws: A Risk-Benefit Analysis in The Context of Domestic and Public International Law, Proceedings of the International Institute of Space Law*, vol. 2018, no. 4, 553–580.

15 Especially in the context of space debris mitigation, see Tapio and Soucek (2019b, pp. 553–590).

16 Principles Relating to Remote Sensing of the Earth from Outer Space, UNGA resolution 41/65 of 1986 (Remote Sensing Principles).

17 Legal Principles Declaration; Principles Governing the Use by States of Artificial Earth Satellites for International Direct Television Broadcasting, UNGA resolution 37/92 of 1982; Remote Sensing Principles; Principles Relevant to the Use of Nuclear Power Sources in Outer Space, UNGA resolution 47/68 of 1992; Declaration on International Cooperation in the Exploration and Use of Outer Space for the Benefit and in the Interest of All States, Taking into Particular Account the Needs of Developing Countries, UNGA resolution 51/122 of 1996 (Space Benefits Declaration).

Additionally, Practice Resolutions[18] have been created to give guidance in determining some important concepts such as the launching State. The most recent examples of "soft law making" are the Guidelines Resolutions[19]. Both the resolutions adopting the Space Debris Mitigation Guidelines and the one welcoming the LTS Guidelines strive toward maintaining outer space as an operational environment for space activities.

The freedom of exploration and use (referred to in Article I of the OST) enables States and private actors to carry out their activities in outer space and receive the benefits. This highlights the significance of having a safe and sustainable operating environment in space, including the near-Earth orbits. Our societies are dependent on services, in whole or in part, by space-based infrastructure, and it is generally agreed that the unavailability of these services would have grave socioeconomic impacts[20]. In this regard, while the non-legally binding instruments go some way in enhancing the interpretation of some of the legally binding principles contained in the UN space treaties, they do not in themselves give rise to principles that would legally bind States to act or not to act in a certain manner, nor do they "constitute a revision, qualification or reinterpretation of the existing principles and norms of international law" (LTS Guidelines, Preamble, paras 14 and 15)[21].

As States moved away from treaty making in the area of space activities toward a tendency to adopt non-legally binding instruments, the law has developed most recently in a marked increase in the adoption of national space legislation (see *supra*, Chapter 2). National laws serve as means of a State meeting its international obligations and implementing these obligations to national rules for the purpose of governing the relationship between the State and its nationals. Arguably, national

18 Recommendations on National Legislation Relevant to the Peaceful Exploration and Use of Outer Space, UNGA resolution 68/74 of 2013 (National Space Law Resolution); Application of the Concept of the "Launching State", UNGA resolution 59/115 of 2005 (Launching State Resolution); and Recommendations on Enhancing the Practice of States and International Intergovernmental Organizations in Registering Space Objects, UNGA resolution 62/101 of 2008 (Registration Practice Resolution).

19 See COPUOS (2007), Space Debris Mitigation Guidelines of the Committee on the Peaceful Uses of Outer Space, A/62/20 (Space Debris Mitigation Guidelines); COPUOS (2019). Guidelines for the Long-term Sustainability of Outer Space Activities of the Committee on the Peaceful Uses of Outer Space, A/74/20 (LTS Guidelines).

20 For an outline of the impacts, see, for example, European Commission (2011).

21 Similarly, see Launching State Resolution, para. 4; Registration Practice resolution, para. 6 and National Space Legislation Resolution, para. 4; The work on the Preamble and 21 LTS Guidelines continues in the Scientific and Technical Subcommittee of COPUOS especially in the frame of the second Working Group on the Long-term Sustainability of Outer Space Activities since 2021 https://www.unoosa.org/oosa/en/ourwork/copuos/working-groups.html.

laws may be also indicative of a State's interpretation of certain international law principles, an example in the space law context being the interpretation of space resource exploitation in certain recent national laws. This way national laws can contribute to the development of space law by providing elements on how the States have understood the international legal framework[22]. One useful example of such contribution is the interpretation of "space activities" in national space laws. Together with the subsequent practice of States, these national interpretations are relevant to determination of the scope and reach of the principles. Accordingly, it can be said in general that national space laws reflect wide acceptance of the principle of exploration and use, which can be inferred from the manner in which national regulations been opted to steer the ways national entities are allowed to "explore" and "use" outer space.

3.3. Article I of the outer space treaty

Article I of the OST is the pinnacle of international law relating to outer space. It promulgates the freedom exploration and use of outer space, and freedom of scientific research by all States. At the same time, it sets certain restrictions on the exercise of this freedom.

3.3.1. *Scope of application*

3.3.1.1. *The subjects and actors*

Although Article I of the OST refers only to States[23], three actors may in reality realize and benefit[24] from the freedom of exploration and use (a) States using their governmental agencies, (b) international intergovernmental organizations and (c) non-governmental entities[25].

22 "Although national space legislation cannot have a direct influence on international law, it might be regarded as an expression of *opinio juris*" (Hobe 2009, p. 31).

23 The first paragraph of the Article refers to "countries", while paragraphs 2 and 3 refer to "States"; States can be in general qualified in international law as entities having permanent population, a defined territory, government and capacity to enter into relations with other States (Warbrick 2006).

24 The realization of the freedom is conditional to Article VI of the OST and the respective national regime concerning space activities.

25 The lack of reference to the other space actors in Article I of the OST has been seen as a sign that international space law has become "obsolete" or "inadequate" (Wassenbergh 1997, p. 173).

The freedom of exploration and use of outer space is granted to all States[26] in the second paragraph of Article I of the OST. The Soviet Union originally proposed that space activities are to be conducted "solely and exclusively" by States[27]. However, no such reference to "States only" was included in the treaty text, opening up the possibility also to other actors, while States remain the predominant actors in space.

In addition to bilateral collaboration, States cooperate in space through international intergovernmental organizations as the activities may require resources that are beyond the capabilities and resources of States[28]. Although international organizations are not mentioned in Article I of the OST, space activities of States acting through such organizations were not meant to be excluded (Jenks 1965)[29]. International organizations are recognized in other provisions of the OST[30]. Accordingly, State participation through such cooperation mechanisms is included and reflects the current practice.

While non-governmental entities, such as legal persons under domestic laws, are not subjects of international law and are not recognized in Article I of the OST as actors, their role in space activities should not be overlooked. Participation by corporations, scientific institutions or even by private persons is permitted, including for commercial purposes[31], as the States retain primary international responsibility for the national activities and require States to authorize and continuously supervise non-governmental space activities. What is more, the multifaceted Article VI of the

---

Non-governmental entities are both natural and legal persons, including inter alia individuals, and companies.

26 Reference is not made only to signatory States, but to all States. The use of "all States" implies a "declaration of a general principle" (Gál 1967, p. 140). On the other hand, reference to States was set out to "avoid the enunciation of a right of adventure in space for individuals, which would go beyond anything which would be thought reasonable even by those that regard private enterprise as having important contribution to make developments in space" (Jenks, 1965, p. 196).

27 Drafting history of Principle 5 of the Legal Principles Declaration, UN Document A/AC.105/C.2/L.1 of 1962, para. 7.

28 See, for example, the first sentence of the Preamble of the Convention for the Establishment of a European Space Agency, 1975, entered into force on October 30, 1980: "The States parties to this Convention, considering that the magnitude of the human, technical and financial resources required for activities in the space field is such that these resources lie beyond the means of any single European country [...]".

29 It has been submitted that space activities by international organizations are considered in the light of the OST as "indirect activities of their member states" (Malanczuk 1997, p. 31).

30 Especially Articles VI and XIII of the OST.

31 The aim of such activity being "to make profit" or at least "to secure a reasonable return on investment" (Traa-Engelman 1993, p. 20). On the permissibility of commercial activities, see, for example, Traa-Engelman (1993, p. 280) and Hobe (2009, p. 35).

OST also sets a limitation for the non-governmental entities to enjoy the freedom of exploration and use: they require an authorization from the appropriate State.

It has also been suggested that the inclusion of "all mankind" in the concept of "the province of all mankind" in Article I of the OST has resulted in 'mankind' being a beneficiary to the freedom of exploration and use, making it "a new subject of law" (Cocca 1984). This line of argument, however, remains subject to debate[32].

3.3.1.2. *Related considerations: Definition and delimitation of outer space*

As a matter of fact, the question on the delimitation between outer space and airspace remains unresolved[33], and there is no agreed definition of outer space or the celestial bodies[34]. In attempt to define limits between the Earth and outer space, the "von Kármán line" has been frequently cited[35]. The *spatial* view advocating that activities taken in outer space as defined would be the determining factor that the legal regime governing activities in outer space should be decisive in a given situation. This approach has been contested by those who advocate for a *functionalist* view, whereby the nature, aim and characteristics of the activities are decisive (von der Dunk 2015, pp. 60–72). Accordingly, reference is made to the types of activities, regardless where they are carried out (Gal 1967, p. 36). The application of this view includes also terrestrial activities if "organically connected" to outer space (Zhukov and Kolosov 1984, p. 4). Even if activities have been left unperformed because of the lack of definition, having no consensus on the delimitation of outer space and airspace may cause practical issues in relation the legal regime to be applied to activities such as sub-orbital flights, and sounding rockets (which do not place a payload on orbit, but reach outer space). It is notable

32 For discussion, see, for example, Malanczuk (1997, pp. 32–34). "The international community as benefactor, however, does not become a subject of international law" (Hobe 2009, p. 32).

33 The issue was identified already in 1959, see Report of the Ad Hoc COPUOS, UN Doc A/4141, Part III, Section III, para. A. The issue remains unsolved and is still debated today in the LSC of the COPUOS under the agenda item: Matters relating to (a) the definition and delimitation of outer space, (b) the character and utilization of the geostationary orbit, including consideration of ways and means to ensure the rational and equitable use of the geostationary orbit without prejudice to the role of the ITU, as well as a dedicated Working Group, see COPUOS (2019c). Annotated provisional agenda (including indicative schedule of work), A/AC.105/C.2/L.308.

34 Celestial bodies are the "result of natural creation of the universe", opposite to artificial space objects (Hobe 2009). The reference to celestial bodies should be understood as "'all land' areas in outer space" (Lachs 1972).

35 Hobe (2019), para.13. Some of the States have included definition of outer space in their national laws, which refer to certain altitudes above sea level (see, e.g. Australia's Space Activities Act No. 123 of December 21, 1998 and Danish Outer Space Act, Act no. 409 of May 11, 2016).

that some States have already begun to legislate on topics such as the regulation of suborbital flights, while international law does not currently provide a clear interpretation. An example of national approach are the provisions of the Space Industry Act enacted in 2018 in the United Kingdom[36].

3.3.2. *The terms of exploration and use of outer space*

There is not universal agreement on the contents the freedoms of exploration, use or scientific research, which are the three main activities, set out in Article I of the OST[37]. These can be categorized as "legally binding behavioral imperatives", which are intended to steer the conduct and character of space activities in a certain direction. While the limitations set out in the Article I of the OST have been discussed in an extensive manner, the definition of the rights have attained lesser attention.

It is notoriously difficult to try to define the clauses of Article I of the OST. Therefore, it is somewhat reassuring that this uncertainty has not caused problems (at least to the author's knowledge) in practice to date. The types of activities that generally take place in outer space — including, for instance, the launch into orbit and on orbit operation of satellites or the maintenance of a permanently inhabited space station on the low earth orbit have been generally accepted to be within the freedom. Today, the satellites in orbit certainly make use of outer space in providing different applications such as communications, navigation and earth observation, such services are essential for various socioeconomic, commercial or even military purposes on Earth.

3.3.2.1. *The freedoms granted in Article I*

"The term 'freedom' means that all entities which are the addressees of these provisions are entitled to use, explore or scientifically investigate in outer space without the need to ask permission from other States or international entity" (Hobe and Chen 2017)[38].

---

36 Space Industry Act of March 15, 2018.

37 It was noted at the time of drafting the OST, that further clarification should be sought (see, e.g. French and Hungarian delegations as cited in Christol 1982, p. 40). It is noteworthy that the term "exploitation" is not mentioned as an activity in Article I, but some include it within the freedom of exploration and use (Christol 1982, p. 39).

38 The quotation reflects the well-founded *Lotus* principle stating that activities that are prohibited are deemed permissible (The S.S. Lotus (France v. Turkey), (1927) PCIJ, A/10 at 18, ibid at 31). The applicability of the *Lotus* principle has been debated: "As correctly stated by

The two nouns, "exploration and use" are commonly referred to together, including in the name of the treaty – Treaty on the Principles Governing the Activities of States in the Exploration and Use of Outer Space, including the Moon and Other Celestial Bodies. However, it seems that the words have distinct and separate meanings, and do not necessarily have to be construed to form a single concept, but have overlapping content (Gorove 1977, p. 53; Hobe and Chen 2017, p. 31). What is more, scientific investigation is not generally referenced together with exploration and use, while it should arguably be treated in a similar way for the purposes of interpreting the scope and contents of Article I of the OST. Importantly, the activities based on these terms, whether referenced together or separately, must comply with the restrictions set out in Article I of the OST, as well as in the other provisions of the OST.

Textual interpretation or distinction of the nouns is challenging due to the possibility of multiple ordinary meanings. Some authors consider the term "exploration" to be self-evident, whereas "use" is less clear (Jenks 1965, p. 196). However, in general parlance, "use" may be interpreted to be the broadest category, including different types of space activities, also for commercial purposes.

What further complicates the discussion is that "use" has, in some circles, been connected with the topic of ownership or property rights in outer space, that is, the rights arising out of "exploitative activities" of natural resources on celestial bodies. The status of such activities is further discussed in other chapters of this book, but it should however be noted that "exploitative activities" may also within the ambit of "use", and not amounting to appropriation or misuse[39]. This is especially as for an activity to be considered as "use" it does not necessarily need to be an activity by a private entity or toward a commercial gain (Hobe and Chen 2017, p. 31).

The activities considered as scientific investigation share many of their characteristics with exploration, most notably the act of discovery, which however can be directed at areas already explored (Hobe and Chen 2017, p. 31). The Moon

M. Lachs, it cannot be implied in international law that everything not prohibited is permitted; the rights of those who act are determined by the rights of those who may be affected by their exercise. Nor is space law an exemption from this rule." (Gál 1967, p. 153). It has been submitted that Article I means that all States have a right to launch objects to space and the exploration and use cannot be prevented by other States (Zhukov and Kolosov 1984, p. 42).

39 French delegate as cited in Christol 1982, p. 39; Developments in discussing governance framework for space resource activities since the research for this Chapter has been made have included various national and international initiatives, perhaps most notably, a Working Group dedicated to 'Legal Aspects on Space Resources' was established in the LSC of COPUOS in 2021 https://www.unoosa.org/oosa/en/ourwork/copuos/lsc/space-resources/index.html.

Agreement elaborates further on the concept of freedom of scientific investigation and expressly seeks to regulate resource utilization through Article 11. As such, the provisions of the Moon Agreement may provide a valid contribution for a future framework, however references to the Moon Agreement could be dismissed as it has not forming, at least today, been widely ratified and its principles are arguably not forming, at least today, part of customary international law.

3.3.2.2. *Related considerations: Customary right of freedom of access*

The question on the legal status of outer space as an international space free from claims of sovereignty, and the activities in that domain are naturally intertwined. Outer space, the Moon and celestial bodies are the object of exploration and use, and the freedom of exploration and use only applies *in space*[40]. The activities of humans physically in outer space remain limited today, and therefore access *to* (and sometimes also *from*) outer space, could be seen as a prerequisite condition for the full exercise of the freedom of exploration and use (Lachs 1972, p. 59).

While the second paragraph Article I of the OST stipulates that "there shall be free access to all areas of celestial bodies"[41], no right of access *to* or *from* outer space, at least *in verbatim* is included. This must however be presumed (Lachs 1972, p. 60; Hobe 2009, p. 33). When this presumption involves passage of a space object[42] traversing *foreign airspace without permission*, it gives rise to a conflict with another rule of international law: States exercise complete and exclusive sovereignty over their airspace[43].

Arguments have been put forward that a customary rule to allow transit of space objects through foreign airspace must exist, as anything else would frustrate the freedom of exploration and use of outer space: the exercise of the freedom necessitates access to outer space[44]. Such presumption favoring the application of space law, however, is argued to exist subject to conditions, and cannot include permanent implied consent (Lachs 1972, p. 61). State practice in support of a

40 Vereshchetin (2006), para.7.

41 The freedom of access to celestial bodies is further elaborated in Article XII permitting access to artificial installations placed on celestial bodies, if certain conditions are met.

42 The quest is further complicated by the absence of an agreed definition for a ""space object".

43 Convention on International Civil Aviation of 7 December 1944, entered into force on April 4, 1947 (Chicago Convention), Art. 1. Article 3 of the Chicago Convention specifies that right of overflight necessitates authorization from the concerned State. In addition to conflict with the freedom of exploration and use, there is a conflicting situation with the jurisdiction and control of the State of registry and the foreign State's sovereignty over its airspace (Hobe 2009, pp. 160–162).

44 On teleological interpretation of Article I, see Lachs (1972, p. 60).

customary rule of freedom of exploration and use, including the right to access, and travel through foreign airspace without permission from the affected State, is not consistent[45]. But as access to space through the airspace of a foreign State, even without explicit consent of the concerned State has not raised objections, a presumption of the existence of a right similar to "innocent passage" in the law of the sea (for landlocked States) has been made (Lachs 1972, p. 59).

The current legal situation seems to be that access to outer space is not "free", but is subject to agreement with the other State(s), unlike the freedom of exploration and use, for which no permission from other States have to be sought. Thus, the provisions of air law curtail the freedom of exploration and use, while the space activity requires overflight through foreign airspace (Hobe 2009, p. 161). Considered as *lege lata*, it may be regarded as hindrance the development of the space sector, for example, in States not having open access to outer space without the need for the rockets to traverse through foreign airspace. As the space services market develops, issues regarding re-entry may also become more pressing for example in the case of reusable rockets or returning aerospace objects. These could act as incentives to seek final clarity in the matter.

3.3.2.3. *Limitations to the freedoms in Article I*

"There can be no doubt that the freedom of action of States in outer space or on celestial bodies is neither unlimited, absolute or unqualified, but is determined by the rights and interests of other States" (Lachs 1972, p. 117).

It is clear that the general freedoms set out in Article I of the OST are limited. What is less clear is the scope of the freedom in relation to its limitations. As for the freedom, the limitations are expressed in conceptual terms, expressing the values considered important in the 1960s Cold War realities. Some of the limitations are included in Article I of the OST, while others can be found in other provisions of the OST, including the critical caveat that the exploration and use of outer space shall be made in accordance with international law[46].

---

45 Lachs notes that States have not, at least in the early days of spaceflight, notified or made agreements with the States whose airspace has been traversed with a space object (Lachs 1972, p. 59). As an example of the consensual State practice, the arrangements made by the United States for an unplanned landing of the Space Shuttle affecting foreign airspaces can be mentioned (Hobe, 2009, p. 161).

46 Article III elaborates the limitation set out in Article I further by stating that the exploration and use of outer space, including the Moon and other celestial Bodies, shall be carried on in accordance with international law, including the UN Charter, in the interest of maintaining international peace and security and promoting international cooperation and understanding.

The limitations to the principle of freedom of exploration and use in Article I of the OST are written in a very general and almost aspirational manner. It can however be submitted that the limitations set out in Article I OST underscore that freedom of exploration and use is not to grant any exclusivity to those States having the capabilities to take up space activities: the interests of also the non-spacefaring States should be taken into account. Attempts have been made to clarify the understanding of "the benefit of all countries" and "the province of mankind" but there are diverging views whether the outcome is sufficient (Benkö and Schrogl 1997, pp. 139–143). A non-legally binding instrument, the Space Benefits Declaration, was adopted in 1996 after discussions in COPUOS to strengthen international cooperation and to further common understanding of the application of the benefits and interests of all countries.

The principle of cooperation is one of the key principles of international space law (OST, Arts. III, IX and X)[47]. Cooperation has been identified as the primary means to take into account the common benefit and interest in space activities. The contents of the principle were clarified in the Space Benefits Declaration, affirming that participation in cooperation is voluntary in nature and the sharing of benefits would not necessarily be free of charge, allowing the parties to agree on "fair, equitable and mutually acceptable contractual terms", and further underlining that exchange of technology and expertise is subject to agreement "on mutually acceptable basis" (Space Benefits Declaration, paras. 2 and 5). It is noteworthy that the principle to share benefits applies to all actors enjoying the freedom of exploration and use, including the non-governmental actors, albeit subject to the provisions Article VI of the OST (Marchisio 2009, p. 174). The Space Benefits Declaration also provides for practical examples on the modalities for voluntary benefits sharing of space activities, through different cooperative arrangements, especially in the scientific domain for the purposes of sharing data, or the provision of capacity building, especially for the needs of the developing countries (for general examples of space benefits, see Christol 1982, p. 43).

The benefit and interest of all countries is closely connected with the idea of outer space as an international space – including no priority or exclusivity rights (Cocca 1984, p. 45). All States are entitled to exercise the freedom of exploration and use irrespective of their technical or financial capabilities (Hobe 2009, p. 33). Further reflections of the common interest can be seen in the limitation referring to the outer space as "the province of mankind", a term understood to mean the "benefit of mankind" (Christol 1982, p. 44); further underlining that there can be no claims of exclusivity (Lachs 1972, p. 117) and that the interest of all States is to be considered (Hobe 2009, pp. 38–39). It has been said that the reference to "mankind"

47 In relation to Article I of the OST, see also the Space Benefits Declaration.

within the provisions of the OST had many important duties: it was inter alia meant to unify the various rights and obligations contained in the treaty instrument, and to reinforce the meaning of "for the benefit and in the interests of all countries" in Article I of the OST (Christol 1982, pp. 44–46).

The obligation to conduct space activities with "due regard to the corresponding interests of all other States" set out in Article IX of the OST reflects the equality of the States, alike Article I, by requiring that space activities of States do not interfere with the activities of others (Marchisio 2009, p. 175). This requirement can also be inferred by logical reasoning: "my freedom of use must not exclude the freedom of others" (Soucek 2016, p. 25; similarly also Zhukov and Kolosov 1984, p. 44). The States have to ensure a "certain standard of care, attention or observance" in the conduct of space activities by governmental or non-governmental entities (Marchisio 2009, pp. 175–176). In case interference is perceived by the State planning to conduct activities, or by others whose activities might be endangered, consultations are to be initiated or requested. It has been suggested that this duty may alleviate the States concerns on the missing definitions in the treaty text (Gál 1967, p. 150). While this may be the case, the ways the obligation of due regard is to be fulfilled, arguably remain unclear. The principles set out in Articles I and IX of the OST are also reflected in the recently adopted LTS Guidelines, stating that in the implementation of the guidelines "States should be guided by the principle of cooperation and mutual assistance and should conduct all their activities in outer space with due regard for the corresponding interests of all other States" (LTS Guidelines, Preamble, para. 14). Arguably, placing an obligation, albeit non-legally binding on the States to reflect the common (corresponding) interest "to meet the needs of the present generations while preserving the outer space environment for future generations" (LTS Guidelines, Preamble, para. 5) at the national level, to ensure that the non-governmental entities carry out their activities in outer space in a way that promotes (long-term) sustainability of the outer space environment (LTS Guidelines, Guideline A.1, para. 4). Moreover, it may be observed that engaging in discussions (deliberating on activities taken, and planned, and their consequences) at international level on the sustainable uses of outer space, for example, in the context of COPUOS, can be seen as one means of consulting the views of others as foreseen in OST.

3.4. Freedom of exploration and use by non-governmental entities – Reflections on State responsibility

3.4.1. *Introduction*

As noted previously, the Soviet Union was at the time of negotiating the OST in the 1960s initially of the opinion that the freedom of exploration and use should be reserved as the prerogative of States. It has been said, that the rationale of this approach was to ensure that "irresponsible private ventures" did not interfere with the activities of States (Gál 1967, p. 142). As private commercial activities were already envisaged by the United States, a compromise solution was accepted: States would bear international responsibility for national activities in outer space, whether those activities were be carried out by governmental or non-governmental entities[48]. In connection, launching States would bear international liability for damages, while compensation for losses resulting from space activities could potentially be claimed under Article VI of the OST[49].

3.4.2. *State responsibility in space law*

One of the special and paramount tenets of international space law is that the States have "an elaborated form" (Traa-Engelman 1993, p. 59) of the general responsibility requiring them to ensure that all national space activities are carried out in compliance with their international legal obligations, in this context particularly with the provisions set forth in the OST. This responsibility applies both to the space activities carried out by the State itself (e.g. through organs/agencies) and those carried out by non-governmental entities. In the case of the non-governmental entities, their activities in outer space are subject to authorization and continuing supervision by the "appropriate State"[50]. In fulfillment of this important obligation, a number of States have taken active steps to develop authorization process, most notably through licensing of space activities, and legal and regulatory frameworks. As noted, another safeguard to ensure the responsible use of outer space by non-governmental entities are the provisions dealing with

48 The profound nature of Article VI is "to ensure that any outer space activity, no matter conducted by whom, shall be carried on in accordance with relevant rules of international law and to bring the consequences of such activity within its ambit" (Lachs 1972, p. 122). "The result is that non-governmental national space activities are assimilated to governmental space activities" (Cheng 1998, p. 14).

49 Compensation under the liability regime is not excluding compensation for damage caused by space activities under the state responsibility (von der Dunk 2015, pp. 52–53).

50 The participation of non-governmental entities is conditional, but in case the State fails to authorize or supervise the activities, this cannot relieve the State from its international responsibility.

liability for damages in Article VII of the OST and further elaborated in the Liability Convention (Liability Convention, Chapter 3)[51].

In a more detailed analysis on the contents of Article VI of the OST, the following separate and independent legal obligations of the States can be inferred. First, the obligation of the State Parties to adhere to the provisions of the OST and international law in their governmental activities is underscored[52]. Second, an obligation is placed on State Parties to "assure" that the non-governmental entities comply with the principles relating to exploration and use of outer space[53]. By assurance, the States assume an objective, direct responsibility, "an air of guarantee" (Cheng 1998, p. 13) that can be described as "obligation of result" rather than a due diligence responsibility (Benkö and Schrogl 2005, p. 107). Third, the compliance of non-governmental entities has to be ensured by authorization and supervision, making the duty to authorize and continuously supervise an independent legal duty of the State[54].

3.4.2.1. *The elements of State responsibility applied in space law*

Under international law, "every internationally wrongful act of a State entails the international responsibility of that State"[55]. For an act (or omission) to be considered a wrongful act, and thus to be the first constitutive element of State responsibility, it must amount to a breach of an international obligation of a State[56]. In space law, an example of a wrongful act could be a failure to comply with the terms of the OST, or those of the subsequent space law treaties and general international law more

---

51 The activity of launching of space objects may today be carried out by private entities, which likely was not thought to be possible in the context of the realities of the time of concluding the Outer Space Treaty. Nevertheless, the Launching State liability remains with the States.

52 "The State Parties to the treaty shall bear international responsibility for national activities in outer space, including the Moon and other celestial bodies, whether such activities are carried on by governmental…" (OST, Art. VI).

53 "…or by non-governmental entities, and for assuring that national activities are carried out in conformity with the provisions set forth in the present Treaty" (OST, Art. VI).

54 "The activities of non-governmental entities in outer space, including the Moon and other celestial bodies, shall require authorization and continuing supervision by the appropriate State Party to the Treaty" (Declaration of Legal Principles, 5).

55 Draft articles on Responsibility of States for Internationally Wrongful Acts of 2001, Supplement No. 10 (A/56/10). Available at: https://www.refworld.org/docid/3ddb8f804.html [Accessed 12 January 2020], Art. 1.

56 Draft articles on Responsibility of States for Internationally Wrongful Acts of 2001, Supplement No. 10 (A/56/10). Available at: https://www.refworld.org/docid/3ddb8f804.html [Accessed 12 January 2020], Art. 2.

broadly[57]. As a second element, this act (or omission), in breach of an international obligation, must be attributable to a State[58]. In space law, "national activities in outer space" are attributed to the appropriate State in accordance with Article VI of the OST. This attribution does not include any standard of "due care" or "due diligence" (Cheng 1998, p. 15), usually relevant in general international law to the element of attribution of private activities to the State. Consequently, subject to further conditions being fulfilled, a breach of an international obligation attributable to a State gives rise to a State's responsibility with no further requirements. The "revolutionary" element in space law is the direct responsibility of States for breach of any provision of the OST, "as it were a breach by the State itself" (Cheng 1998)[59].

As noted previously, in space law, State responsibility arises not only in relation to the space activities carried out by the State itself, but also in relation to those "national activities" by non-governmental entities. The concept of activities qualifying as "national space activities" for which the State is responsible is not defined in Article VI of the OST. The underlying rationale is that States should be in the position to control activities qualifying as "national", including those of the non-governmental entities (Gerhard 2009, p. 106). Traditionally, a State has jurisdiction over its territory and its nationals, and, more broadly, over activities of nationally registered objects (Gerhard 2009, pp. 112–114; Cheng 1998, p. 24). Consequently, those activities that overcome the threshold can be characterized as activities that "could be *linked* or *connected* to a State, or its nationals, territory or facility" (Jakhu and Freeland 2013, p. 384)[60]. This is also the formula to determine the "appropriate State" having the duty to authorize and supervise the non-governmental entities. "The duty to authorize and supervise is an independent legal duty of a State. Subject to their national legal frameworks, States may exercise a "freedom of authorization" by refusing to grant a license and stipulating conditions

57 Especially Articles I, III and VI. The inclusion, and subsequent adherence to private law obligations, afforded by wide interpretation of Article VI, would not be generally included in the scope (Cheng 1995, p. 302).

58 Draft articles on Responsibility of States for Internationally Wrongful Acts of 2001, Supplement No. 10 (A/56/10) [Online]. Available at: https://www.refworld.org/docid/3ddb8f804.html [Accessed 12 January 2020], Art. 2.

59 States would retain indirect responsibility for acts of their nationals under their effective territorial jurisdiction for issues such as "transboundary harm" (Cheng 1998, p. 15).

60 Also making a link between the responsibility and liability regimes by stating that liability is included in the definition of international responsibility as submitted in the Draft Articles "except where modified by international space law" (Jakhu and Freeland 2013).

to it – authorization is not a *fait accompli*"[61]. Arguably, space activities taken by non-governmental entities, even without authorization, are nevertheless subject to the requirement of continuous supervision and State's international responsibility over such national activities[62].

3.4.3. *Responsibility for activities carried on by non-governmental entities: the modalities of authorization and continuous supervision*

As noted previously, States have a legal obligation to authorize and continuously supervise the space activities of non-governmental entities in outer space. The precise modalities and instruments to effect this requirement are not set out in the OST. The discretion of States to organize their internal matters is also reflected in the National Space Law Resolution, calling for "appropriate action at national level" to organize authorization and supervision of non-governmental entities (National Space Law Resolution, Preamble). States have to consider various issues in developing and implementing their space activities regulation. The regulatory framework, including processes for authorization and supervision, should be made flexible enough to enable new technologies and activities, but at the same time it should provide for legal certainty for the operators in the form of clear and unequivocal licensing requirements. The national regulator has a challenging task at the confluence point of national and international law: meeting all the requirements so that they are effective and work in an interoperable manner, which is not arguably an easy task[63].

Authorization is a subset of supervision: the purpose is to ensure that the obligation of continuous supervision can be effectively carried out (Jenks 1965, p. 211). Requiring non-governmental entities to seek permission for their space activities, States have an ex ante possibility to ensure the safety of the planned

61 Tapio, J,. Soucek A. (2022) The European Space Agency's Contribution to National Space Law. In Karjalainen, K., Tornberg, I, Pursiainen, A. (eds) International Actors and the Formation of Laws. Springer, Cham, p. 130.

62 A U.S company was consequently fined for launching small satellites despite the US government's refusal to grant a license (see Henry 2018). This issue of unauthorized launch was noted in the Report of the LSC on its 57 session, held in Vienna from April 9 to 20, 2018, summarizing a statement by a delegation expressing that "unauthorized launches and operations in space posed a fundamental threat to the existing space governance system and that States and the international community as a whole should therefore work to ensure that space activities would be conducted in accordance with international law" (para. 38).

63 For further elaboration, see Tapio, J, Soucek A. (2022). especially Chapters 2.2, 2.3 and 5.1.

activities. Consequently, conditions of a space activities license usually require the operator to submit detailed information in particular about the technical and financial resources of the operator as well as the technical and operational features of the planned mission. Authorities are also encouraged to take action to impose requirement promoting sustainability and safety (LTS Guidelines, especially Guideline A.3). One of the elements is to require the operator to submit information for the purposes of registering the space object in the national registry, subsequently to be notified to the UN (OST, Art. VIII; Registration Convention; registration Practice Resolution; LTS Guidelines, especially Guideline A.5). The physical realities mean that space objects may remain in outer space for substantial time, and many things cannot be changed during the mission, making it vital to ensure that the operator is organized in a responsible manner and adequate safeguards are put in place to ensure resources for the entire mission exist.

The main purpose of the requirement of continuous supervision is to assure that the activities of the non-governmental entities are carried out in accordance with international law throughout the mission. To effectively supervise this, States often need to establish some mechanisms at domestic level to ensure that the authorities responsible for the supervision of private space activities receive accurate information to make informed decisions when needed, in compliance with the State's obligations at national and international level. Due to the dynamics of today's space activities, and actors, spreading across jurisdictions necessitate that States work together in supervising the activities by actively exchanging information and making agreements in division of the responsibilities where possible. While the State responsibility under space law is arguably not collective responsibility, there may be multiple candidates to assume that role.

3.5. Concluding remarks

The rapid development of space activities, the growing number and diversity of actors and the increasing complexity of those activities underline the need to discuss, strengthen and possibly evolve the related regulation of space activities. The principles set out in the OST provide the responsible and sustainable ways to exercise the freedom of exploration and use of outer space, while in the light of the changes of today, further evolution of these principles is necessary. The States remain the primary guardians of this process, additionally there are initiatives by the private industry to contribute to the development. These viewpoints can be reflected for example through national laws or in the practice.

The direct attribution of the activities by non-governmental entities to the State may overcome various problems relating to the identification of the responsible entity and ensure compliance and responsibility. Taking into account the victim-oriented approach of space law, and the dangerous nature of space activities, it seems reassuring that States ultimately will assume international responsibility for the activities, while considering at a national level (e.g. through enactment of their own domestic space laws) the best ways to seek recourse, if possible.

In parallel, the development of new norms and the clarification of the existing ones at international level, while reinforcing cooperation and information sharing between States on national space laws and processes, are mechanisms to support the coherent and unified development of space law. The application of the concepts created in the 1960s has not become any easier in today's context in which private space activities do not necessarily depend on a State's involvement, and the traditional elements of control – territoriality or nationality – may at times be more difficult to determine. The emergence of low cost small satellites and increased private space transportation capabilities have made it relatively easy and affordable to procure a space object and to have it launched into outer space – both activities possibly involving entities from different States. Therefore, the ability to retain the element of control requires greater monitoring and cooperation by the States – as well as active promulgation of the relevant rules concerning private entities to seek authorization and the content of these requirements. Clarity of the legal requirements at the international and national level is also a condition to secure continued interest in space activities and receipt of the benefits.

3.6. References

Benkö, M. and Schrogl, K.-U. (1997). History and impact of the 1996 UN declaration on "space benefits". *Space Policy*, 13(2), 139–143.

Benkö, M. and Schrogl K.-U. (2005). *Space Law: Current Problems and Perspectives for Future Regulation*. Eleven International Publishing, Utrecht.

Brünner, C. and Soucek, A. (eds) (2011). *Outer Space in Society, Politics and Law*. Springer-Verlag, Vienna.

Cheng, B. (1995). International responsibility and liability for launch activities. *Air & Space L*, XX(6), 297–310.

Cheng, B. (1998). Article VI of the 1967 space treaty revisited: "International responsibility", "national activities", and "the appropriate state". *Journal of Space Law*, 26(1), 7–32.

Christol, C.Q. (1982). *The Modern International Law of Outer Space*. Pergamon Press, New York.

Cocca, A.A. (1984). Historical precedents for demilitarization. In *Maintaining Outer Space for Peaceful Uses, Proceedings of a Symposium Held in the Hague, 12–15 March 1984*, Jasentuliyana, N. (ed.). The United Nations University, Tokyo.

von der Dunk, F.G. (2015). International space law. In *Handbook of Space Law*, von der Dunk, F.G. and Tronchetti, F. (eds). Elgar Publishing, Cheltenham.

European Commission (2011). Towards a space strategy for the European Union that benefits its citizens. Report, COM(2011) 152 final.

Gál, G. (1969). *Space Law.* A.W. Sijthoff, Leiden.

Gerhard, M. (2009). Article VI of the outer space treaty. In *Cologne Commentary on Space Law: In Three Volumes*, Vol. I, Hobe, S., Schmidt-Tedd, B., Schrogl, K.-U. (eds). Carl Heymanns Verlag, Cologne.

Gorove, S. (1977). *Studies in Space Law: Its Challenges and Prospects.* A.W. Sijthoff, Leiden.

Henry, C. (2018). FCC fines Swarm $900,000 for unauthorized smallest launch [Online]. Available at: https://spacenews.com/fcc-fines-swarm-900000-for-unauthorized-smallsat-launch/ [Accessed 5 January 2020].

Hobe, S. (2009). Article I of the outer space treaty. In *Cologne Commentary on Space Law: In Three Volumes*, Vol. I, Hobe, S., Schmidt-Tedd, B., Schrogl, K.-U. (eds.). Carl Heymanns Verlag, Cologne.

Hobe, S. (2019). Airspace. In *Max Planck Encyclopedia of Public International Law*, Rüdiger, W. (ed.). Article last updated May 2019 [Online]. Available at: https://opil.ouplaw.com/home/MPIL [Accessed 19 April 2022].

Hobe, S. and Chen, K.-U. (2017). Legal status of outer space and celestial bodies. In *Routledge Handbook of Space Law*, Jakhu, R.S. and Dempsey, P.S. (eds). Routledge, London.

Hobe, S. and Hedman, N. (2009). [The] Preamble. In *Cologne Commentary on Space Law: In Three Volumes*, Vol. I, Hobe, S., Schmidt-Tedd, B., Schrogl, K.-U. (eds). Carl Heymanns Verlag, Cologne.

Jakhu, R.S. and Freeland, S. (2013). The relationship between the United Nations space treaties and Vienna convention on the law of treaties. In *Proceedings of the International Institute of Space Law 2012*, Jorgenson, C.M. (ed.). Eleven International Publishing, The Hague.

Jenks, C.W. (1965). *Space Law.* Stevens & Sons, London.

Kish, J. (1973). *The Law of International Spaces.* A.W. Sijthoff, Leiden.

Lachs, M. (1972). *The Law of Outer Space.* A.W. Sijthoff, Leiden.

Lyall, F. and Larsen, P. (2016). *Space Law. A Treatise.* Routledge, London.

Malanczuk, P. (1997). Actors: States, international organizations, private entities. In *Outlook on Space Law over the Next 30 Years*, Laffaranderie, G. and Crowther, D. (eds). Kluwer Law International, The Hague.

Marboe, I. (2012). *Soft Law in Outer Space: The Function of Non-binding Norms in International Space Law.* Böhlau Verlag, Vienna.

Marchisio, S. (2009). Article IX of the outer space treaty. In *Cologne Commentary on Space Law: In Three Volumes*, Vol. I, Hobe, S., Schmidt-Tedd, B., Schrogl, K.-U. (eds). Carl Heymanns Verlag, Cologne.

Markov, M.G. (1963). General principles of international law in outer space. In *Proceedings of the Fifth Colloquium on the Law of Outer Space 1962.* Andrew G. Haley, Washington.

Palmroth, M., Tapio, J., Soucek, A., Perrels, A., Jah, M., Lönnqvist, M., Nikulainen, M., Piaulokaite, V., Seppälä, T., Virtanen, J. (2021). Toward sustainable use of space: Economic, technological, and legal perspectives. *Space Policy*, 57.

Soucek, A. (2016). *Space Law Essentials Volume I: Textbook.* Neuer Wissenschftlicher Verlag, Vienna.

Soucek, A. and Tapio, J. (2019a). National implementation of non-legally binding instruments: Managing uncertainty in space law. *Air & Space Law*, 44(6), 565–582.

Soucek, A. and Tapio, J. (2019b). Normative references to non-legally binding instruments in national space laws: A risk-benefit analysis in the context of domestic and public international law. In *Proceedings of the International Institute of Space Law 2018*, Jorgenson, C.M. (ed.). Eleven International Publishing, The Hague.

Soucek, A. and Tapio, J. (2020). Does the end justify the means? A legal study on the role and consequences of normative pluralism in international space governance, *Proceedings of the International Institute of Space Law*, 6, 399–418.

Tapio, J. and Soucek, A. (2021). Standardization as an instrument of cooperation: A silver lining for harvesting common benefits on the way back to the moon?, *Proceedings of the International Institute for Space Law*, Eleven.

Tapio, J. and Soucek, A. (2022). The European Space Agency's contribution to National Space Law. In *International Actors and the Formation of Laws*, Karjalainen, K., Tornberg, I., Pursiainen, A. (eds). Springer, Cham

Traa-Engelman, H.L. (1993). *Commercial Utilization of Outer Space.* Martinus Nijhoff Publishers, Dordrecht.

UN Office for Outer Space Affairs (n.d.). Status of international agreements relating to activities in outer space [Online]. Available at: https://www.unoosa.org/oosa/en/ourwork/spacelaw/treaties/status/index.html [Accessed 2 January 2020][64].

64 The list of acceptance of the UN space treaties is updated on this page every year, the most recent list is found at this link.

Vereshchetin, V.S. (2006). Outer space. In *Max Planck Encyclopedia of Public International Law*, Rüdiger, W. (ed.). Oxford University Press [Online]. Available at: https://opil.ouplaw.com/home/MPIL [Accessed 2 January 2021][65].

Warbrick, C. (2006). States and recognition in international law. In *International Law*, 2nd edition, Evans, M.D. (ed.). Oxford University Press, Oxford.

Wassenbergh, H.A. (1997). The law of commercial space activities. In *Outlook on Space Law over the Next 30 Years*, Laffaranderie, G. and Crowther, D. (eds). Kluwer Law International, The Hague.

Zhukov, G. and Kolosov, Y. (1984). *International Space Law.* Novosti Press Agency, Moscow.

65 Note from the editor, found on the website: "A different entry under the title 'Outer Space' written in June 2006 by Judge Vladlen S. Vereschetin appeared as part of the online Max Planck Encyclopedia of Public International Law until the publication of this entry in 2021. This entry has been published at the same web address as the original entry…".

4

The Principle of Exploration and Utilization of Outer Space for the Benefit of All Countries

Olavo DE O. BITTENCOURT NETO and Daniel FREIRE E ALMEIDA
Catholic University of Santos, Brazil

4.1. Introduction

The fundamental principles of space law are provided by the Treaty on Principles Governing the Activities of States in the Exploration and Use of Outer Space, including the Moon and Other Celestial Bodies (OSTs). Opened for signature on January 27, 1967, the OST has enjoyed great reception by the international community. Resulting from intensive discussions at the Legal Subcommittee (LSC) of the United Nations (UN) Committee on the Peaceful Uses of Outer Space (COPUOS), the OST represents the cornerstone of space law. Throughout its 17 articles, norms related to the exploration and use of outer space were codified, including provisions introduced by the UN General Assembly (GA) Resolution 1962 (XVII) of 1963.

The OST may be regarded as an example of umbrella treaty, since it contemplates general legal principles which, in due time, can and have been further developed by other international instruments, including treaties, resolutions and guidelines accorded at the UN (Cheng 1997). Regarded as the *Magna Carta* of space law (Gorove 1992, p. 46), the OST consolidated important provisions and presented innovative features, reflecting outstanding compromises between major space faring nations during a difficult period of the Cold War.

Space Law,
coordinated by Thomas LECLERC.

At the first paragraph of its introductory Article, the OST solemnly states:

> The exploration and use of outer space, including the moon and other celestial bodies, shall be carried out for the benefit and in the interests of all countries, irrespective of their degree of economic or scientific development, and shall be the province of all mankind. (OST, Art. I)

Wide in its scope and comprehensive in its repercussions, the principle of exploration and utilization of outer space for the benefit of all countries deserves to be revisited, considering its motivation, legal content and political repercussions.

4.2. Drafting

To properly comprehend the international relevance of the principle of exploration and utilization of outer space for the benefit of all countries, one should first revisit its drafting history. At multiple drafting sessions at the LSC, different perspectives were advanced by delegates, regarding its content and proper location at the OST final text. In its essence, Article I, paragraph 1 of the OST reflects the first principle of the UNGA Resolution 1962 (XVIII) of 1963, which reads as follows: "The exploration and use of outer space shall be carried on for the benefit and in the interests of all mankind". Based on the referred provision, the OST Soviet Draft I of 1963 suggested similar rule, adopting language somewhat different from the drafts prepared by the United States and the United Kingdom.

At COPUOS, some delegations contested the legal status of the principle, as suggested by the Soviet Union. For instance, it was questioned if any rights were to be derived of such a proposition. Reference on that regard should be made to statements by India, Italy and France, suggesting that the text should be moved to the preamble instead. On the other hand, the Brazilian delegation supported that the common interest principle should be included in the operative part of the Treaty: "the Brazilian proposal ensured the recognition of outer space and the celestial bodies as a 'global commons', a 'public good', and placed inclusive interests of the international community over possible claims by individual States" (Jakhu 2005, p. 5). Subsequently, the US Draft Proposal of 1966 reiterated the principle of freedom and exploration and use by all States, specifically stating that such provision should be construed on the basis of equality and in accordance with international law. The 1966 Draft Proposal provided the groundwork of Article I, paragraph 1 of the OST. The wording would be accepted at a COPUOS' working group on July 29, 1966, and finally accorded and contemplated by UNGA Resolution 2222 (XXI) (Hobe 2007, pp. 182–183).

One should recognize that, in the discussions leading to the provision, developing and non-space faring nations played a relevant part. Working together in what has since been recognized as the Group of 77, they were able to present their perspectives and have their voices heard. For that reason, Article I, paragraph 1 of the OST specifically calls into attention the interests of all countries, "irrespective of their degree of economic or scientific development". The wording makes it clear that space faring nations must explore and use outer space taking into account their global benefits and repercussions. Therefore, all nations, whether capable or not of developing autonomous space activities, shall be entitled to participate in the space age. Equal opportunities, as well as the importance of sharing benefits, are acknowledged as fundamental to peaceful exploration and use of outer space. As explained by Jasentuliyana (1989):

> The language of this provision makes it clear that space exploration is not only a concern of a small number of States that actually carry out such activities, but of all States, including the developing countries. (p. 129)

The treaty text makes it clear that it would be insufficient for the particular exploration and use to be not for the detriment of other people (Gorove 1971, p. 100). "On the contrary", he ascertains: "such exploration and use must be constructively beneficial".

It is appropriate to address, at this point, that Article 31, 2 of the Vienna Convention on the Law of Treaties (VCLT) establishes that

> 2. The context for the purpose of the interpretation of a treaty shall comprise, in addition to the text, including its preamble and annexes:
>
> (a) any agreement relating to the treaty which was made between all the parties in connection with the conclusion of the treaty;
>
> (b) any instrument which was made by one or more parties in connection with the conclusion of the treaty and accepted by the other parties as an instrument related to the treaty. (VCLT, Art. 31, 2)

Due to the above criteria, the initiatives mentioned, involving the construction of the first article of the OST, lead to the conclusion that, in fact, the exploration and use of outer space should benefit humankind, irrespective of the fact that those activities may be carried out by nations with a comparatively more advanced development level.

Through a systematic approach, one may identify a clear link between Article I, paragraph 1 of the OST and other treaty provisions. Indeed, the principle of exploration and use of outer space for the benefit of all humankind should be considered alongside the principles of freedom of access, non-appropriation of outer space and due regard for the corresponding interests of all States Parties to the Treaty.

4.3. Legal content

Article I, paragraph 1 of the OST recognizes the global impact of space activities and takes into consideration desires and demands of all nations on that regard. In particular, the perspective of non-space faring nations is taken into account, something quite relevant in the late 1960s when the treaty was drafted, a time when space activities were reserved to major international powers.

There seems to be no valid reason why the text should not be treated as stating a binding legal obligation:

> While opinion may differ on what constitutes 'the interests of all countries' and compliance with this part of the obligation might have to rely largely on the good faith of the contracting States, the rest of the sentence is much more concrete in meaning. It strongly suggests that 'all States' have a legal right to the fruits of space exploration and use. (Cheng 1997, p. 15)

The most important legal implication of Article I, paragraph 1 of the OST is that it initiated the principle of public interest in outer space (Jakhu 2005, p. 8). Consequently, the interests of all States must be taken into account as far as the exploration and use of outer space are concerned. The common good served as a north star to the international delegates while drafting the OST, considering the vast strategic implications related to space activities. Such standard provides nations, international organizations and the public opinion with a virtuous paradigm upon which to judge space activities (Monserrat Filho 2007, p. 31).

International practice has shed some light in relation to the overall reach of the concern regarding the benefit and interests of all countries vis-à-vis space activities. As provided by the OST, this orientation is applicable irrespective of each nation degree of economic and scientific development. Since 1967, the benefits of space activities to everyday life have become abundantly clear. From the Earth orbit, telecommunication has made it possible, in roughly half a century, to interconnect the whole world. Globalization, characterized by the constant and irresistible flux of information across the globe, has been enabled by orbital satellites. Remote sensing has provided live information about the planet, with vast repercussions from agriculture to national security. Global positional systems are key to navigation and have revolutionized domestic and international transport. Weather has been monitored with increased precision, providing fundamental information to strategic planning and early warning initiatives (Lyall and Larsen 2018, p. 1).

Our world of today relies upon, and is enabled by, space infrastructure. Thus, outer space may be considered as a strategic asset (Johnson-Freese 2007). Nevertheless, access to outer space remains not universal, specially to developing nations. Global collaboration is required to assure that both the exploration and use of outer space acknowledge the common good.

The growth of a global space economy is evident, based on the participation of new space actors, more specifically in the private sector. Nevertheless, international space cooperation is still driven by strategic, national perspectives, as represented by domestic export controls mechanisms. "Hence the call for improved mechanisms of cooperation" (Lachs 1992, p. 300).

The legal status of the norm stablished in accordance with Article I, paragraph 1 of the OST is of vast relevance. By conjugating international recognition and practice, the provision has attained the status of international custom, applicable *erga omnes*.

As per Article 38, 1, b of the Statute of the International Court of Justice, customs represent an "evidence of a general practice accepted as law". This source of international law has two compulsory elements: first, the *consuetudo,* that is, a material fact of behavior, reflecting a consistent and general practice; second, the *opinio iuris*, understood as the subjective belief that such practice is required by law (Brownlie 2003, pp. 6–12).

In this line, "space law also includes its own customary international law. In the history of the development of space law, international treaties and customs have been the primary sources of legally binding international law" (Xinmin 2014, p. 2). Christol (1994) goes further, by suggesting that Article I, paragraph 1 of the OST meets international criteria for *jus cogens* status (p. 43). Peremptory by nature, *jus cogens* norms enjoy such a relevance to the international legal system that they remain binding notwithstanding any agreement to the contrary. In accordance with Article 53 of the VCLT,

> a peremptory norm of general international law is a norm accepted and recognized by the international community of States as a whole as

> a norm from which no derogation is permitted and which can be modified only by a subsequent norm of general international law having the same character. (VCLT, Art. 53)

Accordingly, Oralova (2015, pp. 1–6) developed an interesting study on *jus cogens* norms in international space law, a topic also researched by Sachdeva (2017).

Subsequent international instruments have contemplated the requirement that the exploration and use of outer space shall be carried out for the benefit and interest of all countries (Lyall and Larsen 2018, pp. 57–59). Developing nations drove international debates at the UN to consolidate the use of space technology toward economic and social advancements. For instance, UNGA resolution 51/122 of 1996 reflected the position of non-space faring nations at COPUOS and their appeal for effective mechanisms on that regard. Titled "Declaration on international cooperation in the exploration in use of outer space for the benefit and in the interests of all states taking into particular account the needs of developing countries", the resolution supports international cooperation on an equitable and mutually accepted basis and encourages exchanges of expertise and technology. As concluded by Jankowitsch, UNGA resolution 51/122 certainly combines the principle of the freedom of exploration and use of outer space with a reminder that space powers shall fulfill their obligation in accordance with Article I, paragraph 1 of the OST:

> Space powers should foster international cooperation on an equitable and mutually acceptable basis. Developing countries newly interested in space activities might thus be motivated to pool their energies so as to develop common strategies towards outer space, that could also include efforts towards better legal regulations. (Jankowitsch 2015, p. 10)

4.4. Province of all mankind

The concept of province of mankind, as contemplated by Article I, paragraph 1 of the OST, has led to relevant international debates among scholars and delegates alike. The treaty, in itself, does not define the provision, nor clarify its normative application throughout the subsequent articles. Nevertheless, the provision establishes a clear link between the exploration and use of outer space and the related benefits to all countries:

> The concept of 'province of all mankind' serves as a limitation on the freedoms of outer space in the sense that neither exploration nor use of outer space shall be undertaken for the sole advantage of one country, but done only for the benefit of the international community. The international community as benefactor, however, does not become a subject of international law. (Hobe 2017, p. 189)

The OST does not raise mankind, neither humankind for that matter, into the category of subjects of international law. A subject of international law is an entity capable of possessing international rights and duties and having the capacity to maintain its rights by bringing international claims (Brownlie 2003, p. 57). The reference to mankind, as included therein, serves to clarify the ample scope of repercussions of space activities to all people, irrespective of their nationality or country of residence.

Reviewing the official versions of the OST in English, French, Russian, Spanish, Russian, Arab and Chinese, one may infer those different interpretations of the term "province" may also be traced back to the specific language adopted. Those readings may identify the provision either related to a certain realm, territorially speaking, or to a desired conduct. The concept of province of mankind constitutes "a general political principle with certain moral overtones, meant to govern rights and duties in outer space" (Diederiks-Verschoor 1999, p. 52). Consequently, the context and purpose of said categorization, as far as outer space is concerned, has legal repercussions, which must be properly acknowledged.

For once, Article I of the OST, when reviewed in conjunction with the remaining treaty, may be construed as classifying outer space as a global common. Such conclusion emerges from the realization that outer space constitutes a resource domain to which all nations have access but to which none has the right to claim sovereignty. Therefore, the use and exploration of outer space requires international cooperation, so as to assure due regard for the corresponding interests of all nations (Bittencourt Neto 2020).

As observed by Jasentulyiana (1995):

> The [Outer Space] treaty was (…) one of the first multilateral treaties in which the centuries-old international law concept of sovereignty of States gave way to the modern concept of internationalization of the Global Commons. The 'province of mankind' principle, though controversial, became the watch-word for international relations thereafter, especially in the fields of space law and law of the seas. (p. 360)

Governing global commons in a coordinated, peaceful and prosperous manner is always a great challenge. Difficulties are related to the interconnectivity and complexity of global commons, not only in terms of the resources located therein, but also regarding the diversity of actors who use those resources while interacting with each other (Freeman 2016, p. 18).

The current concerns regarding the long-term sustainability of outer space may also find resonance in the concept of "province of mankind". By considering that the exploration and utilization of outer space shall be conducted for the benefit of all countries, one may infer a leaning toward inter-generation equity.

A link between the concept of province of mankind and environmental protection is unmistakeable. The province of mankind principle can be reasonably seen to include the protection of the outer space environment, including the mitigation and remediation of space debris" (Escolar 2011, p. 361).

Therefore, the Guidelines for the Long-Term Sustainability of Outer Space Activities adopted by the COPUOS in June 2019 should be read in conjunction with Article I, Paragraph 1 of the OST, taking into account the province of mankind rationale.

4.5. Current concerns

State practice in relation to Article I, paragraph 1 of the OST may indicate that the freedom of use and exploration of outer space has achieved prominence, when compared to benefit sharing mechanisms related to space activities. Nevertheless, said provision produces relevant legal and political repercussions as far as the governance of outer space is concerned.

Every nation has the right of access to outer space, in accordance with international law. For instance, as provided by Article I of the OST,

> outer space, including the moon and other celestial bodies, shall be free for exploration and use by all States without discrimination of any kind, on a basis of equality and in accordance with international law, and there shall be free access to all areas of celestial bodies. (OST, Art. I)

Therefore, irrespective of geopolitical considerations and substantive inequalities among space actors, outer space must always be considered as the province of mankind and, hence, shall remain open for use and exploration by all States.

This legal prerogative, nevertheless, implies a necessary limitation: while exercising its right of access to outer space, no State shall deny to others the possibility of doing the same. "The principle of abuse of rights is an important aspect of the international rule as it is directed against arbitrariness in the exercise of sovereign powers of States" (Aust 2011, p. 76).

The rapid growth of space activities and the diversification of space actors have opened opportunities while raising challenges regarding the exploration and use of outer space for the benefit of all humankind. The rapidly transforming space sector and rising global space economy have enabled a wide range of satellite applications and services, while outer space and orbital slots have become more congested. At the same time, commercialization of space activities highlights a growing interest in non-governmental uses of outer space.

As a result, freedom of use and exploration of outer space is not unlimited. Due regard to the interests of other countries and humankind as a whole is required, since outer space is construed as global commons.

The freedom of use of outer space does not include its misuse or abuse, or otherwise would be tantamount to an abuse of rights (Jakhu 2005, p. 13). Accordingly, abuse of rights during use and exploration of outer space may lead to international responsibility (Jasper and Moreland 2012, p. 21).

The prohibition of abuse of rights constitutes a fundamental tenet of international law and, as thus, is applicable to space law, in accordance with Article III of the OST:

> States Parties to the Treaty shall carry on activities in the exploration and use of outer space, including the moon and other celestial bodies, in accordance with international law, including the Charter of the United Nations, in the interest of maintaining international peace and security and promoting international co-operation and understanding. (OST, Art. III)

Such doctrine represents a manifestation of good faith in international relations. "A State is bound to exercise its international rights in good faith. A failure to do so constitutes an abuse of rights" (Kuokkanen 2002, pp. 59–60).

Bindschedler and Bernhardt (1984) further explain that

> the abuse of rights refers to a State exercising a right either in a way which result, for example, from the unconsidered use of a shared

> natural resource (...). Here, the States sharing the same natural resource suffer a reduction in their enjoinment of the resource to which they are entitled. (p. 1)

States are international responsible for their national space activities, both governmental and non-governmental (Article VI of the OST). Whenever the corresponding interests of all nations are disregarded by space faring nations, international responsibility naturally ensue. Failure to do so arguably represents an example of abuse of rights.

4.6. References

Aust, H. (2011). *Complicity and the Law of State Responsibility*. Cambridge University Press, Cambridge.

Bindschedler, R. and Bernhardt, R. (1984). *Encyclopaedia of Public International Law*, Vol. 7. North-Holland, Amsterdam.

Bittencourt Neto, O. (2020). Outer space as global commons and the role of space law. In *A Research Agenda for Space Policy*, Schrogl, K.-U. (ed.). Edward Elgar, Cheltenham.

Brownlie, I. (2003). *Principles of Public International Law*, 6th edition. Oxford University Press, Oxford.

Cheng. B. (1997). *Studies in International Space Law*. Clarendon Press, Oxford.

Christol, C. (1994). Judge Manfred Lachs and the principle of *jus cogens*. *Journal of Space Law*, 22(1–2), 33–46.

Diederiks-Verschoor, I.H.P. (1999). *An Introduction to Space Law*, 2nd edition. Kluwer Academic Publishers, The Hague.

Escolar, G. (2011). The view from the oriels of the stars: The legal obligation to mitigate and remediate space debris in earth orbit. In *Contemporary Issues and Future Challenges in Air and Space Law*, Singamsetty S.S.P., Mendes De Leon, P.M.J., Masson-Zwaan, T.L. (eds). Air and Space Law Books, Utrecht.

Freeman, C. (2016). The fragile global commons in a world in transition. *SAIS Review of International Affairs*, 36(1), 17–28.

Gorove, S. (1971). Freedom of exploration and use in the outer space treaty: A textual analysis and interpretation. In *Proceedings of the 30th Colloquium on the Law of Outer Space*, Shwartz, M.D. (ed). Fred B. Rothman & Co, South Hackensack.

Gorove, S. (1992). Sources and principles of space law. In *Space Law: Development and Scope*, Jasentuliyana, N. (ed.). Praeger Publishers, Westport.

Hobe, S. (2007). Outer space as the province of mankind – An assessment of 40 years of development. In *Proceedings of the 50th Colloquium on the Law of Outer Space*. American Institute of Aeronautics & Astronautics, Reston.

Hobe, S. (2017). Article I. In *Cologne Commentary of Space Law: In Three Volumes,* Vol. I. Hobe, S., Schmidt-Tedd, B., Schrogl, K.-U. (eds). Carl Heymanns Verlag, Cologne.

Jakhu, R. (2005). Legal issues relating to the global public interest in outer space. *Journal of Space Law*, 32, 31–110.

Jankowitsch, P. (2015). The background and history of space law. In *Handbook of Space Law*, von der Dunk, F.G., Tronchetti, F. (eds). Edward Elgar, Cheltenham.

Jasentuliyana, N. (1989). Article I of the outer space treaty revisited. *Journal of Space Law*, 17(2), 129–144.

Jasentulyiana, N. (1995). *Perspectives in International Law*. Kluwer, The Hague.

Jasper, S. and Moreland, S. (2012). *Conflict and Cooperation in the Global Commons: A Comprehensive Approach for International Security*. Georgetown University Press, Georgetown.

Johnson-Freese, J. (2007). *Space as a Strategic Asset*. Columbia University Press, New York.

Kuokkanen, T. (2002). *International Law and the Environment*. Kluwer, The Hague.

Lachs, M. (1992). The treaty on principles of the law of outer space, 1961–1992. *Netherlands International Law Review*, 39(3), 291–302.

Lyall, F. and Larsen, P. (2018). *Space Law: A Treatise*, 2nd edition. Routledge, London.

Monserrat Filho, J. (2007). *Direito e Política na Era Espacial*. Vieira & Lent, Rio de Janeiro.

Oralova, Y. (2015). Jus Cogens norms in international space law. *Mediterranean Journal of Social Sciences*, 6(6), 421–427.

Sachdeva G.S. (2017). Select tenets of space law as Jus Cogen. In *Recent Developments in Space Law*, Rao, R., Gopalakrishnan, V., Abhijeet, K. (eds). Springer, Singapore.

Xinmin, M. (2014). The development of space law: Framework, objectives and orientations [Online]. Available at: https://www.unoosa.org/documents/pdf/spacelaw/activities/2014/splaw2014-keynote.pdf [Accessed 27 September 2021].

5

Non-Appropriation of Outer Space

Philip DE MAN
Centre for Global Governance Studies, University of Leuven, Belgium

5.1. Concept and origins

Outer space, including the Moon and other celestial bodies, is not subject to national appropriation by claim of sovereignty, by means of use or occupation, or by any other means. This principle, also called the non-appropriation principle, is one of the most fundamental rules distinguishing the regulation of the realm of outer space from the subjacent airspace. The latter is, by contrast, subjected to the complete and exclusive sovereignty of the States above whose territory it is situated, as enshrined in Article 1 of the Convention on International Civil Aviation of December 7, 1944, entered into force on April 4, 1947 (Chicago Convention). The legal basis for the non-appropriation principle is provided by Article II of the Treaty on Principles Governing the Activities of States in the Exploration and Use of Outer Space, including the Moon and other celestial bodies (OST). However, the rule finds its origins in United Nations (UN) General Assembly (GA) Declaration of Legal Principles of 1963, which in turn was based on UNGA Resolution 1721 (XVI) of 1961. The principle applies to the entirety of outer space, including the Moon and other celestial bodies, as more recently reiterated in Article 11 (2) of the Agreement Governing the Activities of States on the Moon and Other Celestial Bodies, entered into force on July 11, 1984 (Moon Agreement). All of these instruments were adopted by consensus without substantial discussion of the principle of non-appropriation, indicating widespread acceptance of the norm, although the Moon Agreement has subsequently only been ratified by a small number of

Space Law,
coordinated by Thomas LECLERC.

countries due to differences in interpretation of, among other things, the non-appropriation principle.

The principle that outer space is not subject to national appropriation is a corollary to a number of other basic rules of space law and, as such, is a crucial element of the balancing act engaged in by the international community when negotiating the regime of outer space during the Cold War. In particular, the ban on national appropriation acts as a crucial safeguard for the principles according to which activities in the exploration and use of outer space, including the Moon and other celestial bodies, (i) shall be carried out for the benefit and in the interests of all countries, irrespective of their degree of economic or scientific development (OST, Art. I, 1), (ii) shall be free for all States on a basis of equality without discrimination of any kind (OST, Art I, 2) and (iii) shall be carried out with due regard to the corresponding interests of all other States (OST, Art. IX). These principles would be deeply compromised if outer space were subject to exclusive claims of national appropriation, which would in effect negate the equal freedom of others to explore and use outer space, in particular when taking into account the wide discrepancies in spacefaring capabilities between States.

The main purpose of the non-appropriation principle is to explicitly reject the customarily accepted means of establishing or acquiring exclusive rights under international law on Earth, such as occupation and sovereignty, for application in outer space, including the Moon and other celestial bodies. In turn, the non-appropriation principle is frequently seen as a norm of customary international law itself, as illustrated by relatively longstanding and uniform State practice in spacefaring going back to the dawn of the space age. As such, no State objected to a violation of its territorial sovereignty as a result of the orbital trajectory of the first object launched into outer space, the former USSR's Sputnik I, pointing to the existence of a right of innocent passage in space. This implies a departure from the legal regime of the airspace below, where States' exclusive territorial sovereignty requires prior permission for overflight by other States. Conversely, no State has argued for the creation of property rights over parts of outer space, including the Moon and other celestial bodies, on the basis of a connection to their territory or as a result of earlier spacefaring activities. For example, it was understood that "[t]he placing of national insignia" such as the American flag on the surface of the Moon on the occasion of the Apollo 11 mission "would not of course constitute a sufficient basis to found a claim of sovereignty over unoccupied land masses"[1].

1 US Department of State (1959). Statement of September 14, 1959, MS DoS file 761.56031A/9-1459.

This is not to say that the continuous use of certain orbits and other areas in space by the same States for prolonged duration has not been criticised as a de facto form of appropriation by non-spacefaring countries. However, the absence of any appropriative intention of the former as well as the very need to resort to *de facto* arguments by the latter support the general rule that no legal rights of property can be obtained over outer space. This was confirmed by the near-unanimous condemnation by the international community of the so-called Bogotá Declaration of 1976. The Bogotá Declaration was an attempt by eight equatorial countries, who at the time had not ratified the OST, to claim sovereignty over certain parts of the geostationary orbit, for the reason that these allegedly depended for their very existence on a direct connection with the territory of the signatory countries[2]. The overwhelming rejection of these claims was based in part on the conviction that the non-appropriation principle is binding on all States, regardless of their ratification of the OST, hence indicating customary status[3].

5.2. Scope of application

5.2.1. *Personal scope*

If the non-appropriation principle as such is considered customary in nature, its personal scope of application remained the subject of contention for some time. Various arguments have been made that the term 'national' appropriation in Article II OST, in combination with the Treaty's general scope of application to States, meant that non-governmental actors, or perhaps international organizations, could establish property rights over space or parts thereof. However, it is now broadly accepted that the principle of non-appropriation not only applies to States, including when they act in concert with other States through international organizations, but also to their nationals, including all private natural and legal persons. Though the text of Article II of the OST specifically imposes a ban on "national" appropriation

2 ITU (1977). Declaration of the First Meeting of Equatorial Countries of December 3, 1979, WARC-BS 81-E.

3 The only State arguably still maintaining its claims over parts of the geostationary orbit is Colombia. In its constitutional provision outlining the territory of the Colombian state, it is declared, after noting that the borders of Colombia are those established in international treaties, that "[a]lso part of Colombia is the subsoil, the territorial sea, the contiguous zone, the continental shelf, the exclusive economic zone, the airspace, the segment of the geostationary orbit, the electromagnetic spectrum and the space in which it operates, in accordance with international law or the laws of Colombia in the absence of international regulations" (Constitution of Colombia, Art. 101, para. 4).

of outer space, the provision should not be read in isolation from the rest of the treaty. Article XIII of the OST notes in this regard that the provisions of the Treaty continue to apply to the activities of States Parties when carried on jointly with other States, "including cases where they are carried on within the framework of international intergovernmental organizations". More importantly, the ban on national appropriation should also be read in conjunction with Article VI of the OST, which provides that all activities in the exploration and use of outer space, including those performed by non-governmental entities, are considered national space activities, for which the States Parties to the treaty are responsible under international law. For this reason, States have to authorize and continuously supervise the actions performed by their non-governmental entities and assure that activities by these entities are carried out in conformity with the provisions of the OST, including the non-appropriation principle[4]. Any claims of property over celestial bodies as occasionally put forward by private persons or their companies are hence in violation of the non-appropriation principle in the OST. The Moon Agreement is more explicit on these points, by providing that "[n]either the surface nor the subsurface of the Moon, nor any part thereof [...] shall become property of any State, international intergovernmental or non-governmental organization, national organization or non-governmental entity or of any natural person" (Moon Agreement, Art. 11, 3)[5]. Although the Moon Agreement has only been ratified, at the time of writing, by 18 States, the above part of the provision clarifying the personal scope of the non-appropriation principle is understood by most States as being in line with the OST.

5.2.2. *Functional scope*

According to the text of the UNGA resolutions and subsequent treaties, outer space, including the Moon and other celestial bodies cannot be appropriated "by claim of sovereignty, by means of use or occupation, or by any other means". Rather than banning only sovereignty over outer space as such, the principle thus prohibits the establishment of any type of exclusive right of appropriation, be it through sovereignty or any other means. Together with the commonly accepted interpretation of the provision's personal scope, international space law hence rejects both the exercise of territorial sovereignty by States over outer space or parts thereof, and any claims of private property by individuals and companies, for the recognition of such claims "would presuppose the existence of a territorial sovereign

4 This interpretation has also been put forward in IISL Board of Directors (2004).

5 The omitted words in this citation refer to natural resources in place, which will be discussed *infra*.

competent to confer such titles of ownership"[6]. The ban on appropriation through sovereignty rather than territorial sovereignty as such distinguishes space law from other, related fields of law governing international areas, which explicitly only outlawed territorial sovereignty. For example, the UN Convention on the Law of the Sea (UNCLOS) provides that "[n]o State may validly purport to subject any part of the high seas to its sovereignty" (UNCLOS, Art. 89). This provision simply confirms the Convention on the High Seas, adopted on April 29, 1958 (Convention on the High Seas), which was in force at the time of the negotiations of the UNGA Resolutions and treaties on outer space. Art. 2 of this Convention provides that, "[t]he high seas being open to all nations, no State may validly purport to subject any part of them to its sovereignty" (Convention on the High Seas, Art. 2). The Antarctic Treaty, signed on December 1, 1959 (Antarctic Treaty) states that "[n]o acts or activities taking place while the present Treaty is in force shall constitute a basis for asserting, supporting or denying a claim to territorial sovereignty in Antarctica or create any rights of sovereignty in Antarctica" (Antarctic Treaty, Art. IV). Considering the importance of comparable, terrestrial regimes for the negotiations of the first rules on outer space as a new international area in the 1960s, clear divergences between these regimes, such as in the formulation of the non-appropriation principle, should be treated as deliberate.

Following the above, it is not the occupation of space as such, but rather the appropriation through occupation that is banned by the non-appropriation principle. This insight brings with it some important legal implications. First, even though physical occupation by a State through the establishment of stations, installations and other equipment in outer space, including on the Moon and other celestial bodies, is possible and recognized in space law (see, implicitly, OST, Art. XII and,

---

6 See IISL Board of Directors Statement, *supra*. The full quote reads as follows: "[s]ince there is no territorial jurisdiction in outer space or on celestial bodies, there can be no private ownership of parts thereof, as this would presuppose the existence of a territorial sovereign competent to confer such titles of ownership". This reading is confirmed by the preparatory works of the OST, in which the Belgian delegate is noted as having "taken note of the interpretation of the term 'non-appropriation' advanced by several delegations apparently without contradiction as covering both the establishment of sovereignty and the creation of titles to property in private law". See COPUOS (1966). Summary Record of the seventy-first meeting of the Legal Subcommittee of the Committee on the Peaceful Uses of Outer Space, A/AC.105/C.2/SR.71 and Add. 1, p. 7. Likewise, the French representative stated that "there was reason to be satisfied that three basic principles were affirmed, [including] the prohibition of any claim to sovereignty or property rights in space". See COPUOS (1966). Summary Record of the fourteen hundred and ninety-second meeting of the First Committee of the General Assembly, A/C.1/SR.1492, p. 429. Finally, see the US Act of 2015 "sense of Congress" provision.

more directly, MA, Art. 8), such factual occupation cannot create a basis for the recognition of any form of sovereignty over the relevant part of outer space by that State. Conversely, since the legal notion of occupation presupposes the existence of a territory to be occupied, this situation, as such, cannot provide a basis for the application of the rules of international humanitarian law in outer space, as territorial sovereignty cannot be exercised over outer space or parts thereof by virtue of Article II of the OST. Finally, the addition of "any other means" at the end of this provision is intended to cover any other historic means of acquiring sovereignty under international law on Earth that would otherwise not be covered, such as discovery, acquisitive prescription or historical consolidation of title. It follows that the mere act of laying one's eyes, naming or stepping foot on a hitherto undiscovered celestial body will not be the basis for any exclusive rights thereon.

5.2.3. *Material scope*

The material scope of application of Article II of the OST extends not only to the entirety of outer space, including the Moon and other celestial bodies, but also to the constituent parts thereof, including areas on celestial bodies as well as their surface and subsurface. For the celestial bodies of our solar system and for those States that have ratified it, this interpretation is explicitly confirmed in Article 11 of the Moon Agreement. The same conclusion is also supported by the application of basic interpretative principles of logic and consistency. First, provisions in international treaties have to be interpreted in a way that gives them effect, even if one is not necessarily obliged to choose the interpretation that is most effective. Since it is physically impossible to appropriate the entirety of outer space, there would be no need to formulate and adopt a non-appropriation principle that only extends to the whole of outer space. Indeed, it is precisely because the individualized activities of single actors could legally result in the appropriation of parts of outer space and the celestial bodies contained therein that there was a need to adopt a principle banning appropriation as a legal measure, which is effective only if it applies to all parts of space covered by these activities of States and their nationals.

Second, the terms of a treaty provision have to be interpreted in light of the internal context of the convention, which includes the preamble as well as all other provisions in the actual text of the treaty. It was noted earlier that the non-appropriation principle as codified in the OST is a corollary to a number of other principles in this convention, most importantly the principle that outer space, including the Moon and other celestial bodies, shall be free to be explored and used by all States. It follows that the term "outer space, including the Moon and other celestial bodies" has to be interpreted consistently in all provisions that use it to describe their material scope. And since the same phrase is used to circumscribe both the non-appropriation provision and the principle of free use of outer space, the freedom to use all parts of outer space, including areas on celestial bodies, including their surface and subsurface, exists only if we accept that these same parts are also not subject to appropriation.

5.2.4. *Space resources*[7]

The above should lead to the conclusion that it is not only unlawful for States Parties to the OST to establish sovereignty or to recognize private property rights over areas in space, but also to appropriate or to allow nationals to appropriate natural resources from celestial bodies, as parts thereof. Nevertheless, it is still contested whether the non-appropriation principle also applies to space resources, or if it is limited to the "non-resource parts" of outer space. Since 2015, various States have adopted national legislation and political instruments that appear to recognize the establishment of property rights over natural resources from asteroids and other celestial bodies based on the conviction that space law only bans appropriation of the celestial bodies on which they are found[8]. However, it is consistently stated in these instruments that the establishment and exercise of such property rights should be done in accordance with international law. International law, however, remains imperfect on this point and appears to allow for different interpretations, although some efforts have been done to clarify the application of the non-appropriation principle to space resources.

The Moon Agreement provides that Article II of the OST applies not only to the surface and the subsurface of celestial bodies in our solar system, but also to "any part thereof or natural resources in place", which hence may not "become property of any State, international intergovernmental or non-governmental organization, national organization or non-governmental entity or of any natural person" (MA, Art. 11, 3). It is sometimes argued that the "in place" qualifier in this provision in

7 This part is only about the appropriation of natural resources in space. For more on the regulation of the exploitation of material space resources, see *infra*, Chapter 14.

8 See United States 2015 Commercial Space Launch Competitiveness Act (Public Law 114–90–Nov. 25, 2015); Luxembourg 2017 Law on the Exploration and Use of Space Resources (A674 Law of 20 July 2017); United Arab Emirates 2019 Law on the Regulation of the Space Sector (Federal Law No. 12 of 19 December 2019); Japan 2021 Act on the Promotion of Business Activities Related to the Exploration and Development of Space Resources (Law No. 83 of June 15, 2021). See also the Artemis Accords concluded as bilateral political agreements supporting the US Artemis Programme between the US and 21 countries + 1 territory since October 13, 2020.

the Moon Agreement *a contrario* allows for the appropriation of natural resources once removed from their place. It is unclear, however, what legal rights could be given to a State or company on the basis of the mere removal of a resource from a celestial body, since the inclusive regime of outer space is entirely founded on the rejection of any permanent rights through singular instances of use. The ITU regime on the use of frequencies and associated orbits in space by satellites clarifies that exclusive rights obtained by a single actor are always based on actual and continued use of these resources. Since the basic legal principles of freedom to use and prohibition to appropriate do not distinguish between outer space and the celestial bodies therein, it is not at all clear how the mere physical removal of a part of a celestial body legally speaking could provide the basis for appropriation, or whether it even amounts to a form of use. Further, it is also unclear whether such an act of removal is sufficient to requalify the excavated part of a celestial body as a natural resource, which for this reason alone would then be subject to appropriation. Such an interpretation would effectively render the non-appropriation principle meaningless as it reduces its scope of application to cover only those parts of outer space that already cannot physically be owned. Finally, if the intention of the drafters of the Moon Agreement through the "in place" qualifier was indeed to allow the appropriation of natural resources once removed, it is rather odd to find that those States that want to authorize their nationals to appropriate celestial resources are among the most ardent opponents of the Agreement for reason that it would allegedly stunt the private space resource sector[9].

As noted, though the Moon Agreement was adopted by consensus, most spacefaring States refused to ratify it. Hence, disagreement over the application of the non-appropriation principle to natural resources in space persists. Since 2016, the issue has therefore been put on the agenda of the Legal Subcommittee of the UN Committee on the Peaceful Uses of Outer Space under the item "General exchange of views on potential legal models for the exploration, exploitation and utilisation of space resources". In 2022, a Working Group on Legal Aspects of Space Resource Activities was established in the Legal Subcommittee, whose 5-year workplan aims to finalize "a set of initial recommended principles" for space resource activities in 2027[10].

9 In this respect, we may note that the other, equally authoritative versions of the Moon Agreement may undercut the interpretation that the phrase "natural resources in place" implies that these resources may be appropriated once extracted. For example, the French version of the Moon Agreement refers to "les ressources naturelles qui s'y trouvent".

10 COPUOS (2022). Report of the Legal Subcommittee on its 61st session, held in Vienna from March 28 to April 8, 2022, A/AC.105/1260, Annex II, Appendix.

5.3. Persistent sovereignty

The fact that no sovereignty or rights of property can be established over outer space, including the Moon and other celestial bodies, does not mean that no element of State sovereignty can be exercised in space, nor that space law rejects all claims of private ownership. The activities of astronauts, personnel of spacecraft and other individuals while in space are guided not only by international regulations but by national legislation as well. The application of such domestic rules in an international area depends on the extension of prescriptive jurisdiction as an element of State sovereignty. In this regard, Article VIII of the OST provides that "[a] State Party to the Treaty on whose registry an object launched into outer space is carried shall retain jurisdiction and control over such object, and over any personnel thereof, while in outer space or on a celestial body". The same provision notes that "[o]wnership of objects launched into outer space, including objects landed or constructed on a celestial body, and of their component parts, is not affected by their presence in outer space or on a celestial body or by their return to the Earth". This provision hence dispels the notion that all ownership rights over a space object would be lost simply because it has been launched into an international area that is itself not subject to appropriation.

A particular application of domestic legislation in the realm outer space on the basis of an extraterritorial extension of prescriptive jurisdiction over objects launched into space is the regulation of intellectual property (IP) rights. Though the scope of their application is territorially circumscribed, IP rights are not affected by the fact that activities resulting in their establishment may occur in outer space. In the absence of national territories in space, the legal regulation of IP matters on board objects in outer space follows the laws of the State that exercises jurisdiction and control over the object. As enshrined in the International Space Station (ISS) Intergovernmental Agreement (IGA), signed on January 29, 1998, "for purposes of intellectual property law, an activity occurring in or on a Space Station flight element shall be deemed to have occurred only in the territory of the Partner State of that element's registry" (ISS IGA, Art. 21, 2)[11].

The nationality of the person involved in an invention on board an object in space does not as such affect the application of intellectual property rights law.

---

11 See further the Agreement relating to the International Telecommunications Satellite Organization "INTELSAT", done on 20 August 1971 (INTELSAT Agreement) and the EUTELSAT agreement. See also national US law on the matter (35 USC § 105, added by the Patents in Space Act 1990 (Pob. L. 101-580, Sec. 1(a), Nov. 15, 1990, 104 Stat. 2863)).

However, according to the ISS IGA, for inventions on board the ISS, the invention by a person who is not a national or resident of the State having jurisdiction over the element in which the invention was made does prevent that State from applying its laws concerning secrecy of inventions (ISS IGA, Art. 21, 3). The continued application of the rules concerning IP matters must however be reconciled with the obligation of States Parties to the OST "to inform the Secretary-General of the United Nations as well as the public and the international scientific community, to the greatest extent feasible and practicable, of the nature, conduct, locations and results of such activities" (OST, Art. XI)[12].

A key issue that may arise in the application of the non-appropriation principle in this regard concerns the integration of natural parts of celestial bodies into artificial objects constructed on these bodies, for example, through 3D printing. The current legal regime primarily focuses on the regulation of objects launched into outer space and landed on celestial bodies from Earth. Some rules currently in force that touch upon issues of integration of artificial objects and natural phenomena in space leave the outcome of any future regime in this regard ambiguous. On the one hand, the Moon Agreement clarifies that "[t]he placement of personnel, space vehicles, equipment, facilities, stations and installations on or below the surface of the Moon, including structures connected with its surface or subsurface, shall not create a right of ownership over the surface or the subsurface of the Moon or any areas thereof" (Moon Agreement, Art. 11, 3). At the same time, it was noted that the OST affirms that ownership of objects constructed on a celestial body is not affected by their presence thereon (OST, Art. VIII). It should thus be left to future legal developments to see how the tension between the non-appropriation principle, which covers all of outer space, including the Moon and other celestial bodies, and the retention of ownership rights over artificial space objects should be resolved in case of hybrid artifacts.

12 The Declaration on International Cooperation in the Exploration and Use of Outer Space for the Benefit and in the Interest of All States, Taking into Particular Account the Needs of Developing Countries of 1996 (Benefits Declaration) appears to indicate in this regard that IP rights will have to be taken into account in the application of information-sharing rules as an implementation of the general obligation of international cooperation: "States are free to determine all aspects of their participation in international cooperation in the exploration and use of outer space on an equitable and mutually acceptable basis. Contractual terms in such cooperative ventures should be fair and reasonable and they should be in full compliance with the legitimate rights and interests of the parties concerned as, for example, with intellectual property rights". (Benefits Declaration, para. 2).

Though certain aspects of sovereignty and private property hence persist *in* outer space, it should be reiterated that their exercise cannot provide a legal basis for the appropriation *of* outer space, including the Moon and other celestial bodies, regardless of the activities of performed by a State through its space objects, including their personnel. This follows from the text of the non-appropriation principle, which bans the national appropriation of space by means of use, in addition to other means such as sovereignty and occupation. The fact that a State retains jurisdiction and control over an object launched into space as well as the personnel thereof cannot provide a basis for so-called quasi-territorial sovereignty over the parts of space where these activities are performed, regardless of the intensity and duration of the activity. Any other finding would not only be in violation of the text of Article II of the OST, but also of the underlying principles of international law concerning jurisdiction that provide the basis for the exercise of limited elements of State sovereignty in space.

As noted, the extension of these aspects of State sovereignty in outer space is based on the exercise of prescriptive jurisdiction by that State, which under international law can only be extended extraterritorially if there is a clear connection between the State and the international situation being regulated. Since the non-appropriation principle clearly rejects any claims of territorial sovereignty, it follows that the application of a State's national rules in space through its activities may not affect, directly or indirectly, the legal status of outer space, including the scope of the non-appropriation principle, as that would also affect the actions of other States relating to the same environment in which their own persons and objects carry out their activities[13]. Such legislative action could not be justified on the basis of any individualised connection with a space object or personnel thereof, over which jurisdiction and control is retained, but could only be taken on the basis of the exercise of territorial sovereignty, which is banned by Article II of the OST. As such, this provision also classifies the realm of outer space, including the Moon and other celestial bodies, as an extra-jurisdictional international area whose legal status should be determined by the international community.

13 See above on the national laws of the United States and other States granting their citizens property rights over resources of asteroids if exercised in accordance with the international obligations of the respective State. What this means in terms of the legality of the part of the provision that allows nationals to exercise property rights remains to be determined. To the extent that the regulation of natural resources would be considered as not covered by current international space law, the absence of any territorial basis for the exercise of national prescriptive jurisdiction pursuant to Article II of the OST requires these regulations to be provided for through an international regime.

5.4. References

Doyle, S.E. (2001). Issues of sovereignty and private property. In *Luft- und Weltraumrecht im 21. Jahrhundert: liber amicorum Karl-Heinz Böckstiegel*, Benkö, M. and Kroll, W. (eds). Carl Heymanns, Cologne.

Freeland, S. and Jakhu, R.S. (2009). Outer space treaty: Article II. In *Cologne Commentary on Space Law. Volume I: Outer Space Treaty*, Hobe, S., Schmidt-Tedd, B., Schrogl, K.-U. (eds). Carl Heymanns, Cologne.

Hobe, S. and De Man, P. (2017). National appropriation of outer space and state jurisdiction to regulate the exploitation, exploration and utilization of space resources. *Zeitschrift für Luft- und Weltraumrecht*, 66(3), 460–475.

IISL Board of Directors (2004). Statement on claims to property rights regarding the Moon and other celestial bodies [Online]. Available at: http://www.iislweb.org/docs/IISL_Outer_Space_Treaty_Statement.pdf [Accessed 15 April 2022].

Lee, R.J. (2004). Article II of the outer space treaty: Prohibition of state sovereignty, private property rights, or both. *Australian Int'l L. J.*, 11, 128–142.

de Man, P. (2016). *Exclusive Use in an Inclusive Environment: The Meaning of the Non-appropriation Principle for Space Resource Exploitation*. Springer, Berlin.

Pershing, A.D. (2019). Interpreting the outer space treaty's non-appropriation principle: Customary international law from 1967 to today. *Yale J. Int'l L.*, 44(1), 149–178.

6

Peaceful Uses of Outer Space

Setsuko AOKI
Keio University Law School, Tokyo, Japan

6.1. Background

Within 2 months of the successful launch of the first satellite, Sputnik I, on October 4, 1957, the first United Nations (UN) General Assembly (GA) resolution referring to the peaceful uses of outer space was adopted. This UNGA resolution 1148 (XII) of 1957, however, did not directly provide that outer space shall be used exclusively for peaceful purposes, but urged States to reach a disarmament agreement that contains the provisions that “the sending of objects through outer space shall be exclusively for peaceful and scientific purposes” so as to avoid the launch of nuclear-tipped ballistic missiles. A missile and rocket were not conceptually distinguished either in 1957 or 2002 when the Hague Code of Conduct against Ballistic Missile Proliferation (HCOC) was adopted. HCOC, 4 (a) (ii)-(iii) specifically refers to space launch vehicle programs, while HCOC aims at preventing the proliferation of ballistic missiles. Space utilization has been closely linked with military use of outer space since the advent of the space age.

The following years, ad hoc and permanent Committee on the Peaceful Uses of Outer Space (COPUOS) were established by UNGA resolution 1348 (XIII) of 1958 and UNGA resolution 1472 (XIV) of 1959, respectively, to study “peaceful uses of outer space” while the scope of “peaceful uses” was not defined in either UNGA resolution. Over the next several years, the meaning of “peaceful” was discussed not necessarily openly in the COPUOS but between the political leaders of the United

Space Law,
coordinated by Thomas LECLERC.

States and the Soviet Union (USSR). Except in its first year of space use, the United States claimed "peaceful" meant "non-aggressive", thus military activities in outer space were legally permissible as far as they were not explicitly prohibited under international law. In contrast, the Soviet position equated "peaceful" with "non-military", and as such, any military activities were not permitted (Vlasic 1991, pp. 38–40). In fact, the USSR proposed the prohibition of "the use of artificial satellites for the collection of intelligence information in the territory of foreign States" in the Legal Subcommittee (LSC) of the COPUOS in 1962[1]. This issue, resolved between the two superpowers by around the mid-1960s as a tacit understanding, was not addressed when the Treaty on Principles Governing the Activities of States in the Exploration and Use of Outer Space, including the Moon and other celestial bodies (OST) was negotiated due, in part, to the dominating role of these two countries to determine the text outside the COPUOS. The interpretation of "peaceful", or the scope of "peaceful uses of outer space" remained vague after the adoption of the OST.

6.2. Article IV of the outer space treaty and unresolved issues

Article IV of the OST reads:

> States Parties to the Treaty undertake not to place in orbit around the Earth any objects carrying nuclear weapons or any other kinds of weapons of mass destruction, install such weapons on celestial bodies, or station such weapons in outer space in any other manner.
>
> The Moon and other celestial bodies shall be used by all States Parties to the Treaty exclusively for peaceful purposes. The establishment of military bases, installations and fortifications, the testing of any type of weapons and the conduct of military maneuvers on celestial bodies shall be forbidden. The use of military personnel for scientific research or for any other peaceful purposes shall not be prohibited. The use of any equipment or facility necessary for peaceful exploration of the Moon and other celestial bodies shall also not be prohibited (OST, Art. IV).

The obligation to use "exclusively for peaceful purposes" is explicitly imposed only on "the Moon and other celestial bodies", not in outer void space. Two questions were mainly asked regarding this point in the space law community at that

1 See COPUOS (1962). USSR: Draft Declaration of the Basic Principles Governing the Activities of States Pertaining to the Exploration and Use of Outer Space. A/AC.105/L.2, para. 8.

time. First, if the obligation to use outer void space "(exclusively) for peaceful purposes" was also implied and second, the meaning of "peaceful" was "non-military" or "non-aggressive". Relating to that second question, also asked was if the adverb "exclusively" added the special meaning and would change the interpretation of the "peaceful purposes". In other words, if "peaceful" is interpreted as "non-aggressive", "exclusively for peaceful purposes" would be construed "non-military" or remains "non-aggressive". These questions are considered below to find the permissible scope of the use of outer void space (6.3) and on the Moon and other celestial bodies (6.4).

6.3. Peaceful uses in outer void space

Concerning the first question, both positions are found. There is an interpretation that "peaceful uses" per se is not a legal obligation in outer void space and only the concrete actions specified in the first paragraph of Article IV of the OST are prohibited. Thus, it is only weapons of mass destruction (WMDs) that are banned if it is placed in orbit around the Earth or otherwise stationed in outer void space. It seems noteworthy to state that it is not the OST but the Treaty Banning Nuclear Weapon Tests in the Atmosphere, in Outer Space and Under Water, generally called the Partial Test Ban Treaty (PTBT), adopted in 1963 which obliges States Parties "not to carry out any nuclear weapon test explosion, or any other nuclear explosion" in outer space (PTBT, Art. I, 1, a). From the text, conventional weapons and military satellites (in some States, a military satellite is regarded as a conventional weapon) placed in orbit around the Earth or otherwise stationed in outer void space are not prohibited. Nor is an intercontinental ballistic missile (ICBM) prohibited as it only navigates some parts of outer void space before entering the Earth surface again.

The other interpretation is that there is an obligation to use outer void space for "peaceful purposes" based on other provisions of the OST including especially Article I (the obligation to explore and use outer space for the benefit and in the interests of all countries) and Article III (observance of international law). If "peaceful" is interpreted "non-military", the scope of the lawful use of outer void space would be different from the position interpreting that "peaceful use" is not an obligation, for any military use is not allowed in this case including military satellites and ICBMs.

Some commentators claimed in 1970s that non-military use of outer void space was a legal obligation, but the weakness exists in this interpretation in that this contradicts normal reading of the text of Article IV, paragraph 1 of the OST and state practice (Markoff 1976, pp. 11–12; Zhukov 1969, pp. 36–38). The obligation of non-military uses of outer void space never became a prevalent interpretation.

In the case where "peaceful" is understood as "non-aggressive", whether or not "peaceful use" is the legal obligation, the scope of the permissible use of outer void space is the same. In either case, except the prohibition of a WMD stationed in outer void space, the permissible scope of the military use is within the limitation of the UN Charter including, inter alia, Article 2, paragraph 4 (prohibition of the threat or use of force) and Article 51 (the inherent right of individual and collective self-defence). The interpretation that "peaceful" means "non-aggressive" is in line with the subsequent state practice as evidenced by the ever-increased use of military satellites (including dual-use satellites) and the fact that a Chinese Anti-satellite (ASAT) test was not condemned as the illegal action under the OST[2].

In conclusion, in outer void space, the permissible scope of the exploration and use includes military activities except in so far as they are explicitly forbidden by the OST and other rules under international law including the UN Charter.

6.4. Peaceful uses on the moon and other celestial bodies

From the interpretation above, it seems that the appropriate interpretation of "peaceful" in the OST is "non-aggressive" rather than "non-military". Therefore, there are views that military activities are possible on the Moon and other celestial bodies as long as such activities are not explicitly prohibited in the text (Bridge 1980, p. 659). However, due to the wide variety of concrete actions expressly prohibited in paragraph 2, even if these are not exemplified lists and only definitive lists[3], substantially only "non-military" use would be permissible on the Moon and other celestial bodies.

In addition, some commentators state that the dichotomy between "non-military" and "non-aggressive" may not be a useful criterion to define the lawful scope of the activities in outer space including the Moon and other celestial bodies, and the scope of the peaceful use should be determined pursuant to the text and context of each paragraph of Article IV (Christol 1982, pp. 28–29). This would in particular apply to paragraph 2 of Article IV, for the adverb "exclusively" implies special meaning (Hobe et al. 2009, pp. 82–83), which should be read in the context of this paragraph where a wide range of actions are considered comparable to the Antarctic Treaty,

---

2 For instance, in the 2007 COPUOS/LSC, no statement was made on the Chinese ASAT test conducted on January 11, 2007, and only Canada, Japan, UK and the United States expressed their concerns on its test in the 2007 COPUOS main committee.

3 Article IV (2) of the OST does not contain the phrase to indicate that specifically prohibited actions are only examples while it is found in the Antarctic Treaty as "inter alia, any measures of a military nature" (Antarctic Treaty, Art. I, 1).

signed on December 1, 1959 (Antarctic Treaty), which provides non-military use thereof (Antarctic Treaty, Art. I).

In conclusion, "peaceful use" on the Moon and other celestial bodies are activities not involving military use.

6.5. References

Bridge, R.L. (1980). International law and military activities in outer space. *Akron Law Review*, 13(4), 649–664.

Christol, C.Q. (1982). *The Modern International Law of Outer Space*. Pergamon Press, Elmsford.

Hobe, S., Schmidt-Tedd, B., Schrogl, K.-U. (eds) (2009). *Cologne Commentary on Space Law: In Three Volumes*, Vol. 1. Carl Heymanns Verlag, Cologne.

Markoff, M.G. (1976). Disarmament and "peaceful purposes" provisions in the 1967 outer space treaty. *Journal of Space Law*, 4, 3–22.

Vlasic, I.A. (1991). The legal aspects of peaceful and non-peaceful uses of outer space. In *Peaceful and Non-Peaceful Uses of Space: Problems of Definition for the Prevention of an Arms Race*, Bhupendra, J. (ed.). Taylor & Francis, New York.

Zhukov, G.P. (1969). On the question of interpretation of the term "peaceful use of outer space" contained in the space treaty. In *Proceedings of the 11th Colloquium on the Law of Outer Space*, Schwartz, M.D. (ed). University of California School of Law.

7

State Responsibility and Liability for Space Activities

Armel KERREST
Université de Bretagne Occidentale, Brest, France

7.1. Introduction

A terminological issue arises at the outset because of the existence of two different terms in American English, both of which are translated by one term in the other official languages of the United Nations (UN). This distinction is also noticeable in the organization of the text of the Treaty on Principles Governing the Activities of States in the Exploration and Use of Outer Space, including the Moon and other celestial bodies, opened for signature on January 27, 1967 (1967 Treaty), since the first term, *responsibility*, is the subject of Article VI, whereas the second, *liability*, is used in Article VII. This distinction is also found in the Convention on International Liability for Damage Caused by Space Objects, opened for signature on March 29, 1972 (1972 Convention). In the other official languages of the UN, only one word is used.

From the beginning, English has been the dominant language in the development of space law texts. It must be admitted here that the distinction has a certain relevance. The term "responsibility" in fact covers two complementary but distinct notions. The term responsibility means "having to answer for" an activity or its consequences, having to control it. The term liability indicates an obligation to compensate possible victims. In the United Nations Convention on the Law of the Sea, opened for signature on December 10, 1982 (UNCLOS), the problem has been solved in a different and probably more satisfactory way. The term "responsibility"

Space Law,
coordinated by Thomas LECLERC.

in the English version has in fact been translated into French as *obligation de contrôle*, whereas the term "liability" is translated as *responsabilité* (Kerrest 1997).

It should also be noted, by way of introduction, that Article VI of the 1967 Treaty refers to "national activities", a notion that deliberately encompasses the activities of States themselves through their national space agencies and the activities of private entities that are their nationals. States must ensure that all comply with the treaty and more generally with international law, including the UN Charter. A violation of these obligations could lead to international liability under international law, through the procedures provided for by that law, but that is not the main purpose of this provision. The liability of the launching State applies, on the other hand, to States that meet one of the four criteria that determine it and that are not necessarily linked to "national activity in space" (1967 Treaty, Art. VI). In this case, it is clearly a question of ensuring compensation for victims; absolute liability may apply, which is not the case for the responsibility of Article VI.

It is important to emphasize this distinction because Articles VI and VII deal with different issues, give rise to different solutions involving potentially different States and apply at different times. In principle, States that lead or control the conduct of activities in outer space are at the same time launching States for the objects launched or used. But this is not always the case. The situation is different in time; the State's launching quality is acquired at the moment of the launch and remains until the return of the object. The "national activity" derives from a reality that is established in time, it is not tied to a specific event like the notion of "launching States", it exists over time. If the space object is transferred while in orbit, the State of national activity, the one that exercises control over the object, is not necessarily one of the launching States. This situation can lead to difficulties particularly because Article 1 of the Convention on Registration of Objects Launched into Outer Space, adopted by the United Nations General Assembly (UNGA) on November 12, 1974 (1974 Convention) links the registration of a space object with the notion of a launching State and, according to Article VIII of the Treaty, registration determines the State that exercises jurisdiction and control over the space object (Kerrest 2017).

7.2. The responsibility of States for their national activities in outer space

The negotiation of the legal framework for space activities during the Cold War pitted the advocates of a regulated economy centered on the role of the State against the defenders of a market economy leaving a great deal of room for private initiatives. In all cases, the strategic importance of space and its uses led to the need to recognize an essential role for the State.

At the time of the drafting of the first text organizing the legal framework of the space activities, the USSR refused any private activity in space; it is considered that they had to be led only and exclusively by States[1]. This position was unacceptable for the United States[2].

At these times, private activities in space were extremely rare. A compromise was necessary. It was the subject of item 5 of the Declaration of Legal Principles Governing the Activities of States in the Exploration and Use of Outer Space (UNGA, Resolution 1962 (XVIII) of 1963) and of Article VI of the 1967 Treaty. States would be responsible for "national activities" in outer space. The notion of "national activities" included the activities of governmental bodies but also of "non-governmental entities". As is most often the case in the drafting of compromises, each party can find some of its requirements. For the Soviets, there was the notion of "national activity" and the responsibility of States. For the Americans, through the notion of "non-governmental entities", there was the possibility of allowing the activities of private persons.

The obligation of States is important since they must authorize and control in a continuous manner all private activities in space when they are conducted by their nationals. These rules apply both in space itself and on celestial bodies. States may authorize private companies to conduct activities in space provided that they have the right to do so themselves[3]. These activities are subject to Article VI of the Treaty because they are "national activities" for which the state is responsible and which must be subject to "authorization and continuing supervision".

Bases may be established on celestial bodies. They can be public or private, but in any case they must meet the obligations of Article VI of the Treaty, they must be authorized and under the continuous supervision of the appropriate State. They engage the responsibility of that State.

7.3. Liability for damage caused by space objects

Since the creation of the Committee on the Peaceful Uses of Outer Space (COPUOS), the question of liability for damage has been a preoccupation of the jurists responsible for developing a legal framework for space activities. It can be

1 COPUOS (1962), A/AC.105/C.2/L.1, para. 7.

2 The private satellite *Telstar* 1 was launched for AT&T on July 10, 1962.

3 In application of the principle *nemo plus juris transfere podest quem juste habet*, a State cannot authorize a private person to do what it cannot do itself. See, *infra*, the question of the appropriation of celestial bodies or their resources.

hypothesized that the very protective and well-organized liability provided for in Article VII of the 1967 Treaty and in the 1972 Convention for damage to victims on the ground has been seen as a counterpart to the freedom of use of outer space.

It is essential at the outset to make a distinction, which was made necessary by the drafting of the 1972 Convention. The first two texts of 1963 and 1967 made no distinction according to the place where the damage takes place. This was not the case with the 1972 Convention. Initially, the drafters took over the provisions of the 1963 and 1967 texts. The liability thus established did not distinguish according to the place of the damage. In all cases, it was an objective liability; the text even referring to "absolute" liability of the launching State, without limit and without any exemption clause. This liability being joint and several if there was more than one launching State. During the discussions, this rule appeared excessive and potentially too generous for the victims of damage caused in orbit to other participants in the space adventure. It was decided to make liability for damage caused in orbit, and thus to other space States, subject to demonstration of fault. The idea was undoubtedly perfectly acceptable but the whole text of the Convention would have had to be revised to take account of this modification and clearly distinguish the legal framework of the two situations. The legal regime of a strict liability system is profoundly different from that for fault-based liability. In the case of liability for fault, the attribution should have been modified, as should the joint and several nature of the liability of the launching States. These provisions are directly linked to the objective character of liability and are not applicable in the case of fault-based liability, where it is the fault that determines the attribution in application of the principle *res ipsa loquitur*.

In the text of the Convention, the situation is not explained properly. Responsibility lies with the launching State which has committed a fault, but what if it is the fault of a third State? In this case, the Convention cannot be applied; ordinary law must be applied, since no launching State is at fault and therefore liable under the Convention. Similarly, how can the joint and several nature of the liability between the launching States be applied without taking into account the possible fault of one of them?

It is therefore appropriate to make the place of the damage the *summa divisio* of this question. Liability for damage caused on Earth will first be studied (section 7.4) before turning to liability for damage in orbit (section 7.5). A final section will be devoted to the question of the mode of settlement of disputes which applies in both cases, a study which will be completed by an examination of the particular case of the liability of launching States for private activities (section 7.6).

7.4. Liability for damage caused on land or to aircraft in flight

7.4.1. *Who is liable?*

The objective is to ensure that, in all cases, an obligation to compensate will allow potential victims to obtain said compensation.

Since the 1963 text, the "launching State" has become the most relevant liable party for damage on the ground or to an aircraft in flight, that is, for damage to persons and third States in relation to the activity. The choice is good. It is first of all a State and not a private operator. It is true that at the time, there was no activity conducted in space by private persons. But the choice remains relevant. First of all, because of the delays, because an object in orbit can stay there for a very long time if the orbit is high enough, but also because of the risk of bankruptcy of the companies that conduct these activities. Moreover, it is important that it is easy to determine who is liable in an environment that is very difficult to access.

Four criteria were adopted as early as 1963: the launching State, the State that causes the launch, the State whose facilities are used, the State whose territory is used (UNGA, Resolution 1962 (XVIII) of 1963). These criteria are alternative; it is sufficient for one of them to be met for the State concerned to be a liable launching State. This possible plurality is very favorable to the victims because all are jointly and severally liable, which means that the victim's State can claim compensation from any of the launching States.

This designation of the launching State as liable is the consequence of the need to designate a liable person in advance and this independently of the case in question. The liability provided for is in fact an objective liability without the need to prove fault. In this case, the usual application of the principle *res ipsa loquitur*, which gives the judge the task of determining who is liable by examining the faults that may have been committed, cannot be applied since no fault is taken into account in matters of objective liability. Since only the causal link has been established, it follows naturally, by virtue of the rule, that the State objectively qualified as the launching State is liable. It is sufficient for a State to meet one of these criteria for it to be considered a liable launching State. Even in the case of a launch from the high seas[4] , the launching State remains the State of nationality of a launching enterprise, the State of which an enterprise causes a launch, or the State where the launch facilities are located.

4 See the launches made by the American company *Sealaunch* from the high seas.

In the 2000s, some jurists, particularly in the Netherlands and Belgium, believe to have to have the possibility to consider that in the case of a private company, the liability of the launching State could be avoided. This was to overlook the consequences of Article VI of the Treaty, which integrates and assimilates private activities into "national activities" under the responsibility of the State. Fortunately, the constant practice of States on this point and, above all, the awareness of the destructive effect that such an interpretation would have had on the mechanism itself – which could go so far as to call it into question completely, for example in the case of a launch from the high seas – led to the abandonment of this interpretation (Kerrest 2005). The fact that no damage was caused by private entities in the context of the application of the 1972 Convention facilitated the rejection of this interpretation, the application of which would have emptied the Convention of its content and placed all liability on the State of the territory, a necessarily State criterion.

7.4.2. *What liability?*

The liability provided for the launching State is a strict liability, that is, without fault. The text of the convention even indicates "absolute liability", which no doubt refers to the fact that it leaves practically no room for possible exoneration and is not subject to any ceiling.

7.4.2.1. *Objective liability*

In the event of damage caused by a space object on earth, the victim's State does not have to prove any fault on the part of the launching State. It is sufficient to prove the status of launching State and the link with the launched object. This is an essential point of the mechanism set up to protect persons not involved in space operations in the event of damage caused by a space object. It would indeed be extremely difficult for the victim to prove fault when the activity takes place in space and is most often conducted in a very confidential manner.

7.4.2.2. *Unlimited liability*

In a rather classical way, when one imposes a strict liability on operators, one also accepts to put a ceiling on this liability in order to allow them to take out insurance. This is, for example, the case in the International Convention on Civil Liability for Oil Pollution Damage, concluded on November 29, 1969 (1969 Brussels Convention). A ceiling was put in place to allow the shipowner who is objectively liable to take out insurance. In this case, this provision was not necessary because it concerns the liability of States, which do not usually take out insurance but have sufficient financial resources to insure themselves. There is no ceiling on the amount of compensation to be paid to the victims, neither in terms of amount nor in terms of time. This provision is important insofar as a space object can remain in orbit for a very long time and thus present a risk over a very long period.

7.4.2.3. *Very limited exemptions*

There is very little that the Launching State can exonerate itself from. Its liability is not limited by force majeure or even by the fault of a third party. It is only and possibly limited by the fault of the victim himself/herself. But, here again, this fault must be "a gross negligence or an intentional fault".

7.4.2.4. *Protection limited to third parties to the activity*

This is the consequence of the rationality of the mechanism put in place. It is intended to protect persons who do not participate in the space activity and do not benefit directly from it. Thus, nationals of the launching States and foreign personnel having taken part in the launch are excluded by Article VIII. This does not mean, of course, that in this case these persons will not be compensated, but simply that they will not be compensated in application of the convention. This is why, when the French law on space activities was drafted in 2008, it was planned to provide special protection for French nationals in order to put them in the same situation as that applicable to foreigners and not to be satisfied with allowing the ordinary law of State liability to apply under domestic administrative law.

7.4.2.5. *The distribution of the compensation obligation among the launching States*

As noted above, the victim's State has a choice as to which launching State it wants to claim compensation from. Naturally, this choice will be dictated by various factors, in particular the relationship between the two States and the chances of obtaining significant compensation, but also the ease of proving the status of launching State outside the State of the launching territory, which is most often very well known[5].

Because of the joint and several nature of the liability, the victim's State may claim the full amount of compensation from the launching State which it decides to claim. Article V, paragraph 2, of the 1972 Convention nevertheless provides for the distribution of the burden of compensation between the launching States. The situation is not simple, since it is not possible to refer to the fault of one or the other.

5 Because the Kourou base is located in French Guiana, France is the launch State for many commercial satellites launched on Ariane. In the event of an accident, it risks being implicated, even though it has very few possibilities to control the satellite's activity.

The text merely urges the launching States to enter into agreements among themselves in order to share the burden of compensation. The problem arises in a significant way because, in the case of private activities, the fact that a State may be considered as a launching State may lead it to want to exercise control over the company and the activity it is conducting. Multiple potential launching States risks leading to a plurality of controls that may complicate the activities of these companies. The solution to avoid this redundancy of control would be to use the route opened up by Article V of the 1972 Convention. One could imagine that the obligation to pay compensation could at least be distributed according to the stage of the activity. Article V rightly specifies that this distribution cannot have any consequence on the rights of the victim States, which retain the possibility of taking action against one or other of the launching States. Such an agreement can only have consequences with regard to the distribution, *in fine*, of the financial burden of compensation to the victim. This question was raised during the drafting of the French law. In order to avoid cumbersome procedures, it was not planned to consider these agreements as a prerequisite to the authorization of launching or conducting space operations. Unfortunately, very few agreements have been put in place in this framework. The fact that the convention did not have to be applied in a significant way undoubtedly diminished their interest.

7.4.2.6. *Consequences for third parties of an accident between several space objects*

Article IV of the 1972 Convention refers to the case of damage that could be caused to third parties by an accident involving two space objects. The situation is somewhat complex, but it is possible that, given the increase in space debris, it will become more frequent. The two initial States are, as provided in Article V, jointly and severally liable to the third State. If the damage occurred on the surface of the Earth or to an aircraft in flight, this liability is objective. If not, it must be proved that there was fault.

In both cases, the distribution of the burden of compensation between the responsible launching States will take into account the possible fault of one or the other. If it is impossible to prove this fault, and we know that this is a major difficulty, the distribution will be made "equally"[6]. This simplifying solution could prove unfair in the case of the use of nuclear energy sources by one or other of the launching States, unless their use is in itself wrongful, which is not currently the positive law. A special provision might be useful to take account of this hypothesis.

6 It is understandable that in this situation, it is not possible to provide for an agreement made in advance, as is the case in Article V of the 1967 Treaty, because the two launching States do not know that such an accident will occur.

7.4.2.7. *The nature of the liability*

"A launching State shall be absolutely liable to pay compensation for damage caused by its space object on the surface of the Earth or to aircraft in flight" (1967 Treaty, Art. II). These provisions require a clarification of the terms used.

What is "object" in the first place? The rules of interpretation laid down in Article 33 of the Vienna Convention on the Law of Treaties, concluded on May 23, 1969 (VCLT), provide that in the case of texts adopted in several languages, the meaning that best reconciles these texts must be retained (VCLT, Art. 33, 2). One can note here that four of the six official languages use the word "object" that etymologically means "thing placed in front of" which seems to indicate a material nature of the object. Can we consider, for example, a laser beam as an object? Physicists would certainly do so. In the common language used by the judge, it is less certain.

It is also necessary for the object to be "spatial", which leads to the well-known problem of the definition, or even the delimitation, of outer space. Confusion on this point has led to complicating an already sensitive situation and this for obvious strategic reasons. First of all, a clear distinction must be made between the definition of outer space and the question of applicable law. Contrary to what is sometimes maintained, one cannot define space by a functional criterion, but one can naturally determine the applicable law by such a criterion. As far as the definition of space is concerned, it is essentially a technical question that can be solved by engineers, because the space environment is profoundly different from the air environment. The conditions of circulation, the very nature of the physical laws which apply to it, make two different environments. It remains that with regard to the delimitation, the question is less simple because there is an intermediate zone which it is in principle only possible to cross but where one cannot normally remain. It is however possible for the international legislator to determine the scope of application of treaties and space law. It does so in the 1972 Convention, for understandable reasons of coherence, extending its application to areas and activities that are not space related. Thus, the first stage of a rocket is considered a space object even though it will never go into space, and the same applies to the entire rocket in the case of an aborted launch, the "attempted launch" being considered a launch.

For the time being, some States, in particular the United States and France, do not want a precise delimitation to be decided by COPUOS and the UNGA. This position is more strategic than technical. The stakes are high, since airspace is subject to the sovereignty of the State overflown, whereas outer space is free. A State that has powerful remote sensing capabilities has an interest in seeing that outer space descends as low as possible in order to be able to exercise its freedom of observation more widely[7].

The space object must also be distinguished from meteorites which, although dangerous, are obviously not concerned by the rules of the Convention.

In the case of the so-called suborbital flights, the part of their flight in space is extremely short and marginal and one can wonder about their real space nature[8]. The reason that leads to recognizing the character of space flight is the commercial necessity and the possible qualification of "astronaut". The consequence of this qualification is, in spite of everything, the application to these flights of the 1972 Convention.

7.4.2.8. *The causal link between the space object and damage*

In the formula "caused by" the space object, it is necessary to prove this link. One could wonder about its nature. For example, should the malfunctioning of the satellite be included in this causal relationship? The problem arose in the case of positioning satellites, in particular for Galileo. An extensive interpretation of the notion of "caused by" could have led to the launching State being held responsible for a satellite that had stopped transmitting or had transmitted a faulty signal. This interpretation has been set aside in favor of a more classical conception. It is the material impact of the satellite that causes the damage, not its malfunction. A reason for this can be found in the rationality of the rule of the Convention. It is fundamentally aimed at protecting victims on the ground who are not involved in the activity. In the case of the use of the satellite signal, one can consider that the user, even if they are located on the ground or on an aircraft in flight, is not really a third party but rather a user of the service. There is therefore a certain logic in the fact that they cannot benefit from the objective liability provided for by the Convention, and that they must therefore rely on contractual liability.

7.4.2.9. *The liability procedure*

Contrary to what was seen in the case of damage caused by oil spills at sea under the 1969 Brussels Convention, there is no "channelling" of liability onto the launching States alone. In the case of the 1969 Brussels Convention, the actors in the

7 In the case of satellites in two parts, we can consider bringing one of their parts below the limit of satelliteization. This is the case of "tether satellites" which could be composed of a part in orbit and a smaller part like a camera much lower.

8 The issue of space decay was raised during the so-called "suborbital" flights of *Virgin Galactic* and Jeff Besos. In this case, the qualification of "space flight" has an obvious financial interest.

maritime field obtained, in exchange for their acceptance of the objective nature of their liability, a channeling of this liability onto the shipowner alone. This choice certainly had the advantage of encouraging these owners to maintain their ships well, but it also allowed other operators, in particular the charterer or the owner of the cargo, to escape any liability. This is not the case in the 1972 Convention where Article XI allows the State of the victim or the victim himself to initiate proceedings before the judicial or administrative authorities of the launching State (1972 Convention, Art. XI, 2). The victim may also bring an action against the operator before the courts of another State, since the Convention does not limit this right. The usual problems of recognition of the court's decisions by the operator's legal system would then arise. To take an example, in the event of damage caused by a French private operator to an American victim, this victim could seek compensation from the American court in application of the classic rules of private international law. This would be an action conducted outside the 1972 Convention. The advantage of such a procedure, for the victim, would be that they would not have to have recourse to the good offices of their State of nationality, with all the risks of politicization of the dispute that such an action entails. The 1-year time limit set in Article X would nevertheless undoubtedly prevent the victim from trying to resort first to the ordinary law procedures before following the Convention procedure.

It can be said that the 1972 Convention is not necessarily intended to be used in all cases. Recourse by the victim himself/herself before a national judge in application of local law would probably be the most effective procedure. The 1972 Convention thus plays the role of a safety net that can intervene in case of difficulties, but it is not intended to be used in all cases.

7.4.3. *What damage?*

The 1972 Convention does not specify what damage is involved. It must be concluded that it is all damage. Some people wanted to exclude environmental damage. This is not the case, and there is no such exclusion. On the other hand, it is true, as in other fields, that in the case of damage caused to the environment of international environments, it may be difficult to engage the liability of the author of the damage. The reason for this is not a possible exclusion of environmental damage, but is due to the difficulty of recognizing a victim who would have legal personality and who could therefore act to claim compensation.

It has also been possible to consider excluding damages considered to be "indirect". However, the Convention does not make any distinction between direct and indirect damage. On the contrary, the principle of *restitutio in integrum* clearly shows that the purpose of compensation is "to restore the person, natural or judicial, State or international organization on whose behalf the claim is presented to the condition which would have existed if the damage had not occurred" (1972 Convention, Art. XII), which would be contrary to a distinction between direct and indirect damage.

7.5. Liability for damage suffered elsewhere, not on the surface of the Earth

As we have seen, the amendment of the draft convention to include a distinction between damage caused on Earth and damage caused elsewhere has profoundly changed the mechanism for establishing the liability of launching States for damage caused by a space object. The introduction of the need to prove "fault of the launching State" complicates it considerably.

Where does this rule apply? Article III of the 1972 Convention uses the phrase "elsewhere than on the surface of the Earth". It refers to space objects of other States or to persons on board such space objects. This limitation, which stems from the state of technology in the 1970s, raises the question of damage caused by a space object on a celestial body. It seems that lunar spacecraft or bases on the Moon could be considered as space objects and that, consequently, the convention could apply.

7.5.1. *Proof of fault?*

Proof of fault would certainly be difficult, even if the victim is an operator or a space state. This is first of all the case because of the often confidential nature of the activities. But it is also due to a legal difficulty. Without the existence of fairly precise rules of behavior in space, it will be difficult to qualify as wrongful such or such behavior of the launching State. Reference to codes of conduct would be useful, but on condition that these rules are sufficiently precise and practical to apply. In the absence of consensus and political will within COPUOS, it is very difficult to envisage the establishment of sufficiently precise rules (IADC 2020)[9] that could enable a judge to qualify conduct as "wrongful" within the meaning of the 1972 Convention and thus likely to engage the liability of a launching State.

9 See, also, COPUOS (2009). Space Debris guidelines of the Scientific and Technical Subcommittee, A/AC.105/890.

7.5.2. *Whose fault is it?*

As we have seen, because of the incomplete wording of the 1972 Convention for damage caused in orbit, liability is borne by the launching State and not by the State at fault on the basis of the principle of *res ipsa loquitur*, which can complicate the procedure.

Moreover, solidarity between launching States cannot apply unless each of them is at fault, which is quite unlikely. The victim's State will have to act against the launching State at fault and not against any of the launching States. If it brings an action against a non-defaulting launching State, its claim will be unsuccessful because of the lack of proof of fault on the part of that launching State or the persons for whom it is responsible, which is required by Article III of the Convention. Similarly, only an action against one of the launching States can be brought under the Convention. If the offending State is not one of the launching States, a legal basis other than the Convention will have to be used. This would be the case, for example, in the event of the sale in orbit of the satellite to a State that is not the launching State of the object.

The text of Article III refers to the fault of the initiating State or "persons for whom it is responsible". Who are these persons? In the absence of precision, it can be considered that they are the "governmental organizations" or "non-governmental entities" referred to in Article VI of the 1967 Treaty.

7.6. Questioning the liability of the launching State

The liability established by the 1972 Convention is a liability between States. It is the liability of the launching State called into question by the victim State or the victim's State. The victim must therefore obtain from the State of which he or she is a national that it take up his or her case. Here, we find the well-known procedure in international public law of diplomatic protection.

Being aware, however, of the difficulties encountered by this procedure in international law, the drafters of the Convention tried to make it more effective. In order to enable a victim to obtain compensation when the State of which he or she is a national does not wish to make a claim, the State in whose territory the injury occurred may intervene, even in the case of foreign nationals. If neither the State of nationality of the victim nor the State in whose territory the injury occurred makes a claim, the State of residence of the victim may do so. One can see in these provisions the desire of the drafters of the text to avoid a victim on the ground being unable to obtain satisfaction because of a blockage by the State of which he or she is a national. It is well known that in general international law, the implementation of State responsibility/liability in the context of diplomatic protection is often made difficult by this limitation.

If there are no diplomatic relations between the victim State and the initiating State, the victim State may request a third State to submit its application and represent its interests. If both States are members of the UN, the victim State may request the Secretary General to submit the application. Significant efforts are also made here to prevent the victim from obtaining compensation. The fact remains that, basically, this is a State-to-State procedure, with all the red tape and risks of political blockages that this entails.

Article XI of the Convention allows the victim State to act without having exhausted local remedies, as is in principle the rule under the general international law of diplomatic protection. This is a permission, not an obligation. The victim State will nevertheless have to choose, since it is not possible for it to claim compensation on the basis of the Convention for an injury which has already been the subject of a claim before the judicial or administrative bodies of a launching State (1972 Convention, Art. XI, 2).

In the case of failure of the diplomatic negotiation, it is provided for the establishment of a claims settlement commission, an arbitration commission composed of three members appointed according to the usual procedure at the rate of one member for each party and of a president designated by the two appointed members. The establishment of this commission is an obligation and a right for the parties. If the two appointed members fail to reach agreement on the appointment of a chairperson within 4 months, either party may request the UN Secretary General to appoint the chairperson. The Commission shall determine its procedure and location. It decides on the merits of the claim and determines the amount of compensation to be paid.

The question of the legal value of the Commission's decision was the subject of much discussion during the drafting of the 1972 Convention. Western countries were in favor of making the decision binding, while the Soviet bloc countries were opposed. This point blocked the negotiations for some time. Finally, a compromise was reached. It was foreseen that the decision of the commission would be obligatory for the States accepting this obligation and would constitute a "final and recommendatory award, which the parties shall consider in good faith" for the others. The solution adopted is rather disappointing.

Since the decision is final, reasoned and public, one can nevertheless hope that a launching State condemned by the commission to pay compensation would be pushed to do so for reasons of national prestige. It should be noted here again that this has never happened. In the only case of damage caused by a space object on the territory of a third State[10], the launching State, the USSR, agreed to pay compensation. Even now, Russia, which is the successor to the USSR on this point, nevertheless maintains, against Canada's interpretation, that the convention did not apply in this case because the USSR agreed to pay *ex gratia* and not because of a decision by a commission that was never set up.

7.7. Launching State liability and private activities

The launching State is liable for the activities of private persons when one of the four criteria is met. As in the case of Arianespace, it may be a launch by a private company of the nationality of the launching State, a private company which launches, private installations or a launch from the territory of a state by a private company. All of these cases have been taken into account when establishing national space laws. Contrary to the obligations provided for in Article VI of the 1967 Treaty, there is no obligation for launching States to control private activities. However, the States, through their national laws, have in general provided themselves with the means to control these activities in order to avoid finding themselves liable as launching States. In general, the States have also provided for the possibility of obtaining reimbursement of the sums that they would have had to pay as a result. They also generally provide for an insurance obligation.

The fact that the 1972 Convention excludes any ceiling in amount and time makes the activities of private persons in space perilous. It is true that it places the responsibility on the launching State. However, if the victim prefers to seek compensation from the private operator, or if the launching State intends to obtain reimbursement from the company, the conduct of private activities in space may become very risky and constitute a kind of "Russian roulette".

It should also be remembered that the victim may decide to claim compensation from the private operator before a national judge or to resort to arbitration[11] . In this case, national law may or may not provide for a liability ceiling. Belgian law, in order to support the activity of their companies, provides, for example, the possibility for the King (the executive) not to demand the full reimbursement.

---

10 The *Cosmos 954* satellite, a Soviet spy satellite, carried a nuclear power source. It crashed over northern Canada on January 24, 1978.

11 The Permanent Court of Arbitration has proposed and had adopted by its members optional rules for the arbitration of disputes relating to space activities.

A slightly different and much more favorable mechanism was put in place by the *Commercial Space Launch* of 1984 and the French law of 2008. These two laws establish a liability cap for their companies, which allows them to purchase insurance, which is mandatory in most cases. In the US law, the amount of the cap is set by NASA for the FAA according to the type of launch and its risk; it is the maximum probable loss of $500 million[12]. In France, the amount is set by the 2008 law. The amount retained is 60 million euros. For each launch, Arianespace is insured for 60 million euros. If the amount of the damage taken as a whole is higher than this sum, it is the French State which will pay the difference. This mechanism is interesting because it allows for an insurance mechanism to be put in place without prejudice to the rights of potential victims to obtain compensation without a ceiling and for companies to conduct potentially dangerous activities without risking considerable compensation. The amount of the ceiling covers all the damage that could be caused, not only that covered by the convention but also the damage that could be caused to the launch facilities and on the territory of the launching State.

For the United States, the guarantee thus given by the State is limited to the launch phase stricto sensu. It does not concern the life phase in orbit of the space object. In the case of France, a distinction is made between the launch phase, for which the guarantee applies fully for launches from French territory and for all damage, and the life-in-orbit phase. For this second phase, the State guarantee applies only for damage on earth[13].

For companies, the mechanism put in place by the United States and France has the significant advantage that the cap applies both in the event of recourse by the victim's State on the basis of the 1972 Convention and if the victim decides to take action before national judges. These two laws thus give effectiveness to the cap that they guarantee, while not undermining the rights of potential victims.

7.8. References

Cheng, B. (1979). International liability for damage caused by space object. In *Manuel of Space Law*, Jasentuliyana, N. and Lee, R.S.K. (eds). Dobbs Ferry and Alphen aan den Rijn, Oceana Publications and Sijthoff & Noordhoff, New York.

Goh, G.M. (2007). *Dispute Settlement in International Space Law*. Brill Academic Publishers, Leiden.

12 The amount of this cap changes with the dollar.

13 This limitation has been criticized by the company Eutelsat because there is no risk of damage on earth for geostationary satellites. It is however essential because of the important risk caused by space debris.

Hobe, S. (2012). The permanent court of arbitration adopts optional rules for arbitration of disputes relating to outer space activities. *Zeitschrift für Luft und Weltraumrecht*, 61, 4–25.

IADC (2020). Space debris mitigation guidelines [Online]. Available at: https://orbitaldebris.jsc.nasa.gov/library/iadc-space-debris-guidelines-revision-2.pdf [Accessed 20 May 2022].

Kerrest, A. (1997). Responsibility and liability of States for damages other than those directly caused by the spacecraft. In *Proceedings of the Colloquium of the International Institute of Space Law*. American Institute of Aeronautics and Astronomics, Reston.

Kerrest, A. (2005). Liability for damage caused by space activities. In *Space Law Current Problems and Perspective for Future Regulation*, Benkö, M. and Schrogl, K.-U. (eds). Eleven, Utrecht.

Kerrest, A. (2009). Actualité du droit de l'espace : la responsabilité des États du fait de la destruction de satellites dans l'espace. *Annuaire français de droit international*, 55, 615–626.

Kerrest, A. (2017). Legal aspects of transfer of ownership and transfer of activities. In *4th Luxembourg Workshop on Space and Telecommunication Law: Ownership of Satellites*, Hofmann, M. and Loukakis, A. (eds). Nomos, Baden-Baden.

Smith, J. and Kerrest, A. (2009). Article VII. In *Cologne Commentary of Space Law: In Three Volumes*, Vol. I, Hobe, S., Schmidt-Tedd, B., Schrogl, K.-U. (eds). Carl Heymanns Verlag, Cologne.

Smith, J. and Kerrest, A. (2013). The liability convention. In *Cologne Commentary of Space Law: In Three Volumes*, Vol. II, Hobe, S., Schmidt-Tedd, B., Schrogl, K.-U. (eds). Carl Heymanns Verlag, Cologne.

PART 3

Diversity of Fields of Application for Space Law

8
Space Traffic Management

Ntorina ANTONI[1], Christina GIANNOPAPA[2] and Kai-Uwe SCHROGL[3]

[1] *Eindhoven University of Technology, The Netherlands*

[2] *European Union Agency for the Space Programme, Prague, Czech Republic*

[3] *German Federal Ministry for Economic Affairs and Energy, Berlin, Germany*

8.1. Introduction

The long-term sustainability of outer space activities has become more relevant than ever due to the ongoing transformation of the space sector. Space has evolved from once being the preserve of a handful of States. Nowadays, more than 70 countries and private companies have increased their level of space activity. The most prominent example is the initiative of SpaceX to launch a constellation of 12,000 small satellites for space-based internet access worldwide (Space.com 2020). Such undertakings challenge to a large extent the current space traffic situation featuring 3,300 operational satellites, more than 34,000 debris objects greater than 10 cm, 900,000 debris objects from greater than 1–10 cm and 128 million debris objects from greater than 1 mm–1 cm (ESA 2022). These figures demonstrate the increasing proliferation of space debris and the massive risk for collisions with operational satellites (Ailor 2020). This situation has been significantly exacerbated with anti-satellite missile tests carried out in 2007 by China, in 2008 by the United States, in 2019 by India and in 2021 by Russia. These behaviors have heightened the concerns of the international community for the peaceful uses of outer space.

Simultaneously space technologies are advancing, and satellite applications are improving by continuously innovating. This is facilitated by the increasing participation of a diverse array of private actors and the rising space budgets from many States, and innovative approaches for the growth of satellite technology and applications. In particular, the expenditures in defense and security space programs have substantially increased worldwide, expecting further growth in spending toward 2028 (Euroconsult 2019). Over the years, the space sector has moved beyond the traditional confines. It has moved into the more entrepreneurial and innovative phase, termed *NewSpace*. The programmatic objectives range from commercial to military, including robotic space exploration, human spaceflight, Earth observation, telecommunications and navigation. In particular, *NewSpace* witnesses a growing development of dual-use technologies such as autonomous rendezvous and docking capabilities. This growth in the commercial space sector raises issues concerning the protection of quasi-military systems and the unclear security implications of global markets for dual-use space capabilities (Hays 2020). Consequently, the increasingly congested and complex orbital space environment combined with the rise in the number and type of actors and civil–military activities have clear implications for the safety of space operations and the long-term sustainability of outer space.

With the world's ever-increasing dependence on space technologies, it is more necessary than ever to ensure uninterrupted access to and safety of operations in and from space. In this context, the development of a comprehensive regulatory approach to managing space traffic has come to the forefront of international discussions on the regulation of space activities. The Legal Subcommittee (LSC) of the United Nations (UN) Committee on the Peaceful Uses of Outer Space (COPUOS) introduced in 2016 "Space Traffic Management" (STM) as a regular agenda item, thereby providing the opportunity for multilateral, equitable and inclusive deliberations and negotiations[1]. In 2019, the Long-Term Sustainability of Outer Space Activities (LTS) Working Group in the Scientific and Technical Subcommittee (STS) of the COPUOS, established in 2010, presented its results on the additional guidelines adopted[2]. Other diplomatic initiatives concern the UN Governmental Group of Experts (GGE) on Outer Space Transparency and Confidence Building Measures (TCBMs), which presented its results in 2013[3], and the Draft International Code of Conduct for Outer Space Activities proposed by the European Union (EU) in 2007, now suspended. STM has also been addressed on governmental, regional and intergovernmental levels. In the United States, the Space

1 COPUOS (2015). Proposal for a Single Issue/Item for discussion at the 55th session of the Legal Subcommittee in 2016 on: "Exchange of views on the concept of Space Traffic Management", A/AC.105/C.2/2015/CRP.23/Rev.1.

2 COPUOS (2019). Report of the Scientific and Technical Subcommittee on its 56th session, held in Vienna from 11 to 22 February 2019, A/AC.105/1202.

3 UNGA (2013). Transparency and confidence-building measures in outer space activities, A/68/189.

Policy Directive-3, National Space Traffic Management Policy (SPD-3) of June 2018 was issued to improve space situational awareness and address STM issues. In Europe, besides the Member States' initiatives, in November 2020, the Council of the EU adopted a set of conclusions devoted particular attention to STM, highlighting the need for a coordinated European approach[4]. These ongoing parallel efforts call for synergies and the possibility to merge under an overarching STM framework. The development of a comprehensive STM regime shall correspond to the space sector dynamics and safeguard the peaceful uses of outer space.

Accordingly, the remainder of this chapter will present the evolution and current status of STM by emphasizing the principle of international cooperation as a foundation for maintaining stability and ensuring the peaceful uses of outer space (section 8.2). Additionally, it will provide an overview of STM's national perspectives (section 8.3). This overview will explain the underlying challenges for establishing STM and will focus on the effect of the emerging civil–military paradigm. Further, it will elaborate on how civil–military cooperation is treated on comparable traffic regimes: air traffic, maritime traffic and frequency spectrum (section 8.4). Finally, the chapter provides considerations for the way forward of STM (section 8.5)

8.2. Evolution and status of space traffic management

International space law embodies the five UN space treaties and five sets of UN principles governing outer space (UNOOSA 2008). Dating back to 1967, the States parties to the Treaty on Principles Governing the Activities of States in the Exploration and Use of Outer Space, including the Moon and other celestial bodies (OST) of December 19, 1966 recognized that it was in the common interest of all mankind to commit to "broad international cooperation in the scientific and legal aspects of the exploration and use of outer space for peaceful purposes" (OST, Preamble). The latter appears in several parts of the body of space law. The Declaration on International Cooperation in the Exploration and Use of Outer Space for the Benefit and in the Interest of All States, Taking into Particular Account the Needs of Developing Countries of 1996 (Space Benefits Declaration) further elaborates on the modes of cooperation considered the most effective and appropriate, including governmental and non-governmental; commercial and non-commercial; and global, multilateral, regional or bilateral and international cooperation.

---

4 EU Council (2020a). Orientations on the European contribution in establishing key principles for the global space economy, 12851/20.

Although not defined under international space law, the STM concept appeared first in the 1980s. STM evolved in discussions of the American Institute of Aeronautics and Astronautics (AIAA). Eventually, it was conceptualized by the 2006 Cosmic Study of the International Academy of Astronautics (IAA), the first in-depth multi-disciplinary study on STM. Accordingly, STM is considered a "concept that provides a framework for the safety, stability, and sustainability of space activities" (Perek 1982; Schrogl et al. 2018). As per its definition in the 2006 Cosmic Study, STM is

> the set of technical and regulatory provisions for promoting safe access into outer space, operations in outer space and, return from outer space to Earth free from physical and radio-frequency interference. (Jorgenson et al. 2006; Schrogl et al. 2018)

The objective of STM to provide for the safety, stability and sustainability of space activities falls under the OTS's peaceful purposes. Applying the principle of international cooperation to STM would entail ensuring the effective and appropriate collaboration between commercial and non-commercial actors, such as commercial/civil and non-commercial/military actors. Yet, international law instruments do not provide sufficient mechanisms dealing with this broad spectrum of space activities (Freeland and Gruttner 2020). The complication lies in distinguishing between military objectives or civil and commercial ones due to space technology's dual-use nature. In this context, the 2018 IAA study (Schrogl et al. 2018) revisited and adjusted the STM topic to address the space sector's increasing challenges and provided a roadmap for its implementation. The study analyses and evaluates the potential of STM to exert influence or shape the evolution of the legal and regulatory framework for space activities (Schrogl et al. 2018; Schrogl 2008).

8.2.1. *The United Nations context*

The LSC of the COPUOS has acknowledged STM as a topic appropriate for formal consideration deliberation at the intergovernmental level. LSC added STM in 2016 as a regular agenda item in its subsequent sessions for debate and negotiation (UNOOSA 2015). The LSC session in 2018, taking note of the 2018 IAA Study, highlighted that "the development of an integrated, harmonized and comprehensive space traffic management system was regarded as an urgent priority". In this direction, it reiterated the need for "complementing the five UN Treaties on outer space by further international agreements containing basic space traffic management rules". The second level of rules and regulations could comprise dynamic standards considering ongoing technological developments[5].

The Scientific and Technical Subcommittee agreed on the first set of guidelines for the long-term sustainability of outer space in 2016. In June 2019, STS approved another a set of 21 voluntary, non-binding guidelines (UNGA 2019). Although not explicitly referring to STM, these guidelines lay the foundation for developing policy and regulatory framework for space activities, the safety of space operations, international cooperation, capacity-building and awareness, and scientific and technical research and development. Moreover, STS expressed the idea that if a working group was to be established, it should consider STM. At the same time, it was reiterated that international cooperation and coordination is a necessary condition for STM (UNGA 2019). Close to this work was also the UN GGE, established by the Secretary-General (SG) of the UN in 2010, with the Report on TCBMs[6].

STM gained even more relevance and prominence at the UNISPACE+50 process prepared by the United Nations Office of Outer Space Affairs (UNOOSA) in 2018[7]. The Space2030 vision aims at strengthening international cooperation in the peaceful uses of outer space through a comprehensive and inclusive long-term strategy for global governance[8]. STM constitutes an essential condition for the sustainability of space activities[9]. Concerning strengthening the global governance of space activities, it is proposed to

> restructure the Committee's agenda to address, in a comprehensive manner, the use and utility of space as a driver of sustainable development and the issues of safety, security and sustainability of outer space activities, including the exchange of information on space

5 COPUOS (2018). Report of the Legal Subcommittee on its 57th session, held in Vienna from 9 to 20 April 2018, A/AC.105/1177.

6 UNODA (2013). Report of the group of governmental experts on transparency and confidence-building measures in outer space activities, A/68/189.

7 UNISPACE+50 was held in Vienna on 20-22 June 2018. Previous UNISPACE conferences were held in Vienna in 1968, 1982 and 1999.

8 COPUOS (2018). Report, on the United" Nations Office for Outer Space Affairs/Holy See Seminar Exploration and Development of Space Opportunities and Issues in the Context of the Sustainable Development Goals (27–28 March 2018), A/A/AC.105/2018/CRP.23.

9 COPUOS (2017). Priorité thématique 2. Le régime juridique de l'espace extra-atmosphérique et la gouvernance mondiale de l'espace : perspectives actuelles et futures, A/AC.105/1169.

objects and events, in-orbit collisions and interferences, space operations and space traffic management[10].

8.2.2. *The European regional context*

In addition to the UN system, there have been many STM-related initiatives at the regional and intergovernmental levels in Europe. Responding to the UN calls for concrete proposals on TCBMs in outer space activities, the EU proposed in 2008 the International Code of Conduct for Outer Space Activities (ICoC) to enhance the safety, security, and sustainability of space activities. Although no consensus was reached, the ICoC represented a distinctive approach to the rules of the road in support of international cooperation and peaceful uses of outer space.

In November 2020, the EU Council adopted a set of conclusions entitled "Orientations on the European contribution in establishing key principles for the global space economy" in preparation of the tenth Space Council meeting. The Council Conclusions devoted particular attention to STM, highlighting the need for a coordinated European approach in order to foster European autonomy, security and resilience (EU Council 2020a). On 16 December 2020, under the German Presidency, the EU Council reached a political agreement with the European Parliament on the regulation establishing the EU space program for 2021–2027 (Eu Council 2020b)[11]. The EU Space program component for Space Situational Awareness (SSA) builds on and extends the current EU Space Surveillance and Tracking (SST) framework to space weather and near-Earth objects[12]. By mid-2015, a Consortium Agreement on the European Space Surveillance and Tracking (SST) Consortium was signed by the national designates entities of the following EU

10 COPUOS (2018). The "Space2030" agenda and the global governance of outer space activities, A/AC.105/1166.

11 EU Council (2020). Proposal for a Regulation of the European Parliament and of the Council establishing the Space Programme of the Union and the European Union Agency for the Space Programme and Repealing Regulations, (EU) No 912/2010, (EU) No 1285/2013, (EU) No 377/2014 and 541/2014/EU.

12 European Commission (2018). Proposal for a Regulation of the European Parliament and of the Council establishing the space programme of the Union and the European Union Agency for the Space Programme and Repealing Regulations (EU.) No. 912/2010, (EU.) No. 1285/2013, (EU.) No. 377/2014 and Decision 541/2014/E.U. After adopting the decision of the EU Council and the European Parliament establishing a framework on SST support in March 2014, the European Commission adopted, in September 2014, the related implementing decision. European Commission (2014). Decision No 541/2014/EU of the European Parliament and of the council of April 16, 2014 establishing a Framework for Space Surveillance and Tracking Support, L158/227.

Member States: France, Germany, Spain, Italy and the United Kingdom. Finally, the 2016 Space Strategy for Europe has also acknowledged the relevance of STM[13].

The EU and the European Space Agency (ESA) signed a Joint Statement on Shared Vision and Goals for the Future of European Space on October 26, 2016. EU and ESA emphasized their intention to reinforce their cooperation in the future, as foreseen in the ESA/EU Framework Agreement of 2004 (ESA 2016). What is more, after proactive US positioning at the International Civil Aviation Organization (ICAO), the European Commission is investigating the so-called new entrants, including balloons, solar planes, supersonic planes, suborbital vehicles and High-Altitude Pseudo Satellites (HAPS). Higher Airspace Traffic management (HATM) is considered on top of the existing Air Traffic Management. There have been several interactions among the European Commission, Eurocontrol, the European Aviation Safety Agency (EASA), ESA, the European Defense Agency (EDA) and the European GNSS Agency (GSA) (Eurocontrol 2019).

Furthermore, ESA has established close collaboration with European institutions dealing with space security/defense, namely the European Commission, the European External Action Service (EEAS), EDA, the European Union Satellite Centre (SatCen) and GSA. ESA has been involved with security-related projects and programs and addresses the security threats to its own activities to secure the Member States' investments (Giannopapa et al. 2018). A document entitled "Elements of ESA's Policy on Space and Security" was presented in 2016 to the ESA council and the ESA Security Committee, elaborating on ESA's involvement in space and security activities (Giannopapa et al. 2018). At the ESA Ministerial Council of 2019, the Ministers agreed that the Space Safety Programme (S2P), as a successor to the SSA Programme, aims to contribute to the protection of Earth, humanity and assets from hazards originating in space shall be conducted within the framework of the agency. S2P will contribute significantly by establishing overall complementarity and maximizing synergies among space safety initiatives in Europe, providing opportunities for cooperation among the Agency, Member States, the EU and other partners to those ends[14]. The innovative and promising approach aims at establishing a long-term framework for in-orbit servicing, starting with active debris removal performed on an ESA space object. This provides ESA the opportunity to apply more efficiently in the coming years its policy and guidelines aimed at reducing the harmful effects of space debris.

---

13 European Commission (2016). Communication from the Commission to the European Parliament, the Council, the European Economic and Social Committee and the Committee of the Regions – Space Strategy for Europe, COM (2016) 705 final.

14 ESA (2019). Resolution on ESA programs: Addressing the challenges ahead, ESA/C-M/CCLXXXVI/Res.3.

8.3. National STM initiatives and the civil–military challenge

The evolution of STM goes hand in hand with the development of the space sector and the exploding opportunities for private, military and commercial actors. Concerning STM overall responsibility for operating SSA is a condition *sine qua non* for STM (Schrogl et al. 2018). This responsibility is typically assigned to military authorities with civilian support (Lal et al. 2018). The civil–military nexus in space activities impacts space technology and the interoperability of civilian and military systems. As such, the military/security aspect gives rise to a significant regulatory challenge for establishing a comprehensive STM system: developing a common ground for further cooperation despite the divergent interests and approaches. The criticality of the civil–military nexus for effective collaboration makes it a significant element to study STM (Schrogl 2008). The selected countries below serve as examples to reflect upon the said civil–military challenge in STM developments. Namely, the Unite States provides the most recent and elaborate initiative, while national positions are also built-in Europe and especially in France, Germany, the United Kingdom, Italy and Spain.

8.3.1. *United States*

The Space Policy Directive-3 (SPD-3), National Space Traffic Management Policy, was issued to provide an interagency approach to STM. The SPD-3 was released as part of a comprehensive national security strategy, including the National Space Council's revival and a broad series of Space Policy Directives. The SPD-3 aims for the US leadership in space. It stipulates the need to

> set priorities for space situational awareness (SSA) and STM innovation in science and technology (S&T), incorporate national security considerations, encourage the growth of the US commercial space sector, and establishes an updated STM architecture, and promote space safety standards and best practices across the international community.

One of the goals stated in Section 4 of SPD-3 as a critical first step is

> to develop US-led minimum safety standards and best practices to coordinate space traffic' and it is further elaborated meaning to include: 'space traffic standards and best practices, including technical guidelines, minimum safety standards, behavioral norms, and orbital conjunction prevention protocols related to pre-launch risk assessment and on-orbit collision avoidance support services.

Further, the SPD-3 attributes the responsibility for providing SSA data for civil use to the Department of Commerce (DoC), while the Department of Defense (DoD) will focus on maintaining access to and freedom of action in space. In particular, the DoC becomes the agency responsible for SSA data sharing and timely warning of collision avoidance, including conjunction assessments and maneuver plans, available to the public through the publicly releasable portion of DoD authoritative catalog. The availability of the data is and will remain to be free of direct user fees. Therefore, the DoD shifts the civilian part of its responsibilities to the DoC and will be in charge of the authoritative catalog of space objects' military aspect. The directive stipulates in this regard the increasing demand for DoD to focus on protecting and defending US space assets and interests (US Department of State 2019). Shall the SDP-3 proposal be approved it does raise the following fundamental question: first, how unusual can the military SSA activities be from the civilian SSA activities and, second, will it be possible for them to integrate under one comprehensive regime? This has implications for the governance of the safety of operations and national security and the exchanges and coordination with other national and international organizations (Hitchens 2019).

In response to the US developments and the SPD-3, which is the most comprehensive national initiative, Europe has taken action at the regional level and with national policies and programs. In this context, ESA, the EU and their respective Member States have strengthened their existing partnerships and have established new ones with security stakeholders in Europe. Many European countries have developed a dedicated space and security/defense strategy or policy with implicit or explicit at times reference to STM elements (Antoni et al. 2020). In the following section, the selected European countries will be presented: France, Germany, UK, Italy and Spain.

8.3.2. *France*

France's announcement for creating a new military space command in September 2019 follows similar moves by the United States. It reflects France's renewed military focus on space. In July 2019, the president of the Republic asked for the drafting of "a defense space strategy" (Financial Times 2019; Le Monde 2019). This new military strategy depicts a doctrine of active defense, allowing France to go on the attack when a hostile act has been detected. Given the dual use of space assets and the importance of space, and the general acceptance of military capabilities for civilian objectives in France, the French *Direction générale de l'armement* (DGA), on behalf of the Ministry of Armed Forces, exercises guardianship over the *Centre national d'études spatiales* (CNES). CNES assesses the French space policy under the responsibility of the Minister for Higher Education, Research and Innovation and has a unique partnership with the Ministry of the Armed Forces for what concerns military space activities. The close cooperation of the two ministries in relation to CNES is reflected in the dual-use research Program 191 (The French Ministry for the Armed Forces 2019).

Concerning the role of STM in particular, the 2017 Defense and National Security Strategic Review emphasizes "the need to develop space situational awareness and to ensure the resilience of space capabilities". It also refers to space as "a provider of essential navigation, communication, meteorological and imagery services, while also a domain of confrontation" (The French Ministry for the Armed Forces 2017). The French dual-use operational SSA activities are organized on three levels. At the programmatic and decision-making level, CNES and the French Ministry of Armed Forces closely cooperate to define policy, capabilities and priorities. At the sensors level, the French Ministry of Armed Forces operates the GRAVES survey radar and several tracking radars that CNES uses on a regular basis (Assemblée Nationale 2016). CNES and the National Centre for Scientific Research (CNRS) also operates three TAROT telescopes for survey and tracking purposes with locations at Calern, Chili and La Réunion. The Operational Center for Military Surveillance of Space Objects (COSMOS) implements space surveillance sensors to manage the national catalog of space objects, detect abnormal behavior, provide space support for military operations and manage space meteorology (Assemblée Nationale 2019; Testé 2020).

8.3.3. *Germany*

In Germany, the Federal Ministry for Economic Affairs and Climate Action (BMWk) is mainly responsible for Germany's civilian space activities. The Ministry of Transport and Digital Infrastructure oversees selected EU flagship programs Galileo and Copernicus. The Foreign Office is also active in space and security realm, contributing to international space law discussions. The Ministry of Defence operates Germany's reconnaissance satellites. At the same time, the Chancellery oversees the intelligence service, and the Ministry of the Interior coordinates the protection of critical infrastructure and national crisis management (Governance Post 2017).

The Federal Government Coordinator of German Aerospace Policy coordinates the Federal Government's measures to strengthen Germany's aerospace sector's international competitiveness in the fields of research and development. The *Deutsches Zentrum für Luft- und Raumfahrt* (DLR) is the national space agency. It reports directly to the BMWk. DLR was founded in 1969 through the merger of several institutions. Acting on behalf of the Federal Government, DLR Space Administration designs and implements Germany's Space Program, integrating all German space activities at the national and European level (BMWk 2021).

In 2016, the German government published its White Paper on German Security Policy and the Future of the Bundeswehr, which set out the new concepts and strategic priorities underlying German security and defense policy, guaranteeing a whole-of-government approach for the development of sustainable security. The 2010 German Space Strategy recognizes the importance of synergies between civil and military space research and stresses the need to build up national competence for SPA through existing resources (Governance Post 2017). This includes STM-related activities and complements Germany's active role in the EU SST Consortium. Within Germany's Air Force General Staff, there is a dedicated Space Office. Part of the Air Force Operations Centre, the German Space Situational Awareness Centre (GSSAC), deals with SST/SSA in collaboration with DLR. The GSSAC is an interagency organization formed by DLR and Air Force entities. The German Air Force's Air Operations Command provides the infrastructure, IT equipment and staff. Simultaneously, the DLR Space Administration's SSA division brings in additional IT equipment for development and testing and staff (Auras 2015). The Centre uses catalog data from its partner nations, the United States and France, and national resources (BMWk 2010). The new German Experimental Space Surveillance and Tracking Radar (GESTRA) – the first German space radar – was developed by Fraunhofer for DLR and was inaugurated on October 13, 2020 (DLR 2020).

8.3.4. *United Kingdom*

The UK Government published the National Security Strategy (NSS) and the Strategic Defense and Security Review (SDSR) on November 23, 2015 (HM Government 2015, 2016). They cover both domestic and overseas elements of security. The NSS focuses on the country's objectives, while the focus of the SDSR is the "ways" and "means" to achieve them (HM Government 2010). Resorting under the Department for Business, Enterprise and Industrial Strategy, the UK Space Agency (UKSA) helps to bring together UK civil and commercial space programs and interests. It is an executive agency of the Department for Business, Energy and

Industrial Strategy and is responsible for all strategic decisions on UK civil space programs[15].

From a military perspective, the Royal Air Force is recognized as the most significant contributor to space operations since it retains most of the UK's military space expertise and manages key UK space capabilities. The UK has three space surveillance systems, two radar systems (Fylingdales and Chilbolton) and one optical system (Starbrook). The primary role of Fylingdales is to maintain an uninterrupted missile warning service, thereby demonstrating that a surprise missile attack could not succeed. The secondary function is space surveillance, which forms an essential element of the primary task (www.raf.mod.uk). Chilbolton is the largest steerable meteorological radar in the world. It can remotely determine the shape and orientation of cloud and precipitation particles in the atmosphere (www.https://archive.ceda.ac.uk/). Starbrook, located in Cyprus, is a ground-based optical sensor system for the surveillance of space. Designed and implemented in 2006, Starbrook has an added experimental survey sensor and is a new class of wide field of view telescopes that scan near-Earth space to discover and catalog objects. The first is part of the US Space Surveillance Network (SSN) and is operated by British armed forces. The second is managed by the Rutherford Appleton Laboratory and is mainly used for atmospheric and ionospheric research. The radar has been upgraded to track down to 10 centimeters of objects at an altitude of 600 km (Space Insight 2014).

8.3.5. *Italy*

In Italy, responsibility for space activities has recently been put directly under the supervision of the Prime Minister. The Italian Parliament approved on December 22, 2017 the new space governance law, mandating an Inter-ministerial Committee for Space and Aerospace Research Policies, and the President of the Italian Space Agency (ASI) reporting directly to the Prime Minister's Office (Instituto Affari Internazionali 2003). The ASI Strategic Vision Document 2016-2025 (ASI 2016-2025) highlights the use of space assets also for security and defense purposes, including Earth observation data for crisis management, security, defense and disaster monitoring. In addition, the document makes reference to SSA, STM, Confidence-building measures and Code of Conduct. ASI will make the sensors located at the Space Geodesy Centre in Matera (optical telescope and laser-tracking

15 The Department for Business, Innovation and Skills was a ministerial department of the UK Government created on June 5, 2009 by the merger of the Department for Innovation, Universities and Skills and the Department for Business, Enterprise and Regulatory Reform. It was disbanded on the creation of the Department for Business, Energy and Industrial Strategy on July 14, 2016.

system) available to the initiative. Its role is that of the Centre of Expertise for Civil Protection in monitoring the uncontrolled re-entry of space debris.

The Ministry of Defense will contribute, with optical telescopes and radar used for detection and tracking, together with the National Operations Centre. The National Institute for Astrophysics (INAF) will contribute with the Sardinia Radio Telescope and the Northern Cross Radio Telescope located in Medicina (Bologna) (ASI 2016-2025). ASI, INAF and the Ministry of Defense have signed a Framework Agreement for the SST program. The agreement – which runs from June 2015 until the end of 2020 – foresees a Steering Committee for Space Surveillance and Tracking activities, responsible for coordinating the national activities in the European SST initiative and directing ASI as National Entity representing Italy within the European SST Consortium (INAF 2015; Telespazio and TAS 2017). The Air Force's 4th Telecommunication and Systems for Air Defense and Flight Assistance Brigade are involved in SSA activities. The ASI 2016-2025 foresaw the establishment of the Italian SST Operation Centre (ISOC) managed by the Italian Military Air Force, as the collector and the agent for "data fusion" national SST sensors. ISOC is the hub for all national SST sensors belonging to the Ministry of Defense, National Institute for Astrophysics and ASI (Italian Air Force 2015; Leonardo 2015; GIAS 2017).

8.3.6. *Spain*

The Spanish Space Surveillance and Tracking (S3T) system provide services with two main objectives: first, it ensures the long-term availability of space infrastructures essential for the safety and security of worldwide citizens; second, it provides the best available information to governmental and civil protection services in the event of uncontrolled re-entries of entire spacecraft or space debris thereof into the Earth's atmosphere. The SST services provided comprises collision risk assessment, the generation of conjunction data messages between objects in space, the detection and characterization of in-orbit fragmentations and collisions, and the characterization and surveillance of uncontrolled re-entries of space objects into the Earth's atmosphere (Gomez et al. 2017). Spain houses two major European sites relevant for SSA/SST: the Space Surveillance Test and Validation Centre (SSTC) located at ESA's European Space Astronomy Centre and the EU Satellite Centre (EU Satcen).

Through the Directorate General of Armaments and Material (DGAM), the Ministry of Defense has participated in the selection of the site of the future advanced surveillance radar SST, as well as data policy and information security. In June 2015, it carried out the Transfer of the Operative Control of the Radar SSA of

Santorcaz from ESA to the Air Force (Ministerio de Defensa 2015). The monostatic breadboard surveillance radar, located at the Santorcaz Naval Radio Communication Station, near Madrid, Spain, was developed within ESA's SSA Program. Following the completion of its development and upon the Spanish government's request, the Radar's operation was transferred on a loan basis to the Spanish Ministry of Defense for use in Spain's national SST activities[16].

8.3.7. *Europe*

Europe finally entered the scene of STM as a unified actor at the EU-ESA Space Council on November 20, 2020. During this concomitant meeting of the EU Competitiveness Council and the ESA Council at ministerial level, the Orientations adopted contain two paragraphs on STM, demonstrating the ambition of Europe in this field:

> 14. NOTES that there are a number of Space Traffic Management initiatives worldwide; UNDERLINES that Europe develops spacecraft collision avoidance, space debris removal techniques, space debris avoidance, operates and develops Space Surveillance and Tracking (SST) capabilities to achieve a higher level of strategic autonomy, in order to manage effectively access and return from outer space and in-orbit operations;
>
> 15. RECOGNISES the need for increased European coordinated approach to space traffic management (including technical and operational), in order to preserve its interests, to protect its private and public investments in space in a sustainable manner; and in this view PROPOSES to all competent actors to start a European dialogue together with academia and industry, including a mapping exercise of current regulatory frameworks in Europe to be discussed in a dedicated European conference. (EU Council 2020a; SpaceWatch Europe 2021a)

The format was carefully chosen, since it is neither the EU and its Commission, nor the ESA Executive, which has an exclusive competence for STM, but the Member States have to lead a triangle of actors. On July 7, 2021, the European STM Conference took place on a virtual format with the aim to build a coherent and

16 ESA (2018). Status report on the implementation of the Agreement between ESA and the Spanish Ministry of Defense regarding the Monostatic Breadboard Radar at Santorcaz – Status as of March 2018, ESA/PB-SSA(2018)15.

comprehensive position for Europe. Member States from the EU and the ESA together with EU institutions and ESA Executive endorsed an Outcome paper entitled "Fostering a European approach on Space Traffic Management" (SpaceWatch Europe 2021b).

8.4. Civil–military cooperation in the air, seas and radio frequency spectrum

Space traffic is comparable to the traffic that concerns air space, the seas and radio frequency (RF) (Schrogl 2008; Schrogl et al. 2018). Air space, seas, radio spectrum/frequency and outer space share a common denominator: they are valuable resources that have become scarce to different degrees. A perspective of these regimes on civil–military cooperation will facilitate to draw analogies for the further refinement and practical negotiation of an STM regime. The latter shall develop "in one coherent way from the overarching principle of guaranteeing safe operations in the space traffic system" (Schrogl 2008).

8.4.1. *Air traffic management*

The close interconnection between civil and military aviation sharing common airspace has led to boundary-setting or segregation that set constraints to airspace access/use. Accordingly, the authorities dealing with air traffic in each country differ. In Germany, the *Deutsche Flugsicherung* has been entrusted by law to control German civil and military air traffic. The law only exempts local military air traffic control services at military aerodromes, including military aerodromes also used by civil aviation (www.dfs.de). In France, the *Direction des Services de la Navigation Aérienne* is in charge of civil air traffic, while the *Direction de la Circulation Aérienne Mililtaire* is in charge of military air traffic.

Yet, the modernization of the air navigation system and new civil and military technologies/aircraft create discrepancies and complications for air traffic at the international level, calling for a harmonized and integrated civil–military approach in civil aviation[17]. ICAO has gathered civil and military communities to improve collaboration in practice and ensure air transport safety and security. ICAO envisages an integrated, harmonized and globally interoperable ATM system, the "Global ATM Operational Concept"[18]. Most recent recommendations call for a

17 ICAO (2011). Civil/Military Cooperation in Air Traffic Management, Cir 330. AN/189 2011.

18 ICAO (2005). Global Air Traffic Management Operational Concept, Doc. 9854, AN/458 2005.

systemic approach based on, collaborative decision making and conceptual interoperability for a future seamless and cost-efficient access to airspace. Such a system allows for performance equivalence and dual use while respecting State sovereignty and national security, defense and law enforcement obligations[19].

8.4.2. *Maritime traffic management*

The UN Convention on the Law of the Sea of December 10, 1982 (UNCLOS) represented unprecedented international negotiations efforts in codifying the rules governing the ocean space. According to the Convention, the safety of navigation and the regulation for maritime traffic in the territorial sea are subject to the coastal State's sovereignty. Realizing the importance of coordinating all maritime actors, Germany has developed a Maritime Development Plan with the Transport Ministry having the lead yet with no direct civil–military interaction. The Netherlands, on the contrary, makes more explicit the need for an integrated approach between military and civilian situational awareness, as noted in the Maritime Vision 2030 (Drent et al. 2013). The challenge in national practices lies in the involvement of multiples actors with distinct competencies.

However, an integrated approach at the international level is essential to achieve maritime security. Accordingly, UNCLOS provides a legal framework that includes countries' military authorities in organizing to ensure safety and security at sea. In particular, concerning maritime traffic regulation, Vessel Traffic Service (VTS) is considered significant for maritime traffic safety and efficiency. Technological developments have made necessary for maritime security and collision avoidance, the development of the Universal Shipborne Automatic Identification System-AIS[20]. The purposes of the AIS system include promoting the safety of navigation, collision avoidance, enabling coastal Countries to obtain information about ships and their cargoes, and as a VTS tool.

8.4.3. *RF spectrum management*

Since the early 1900s and radio transmission development, the International Telecommunication Union (ITU) has managed RF spectrum and the regulation of its use via the radio regulations (RR). Most vital industry and military sectors are based

---

19 ICAO (2014). Assembly Resolutions in Force (as of 4 October 2013), Doc 10022.

20 IMO (2015). Guidelines for the onboard operational use of shipborne Automatic Identification Systems (AIS), Resolution A.917 (22), Revised on 2 December 2015, A 29/Res.1106, 2015.

on access to radio frequencies: mobile communications, audio and television broadcasting, satellites, radiolocation, transportation and the Internet of Things (Mazar 2016). Hence, a particular concern in radio spectrum management is the military's use and need to ensure that critical missions are interference free. The multi-jurisdictional process separates agencies managing spectrum in civilian and military services, with an intermediary committee having the coordination role. This applies in most countries, including the UK and the United States. The single authority approach assigns one single agency to manage all the RF spectrum. The latter is not the usual model, with Malaysia and New Zealand being a few countries adopting it (ITU 2016). Therefore, defense systems can be subject to the discretion of each country and its sovereign right to regulate the military RF spectrum, leading to fragmented allocation.

A harmonization approach that delivers these benefits considers accommodating the military, government and commercial users' spectrum needs. In this direction, the ITU International working groups in collaboration with the European Conference of Postal and Telecommunications Administrations (CEPT) discuss the industry's requirements to provide frequency regulation as target oriented as possible. Although these discussions so far are limited to civil use frequency, ITU recognizes the importance of RR for defense systems (ITU 2016b). The discussions in these working groups with industry involvement are a crucial element that enables ITU to help countries overcome the various users' conflicting interests. These jointly developed principles then serve to allow internationally harmonized use of frequency resources.

8.5. The way forward for STM

The space sector witnesses a transformation resulting, among other things, from the increasing technological advances, complex interdependencies between diverse actors and the emergence of the civil–military nexus leading to conflicting interests. These circumstances call for a comprehensive regulatory solution at the international level. In this regard, establishing a STM regime is essential for the safety and sustainability of space activities. Accordingly, this chapter looked into the evolution and current status, including the role of the principle of international cooperation. Also, STM initiatives were presented at the international, intergovernmental and national levels, with an emphasis placed on the civil–military challenge. The aim was to examine how the principle of cooperation between civil and military actors is treated in comparable traffic regimes with established practices at the international level – namely, air traffic, maritime traffic and RF spectrum management.

STM has acquired a clear and prominent role in the UN Committee's agenda on the Peaceful Uses of Outer Space as a comprehensive solution for the safety, stability and sustainability of space activities in conformity with the peaceful purposes of the OTS. Simultaneously, STM is emerging as a relevant policy objective in national and regional contexts such as in the United States and Europe. Subsequently, the divergent expectations and fragmented regulatory landscape cannot be addressed under the OTS. COPUOS, in line with the 2018 IAA Study on STM, considers a two-way approach for the future establishment of STM. First, through international agreements containing basic STM rules in the long-term, and second, through international administrative rules and regulations comprising dynamic standards that would need to be easily amendable and would consider ongoing technological developments. The principle of international cooperation through effective and appropriate approaches between commercial and non-commercial actors, civil and military, is therefore fundamental for establishing STM.

To further explore the operationalization of the cooperation principle, this chapter looked at how the civil–military nexus affects cooperation in other traffic regimes: air traffic, maritime traffic and frequency management systems. Despite the differences between traffic regimes, there is one common denominator for all of them, ensuring the safety of traffic, thus navigation of aircraft, vessel, radiofrequency and spacecraft. Safe navigation cannot be guaranteed unless all actors are involved at the international level to negotiate the best approach to reach the objective. In this regard, it can be relevant to look at these analogies and implement practices to help overcome the challenges in developing an STM regime. The role of ICAO, IMO and ITU has been crucial in dealing with the transition from fragmented approaches at the national level in civil–military cooperation to more harmonized and integrated traffic management regimes that respond to the technological advances in the needs of all users. The evolution of these traffic regimes has not been a simple, linear and straightforward one. The actors involved have encountered challenges in developing a comprehensive and inclusive approach at the international level. It took 14 years of long and complicated negotiations to establish the UNCLOS governing the ocean space. Still, IMO is adapting to the new technologies – including spa-based systems – to enhance maritime security. ICAO's recommendations are not binding norms yet are accepted by States, offering gradual adherence to a changing air traffic management system. ITU has developed a system since the 1900s to reach the level of today's normative flexibility for the governance of RF spectrum, including wireless defense systems. Drawing on the fundamental principle of international cooperation and essential elements of these comparable traffic regimes can further refine international negotiations on STM.

8.6. References

Ailor, W. (2020). Evolution of space traffic and space traffic management. In *Handbook of Space Security. Policies, Applications and Programmes*, 2 Volumes, 2nd edition, Schrogl, K.-U., Adriaensen, M., Giannopapa, C., Hays, P.L., Robinson, J., Antoni, N. (eds). Springer, New York.

Antoni, N., Adriaensen, M., Giannopapa, C. (2020). Strategic overview of European space and security governance. In *Handbook of Space Security. Policies, Applications and Programmes*, 2 Volumes, 2nd edition, Schrogl, K.-U., Adriaensen, M., Giannopapa, C., Hays, P.L., Robinson, J., Antoni, N. (eds). Springer, New York.

Assemblée Nationale (2016). Compte rendu n°48 de la Commission de la défense nationale et des forces armées [Online]. Available at: https://www.assemblee-nationale.fr/14/cr-cdef/15-16/c1516048.asp [Accessed 20 April 2022].

Assemblée Nationale (2019). Compte rendu n°21 de la Commission de la défense nationale et des forces armées [Online]. Available at: https://www.assemblee-nationale.fr/dyn/15/comptes-rendus/cion_def/l15cion_def1920021_compte-rendu [Accessed 20 April 2022].

Auras, G.S. (2015). The German interagency approach to SSA. Presentation, 2015 Japan Space Forum SSA Symposium [Online]. Available at: http://www.jsforum.or.jp/debrisympo/2015/pdf/19%20The_German_interagency_approach_to_SSA_LTC_Auras_Script.pdf [Accessed 5 January 2021].

BMWk (2010). Aerospace policy [Online]. Available at: https://www.bmwi.de/Redaktion/EN/Dossier/aerospace-policy.html [Accessed 5 January 2021].

BMWk (2021). Making Germany's space sector fit for the future – The space strategy of the German Federal Government [Online]. Available at: https://www.dlr.de/rd/en/Portaldata/28/Resources/dokumente/Raumfahrtstrategie_en.pdf [Accessed 20 April 2022].

DLR (2020). Improved safety in space – GESTRA space radar ready to begin operations [Online]. Available at: https://www.dlr.de/content/en/articles/news/2020/04/20201013_space-radar-gestra-begin-operation.html [Accessed 5 January 2021].

Drent, M., Homan, K., Zandee, D. (eds) (2013). Civil-military capacities for European security [Online]. Available at: https://www.clingendael.org/sites/default/files/2016-02/Study-Civil-Military-Capacities-European-Security.pdf [Accessed 20 April 2022].

ESA (2016). Joint statement on shared vision and goals for the future of Europe in space by the EU and ESA [Online]. Available at: https://m.esa.int/About_Us/Welcome_to_ESA/Shared_vision_and_goals_for_the_future_of_Europe_in_space [Accessed 5 January 2021].

ESA (2022). Space debris by the numbers [Online]. Available at: https://www.esa.int/Our_Activities/Space_Safety/Space_Debris/Space_debris_by_the_numbers [Accessed 5 January 2021].

Euroconsult (2019). Government space programs: Benchmarks, profiles & forecasts to 2028 [Online]. Available at: https://www.euroconsult-ec.com/shop/?id_product=115&controller=product [Accessed 20 April 2022].

Financial Times (2019). France follows US to set up military space command [Online]. Available at: https://www.ft.com/content/a479bcb6-a628-11e9-984c-fac8325aaa04 [Accessed 20 April 2022].

Freeland, S. and Gruttner, E. (2020). The laws of war in outer space. In *Handbook of Space Security. Policies, Applications and Programmes*, 2 Volumes, 2nd edition, Schrogl, K.-U., Adriaensen, M., Giannopapa, C., Hays, P.L., Robinson, J., Antoni, N. (eds). Springer, New York.

Giannopapa, C. (2018). Strategy for space safety and security, 2018 EISC Workshop [Online]. Available at: https://eisc-europa.eu/images/stories/2018/Workshop/Final_upload/EISC_Presentations/Strategy_for_space_safety_and_security_-_Christina_Giannopapa.pdf [Accessed 5 January 2021].

GIAS (2017). Aero-space system engineering group (GIAS) and Italian SST operational center (ISOC) [Online]. Available at: http://www.afceanaples.it/wp-content/uploads/2017/06/02.-Space-and-Aerospace-at-the-Flight-Test-Wing.pdf [Accessed 5 January 2021].

Gómez, I.A., Vildarraz, S.A., García, C., Garnica, J. M.H., Hernandez, C.P., Prada, M.A.R., Ortiz, N. (2017). Description of the architecture of the Spanish space surveillance and tracking system. In *Proceedings of the 7th European Conference on Space Debris*.

Governance Post (2017). After the election, Germany's space policy must take shape [Online]. Available at: https://www.hertie-school.org/the-governance-post/2017/12/election-germanys-space-policy-must-take-shape/ [Accessed 5 January 2021].

Hays, P.L. (2020). Spacepower theory and organizational structures. In *Handbook of Space Security. Policies, Applications and Programmes*, 2 Volumes, 2nd edition, Schrogl, K.-U., Adriaensen, M., Giannopapa, C., Hays, P.L., Robinson, J., Antoni, N. (eds). Springer, New York.

Hitchens, T. (2019). Space traffic management: US military considerations for the future. *Journal of Space Safety Engineering*, 6(2), 108–112.

HM Government (2010). Fact sheet 1: Our approach to the national security strategy [Online]. Available at: https://www.gov.uk/government/uploads/system/uploads/attachment_data/file/62483/Factsheet1-Our-Approach-National-Security-Strategy.pdf [Accessed 5 January 2021].

HM Government (2015). National security strategy and strategic defense and security review 2015: A secure and prosperous United Kingdom. Report, Cm 9161.

HM Government (2016). The 2015 strategic defense and security review. Briefing paper 7462.

INAF (2015). Costituzione dell'organismo di coordinamento e di indirizzo relativo all'iniziativa space surveillance and tracking della commissione europea [Online]. Available at: http://www.inaf.it/it/sedi/sede-centrale-nuova/consiglio-di-amministrazione/delibere/archivio_delibere/delibere-2015/Delibera%2015-15.pdf [Accessed 5 January 2021].

Instituto Affari Internazionali (2003). Space and security policy in Europe [Online]. Available at: https://www.iai.it/sites/default/files/2003_space-and-security-in-europe.pdf [Accessed 20 April 2022].

Italian Air Force (2015). Italian Air Force radar and optical sensor experiments for the detection of space objects in LEO orbit [Online]. Available at: http://www.amostech.com/TechnicalPapers/2015/Adaptive_Optics_Imaging/DelGenio.pdf [Accessed 5 January 2021].

ITU (2016a). West African common market project: Harmonization of policies governing the ICT market in the UEMOA-ECOWAS space radio spectrum management techniques [Online]. Available at: https://www.itu.int/ITU-D/treg/projects/itu-ec/Ghana/modules/Final Documents/Spectrum.pdf [Accessed 5 January 2021].

ITU (2016b). Celebrating the radio regulations [Online]. Available at: https://www.itu.int/en/itunews/Documents/2016-05/2016_ITUNews05-en.pdf [Accessed 5 January 2021].

Jorgenson, C., Lála, P., Schrogl, K.-U. (eds) (2006). *Cosmic Study on Space Traffic Management*. International Academy of Astronautics, Paris [Online]. Available at: https://www.black-holes.eu/resources/IAA_spacetrafficmanagement.pdf [Accessed 13 June 2022].

Lal, B., Balakrishnan, A., Caldwell, B.M., Buenconsejo, R.S., Carioscia, S.A. (2018). *Global Trends in Space Situational Awareness (SSA) and Space Traffic Management (STM)*. IDA Science and Technology Policy Institute, Washington [Online]. Available at: https://apps.dtic.mil/sti/pdfs/AD1123106.pdf [Accessed 13 June 2022].

Le Monde (2019). La France va préciser sa nouvelle stratégie spatiale militaire command [Online]. Available at: https://www.lemonde.fr/international/article/2019/07/15/la-france-va-preciser-sa-nouvelle-strategie-spatiale-militaire_5489589_3210.html [Accessed 20 April 2022].

Leonardo (2015). SELEX ES radars key players of the space surveillance and tracking program [Online]. Available at: http://www.leonardocompany.com/en/-/selex-es-am-space-junk [Accessed 5 January 2021].

Mazar, H. (2016). *Radio Spectrum Management, Policies Regulations and Techniques*. John Wiley & Sons, New York.

Ministerio de Defensa (2015). Plan Director de Sistemas Espaciales [Online]. Available at: https://publicaciones.defensa.gob.es/plan-director-de-sistemas-espaciales.html [Accessed 10 April 2022].

Perek, L. (1982). Traffic rules for outer space. In *Proceedings of the Twenty-Fifth Colloquium on the Law of Outer Space*. American Institute of Aeronautics and Astronautics, Reston.

Schrogl, K.-U. (2008). Space traffic management: The new comprehensive approach for regulating the use of outer space – Results from the 2006 IAA cosmic study. *Acta Astronautica*, 62(2–3), 272–276.

Schrogl, K.-U., Jorgenson, C., Robinson, J., Soucek, A. (eds) (2018). *Space Traffic Management – Towards a Roadmap for Implementation*. International Academy of Astronautics, Paris.

Space Insight (2014). An introduction to Space Insight's Starbrook systems for the surveillance of space [Online]. Available at: http://www.spaceinsight.co.uk/assets/uploads/SpaceInsight_Starbrook.pdf [Accessed 5 January 2021].

Space.com (2020). SpaceX launches 60 Starlink satellites, lands rocket at sea in 2nd flight this week [Online]. Available at: https://www.space.com/spacex-launches-60-starlink-15-satellites [Accessed 5 January 2021].

SpaceWatch Europe (2021a). Interviews: Kai-Uwe Schrogl "A challenging presidency – A bright perspective for Europe in space" [Online]. Available at: https://spacewatch.global/2021/01/spacewatchgl-interviews-kai-uwe-schrogl-a-challenging-presidency-a-bright-perspective-for-europe-in-space/ [Accessed 5 January 2021].

SpaceWatch Europe (2021b). Interviews: Kai-Uwe Schrogl "Europe enters a center stage in space traffic management" [Online]. Available at: https://spacewatch.global/2021/07/spacewatchgl-interviews-kai-uwe-schrogl-europe-enters-centre-stage-in-space-traffic-management/ [Accessed 5 July 2022].

Telespazio and TAS (2017). Space situation awareness & space surveillance tracking [Online]. Available at: http://www.cesmamil.org/wordpress/wp-content/uploads/2017/05/9-_-Matarazzo-Brancati-_-Thales-Telespazio.pdf [Accessed 20 April 2022].

Testé, J.D. (2020). Future of French space security. In *Handbook of Space Security. Policies, Applications and Programmes*, 2 Volumes, 2nd edition, Schrogl, K.-U., Adriaensen, M., Giannopapa, C., Hays, P.L., Robinson, J., Antoni, N. (eds). Springer, New York.

The French Ministry for the Armed Forces (2017). Defence and national security strategic review [Online]. Available at: https://www.diplomatie.gouv.fr/IMG/pdf/strategic_review_of_defense_and_national_security_2017_-_key_points_cle0d61fa.pdf [Accessed 20 April 2022].

The French Ministry for the Armed Forces (2019). Space defense strategy [Online]. Available at: https://www.gouvernement.fr/sites/default/files/locale/piece-jointe/2020/08/france_space_defence_strategy_2019.pdf [Accessed 20 April 2022].

UNOOSA (2008). United Nations treaties and principles on outer space, including related General Assembly resolutions and other documents [Online]. Available at: https://www.unoosa.org/pdf/publications/st_space_11rev2E.pdf [Accessed 20 April 2022].

US Department of State (2019). Events – Office of Space and Advanced Technology [Online]. Available at: https://www.state.gov/spaceentsummit [Accessed 5 January 2021].

9

ITU Regulatory Regime Related to Non-GSO Satellite Systems

Yvon HENRI[1] and Attila MATAS[2]

[1] *ITU Radio Regulations Board, Échenevex, France*

[2] *Senior Satellite Regulatory Consultant, Geneva, Switzerland*

9.1. Introduction

This chapter explains the relevant regulatory issues and activities related to non-geostationary satellite orbit (non-GSO) Fixed Satellite System, non-GSO Radio Navigation Satellite Systems and non-GSO satellite systems with short-duration missions (SDM) at the international level.

The radio-frequency spectrum and associated orbits for space systems are limited to natural resources essential to most of the applications that our world has come to rely on for its economic and societal development since the end of the 1950s. As all scarce resources, they must be used rationally, efficiently and economically so that all States or groups of States may have equitable access to those frequencies and associated satellite orbits. These resources are managed through a system of international regulations and standards adopted by the International Telecommunication Union (ITU), the United Nations (UN) specialized agency for Information and Communication Technologies (ICTs), whose role is to facilitate international connectivity between communication networks by allocating global radio spectrum and satellite orbits and developing the technical standards that ensure

Space Law,
coordinated by Thomas LECLERC.

networks and technologies seamlessly interconnect, and to strive to improve access to ICTs to underserved communities worldwide. ITU's global membership includes 193 Member States as well as over 900 companies representing the world's leading ICT operators, equipment manufacturers, software developers, service providers, R&D organizations, universities and local, regional and international ICT bodies that seek to reach regional and worldwide consensus about the optimal way to manage the spectrum by modifying and enhancing national and international laws and regulations so as to adapt them to the rapid evolution of technology and uses without jeopardizing the investments already made.

The ITU spectrum and satellite orbit regulatory regime started with the first ever World Administrative Radio Conference in 1963 (WARC-63) that included in the Radio Regulations, first edition in 1906, the basic articles related to space services. Up to the last World Radiocommunication Conference, in 2019 (WRC-19), many ITU conferences have addressed, reviewed and improved the regulation of spectrum and orbit usage by stations of the space radiocommunication services. This legal regime, agreed upon by the ITU Member States at WRCs is codified in the ITU Constitution of 1992 (ITU Constitution; ITU 1992), including the ITU Radio Regulations of 2020 (RR; ITU 2020a).

These instruments contain the main principles and lay down the specific regulations governing the following major elements:

– frequency spectrum allocations to different categories of radiocommunication services;

– rights and obligations of Member administrations in respect of their own and other administrations' spectrum and orbit resources; in particular the international recognition derived from the recording in the ITU Master International Frequency Register (MIFR) of frequency assignments and, as appropriate, orbital information for a space station on-board a geostationary-satellite (GEO) or for space station(s) on-board non-geostationary satellite(s) (non-GEO), used or intended to be used or from their conformity, where appropriate, with a Space Plan.

Although the GSO networks have imposed themselves to provide commercial telecommunication services since the 1960s, technological progress and the ITU's decisions from the last thirty years have set the conditions for launching constellations from a few dozens to thousands of satellites for many different applications from broadband Internet applications to countless narrow band applications via specific smaller non-GSO satellite systems with SDM, including the exponential development of radionavigation satellite system applications.

9.2. International regulations applying to the use of frequencies and orbits particularly by non-GSO FSS satellite systems

Although the first launched satellites at the end of the 1950s and early 1960s were operating on non-GSOs, GSO Fixed Satellite Service (FSS) networks, as defined in the RR No. 1.21[1], quickly became the main means to provide commercial telecommunication services for economic reasons. Indeed, the GSO, a geosynchronous equatorial orbit at 35,786 km above Earth's equator and following the direction of Earth's rotation, allows a satellite to remain fixed in relation to Earth and therefore only requires a single antenna pointed in a fixed direction in space, whereas in the case of a non-GSO system several antennas are required that must constantly change pointing direction to ensure seamless tracking.

The predominance of GSO satellite network use was translated in the ITU regime by RR No. 22.2 specifying that non-GSO systems shall not cause unacceptable interference to nor claim protection from GSO networks in the FSS and the broadcasting-satellite service (BSS), and also a resolution that went even further by prohibiting any satellite other than GSO in the bands allocated to BSS (RR, Resolution 506 [REV.WRC-97]).

With the notable exception of satellites used for scientific purposes and mobile services for which omnidirectional earth stations could operate with non-GSO satellites and systems using a strongly elliptical orbit to provide services to the polar areas, GSO systems were prevalent and the international ITU regulations were based on such systems.

The specific procedures setting out the rights and obligations of ITU Member States for the management of spectrum and orbits are based on two main principles: efficient use and equitable access to the resources with the view to achieve interference-free radiocommunication operation.

These principles are embedded in two major mechanisms for the sharing of spectrum and orbit resources (Henri and Matas 2019):

1 "A radiocommunication service between earth stations at given positions, when one or more satellites are used, the given position may be a specified fixed point or any fixed point within specified areas; in some cases, this service includes satellite-to-satellite links, which may also be operated in the inter-satellite service; the fixed-satellite service may also include feeder links for other space radiocommunication services" (RR No. 1.21).

A priori planning procedures (guaranteeing equitable access to spectrum and orbit resources for *future use*) specifically FSS and BSS frequency ranges (RR, Appendix 30B, 30 and 30A [the associated Plan for feeder links]).

Coordination procedures (with the aim of efficiency of spectrum and orbit use and interference-free operation satisfying *actual requirements*), for satellite systems in all other GSO FSS and BSS frequency bands and space services, including non-GSO systems specifically services and bands (RR, Arts. 9 and 11).

In the early 1990s, the first non-GSO satellite constellation projects started, which capitalized on the economies of scale that a non-GSO system covering the whole planet from the outset would make possible by providing the equivalent of a cellular mobile service with cell phones, then at the dawn of its development.

As a result, the ITU's WRCs, successively from 1992 to 1997, reviewed the regulatory framework for non-GSO systems, assigned a few hundred megahertz to these new applications and defined procedures for access to the radio-frequency spectrum based on the coordination procedure that was at that time included in RR Resolution 46, and then embedded in Article 9 under RR No. 9.11A ("first come, first served" coordination principle vis-à-vis GSO systems, other non-GSO systems and terrestrial services), in other words, the royal path for accessing spectrum, international recognition and gaining a protection conditioned by the successful completion of the coordination procedures.

By 1995, plans for systems with hundreds of non-GSO satellites were being made to offer all countries on the planet access to the Internet. The WRCs in 1997 and 2000 satisfied these requests by allocating several gigahertz spectrum and by adopting (after surmounting major difficulties) a global approach that replaced RR No. 22.2 with a system that had verifiable, mandatory limits (equivalent power flux density (epfd) limits)[2] in order to protect the full spectrum shared with GSO

2 More precisely, the epfd limits of Article 22 of the RR, radiated cumulatively (1) by all the satellites in the constellation and verified at any point on Earth surface, in a set of typical GSO network earth station antennas pointed to any point in the GSO, and for given percentages of time (epfd ↓), (2) by all earth stations of the non-GSO constellation at any point in the GSO (epfd ↑), and (3) by all the satellites of the non-GSO constellation at any point of the GSO (epfd). The calculation necessary to verify the compliance of a non-GSO system with these limits requires modeling the entire non-GSO constellation. For this calculation to be feasible, it is assumed that each satellite in the constellation permanently radiates at maximum capacity, which makes the result more protective for GSO systems.

networks and also deleting Resolution 506 for all bands of the BSS other than those in the Plans of Appendix 30 of the RR. Full details of epfd limit calculations for determination of conformity of non-GSO FSS system networks with the limits contained in Article 22 are described in ITU-R Recommendation S.1503 (ITU 2018).

In the first years of the 21st century, the dot-com bubble overwhelmed most of the planned non-GSO systems for fixed and broadcasting satellite services. However, since 2015, with the availability of launch vehicles capable of supporting multiple satellite launches, mega-constellations consisting of dozens to thousands of spacecrafts are becoming a popular solution again for the provision of Internet in areas not covered by fixed or mobile networks, which still represent most of the inhabited earth's surface.

Recognizing that the regulatory sharing solution that had been adopted by WRC-95 was not suitable in the long-term, being designed on non-GSO technologies of the late 1990s and considered today as quite restrictive for actual non-GSO operation, WRC-19 (while continuing to endorse the existing epfd approach for non-GSO FSS systems in KU and Ka bands[3]) chose new methods of sharing between GSO and non-GSO systems in the Q/V bands (RR No. 22.5L and 22.5M; RR Resolutions 769 and 770) based on a single entry unavailability threshold with RR No. 22.2 continuing to apply. Along with the methodology to be used for the calculation, the conference provided a set of technical characteristics of generic GSO reference links parameters representative of the GSO networks against which the compliance of a non-GSO system into GSO networks with the single entry permissible interference will be validated; notifying administrations of non-GSO constellations were also asked to hold regular meetings to ensure that the potential interference from all non-GSO systems remain below agreed aggregate limits.

3 Ku-band uses frequencies in the 12–18 GHz range, while Ka-band uses frequencies in the 18–31 GHz range. FSS allocations in the Ku and Ka bands have been and are intensively used by GSO networks for FSS applications (in addition to C-band [4–8 GHz] that was the first frequency range used for FSS in the 1970s). To protect those existing and future GSO applications, epfd approach has been introduced as hard power limits to be respected by NGSO FSS systems in these bands. As mentioned in the article, those epfd limits have been agreed upon back in the 1997–2003 time frame based on the NGSO technology at that time…and the fact that none of the foreseen NGSO FSS project materialized then. With the new 2020 technology and experience with real FSS NGSO systems, the NGSO community felt that those epfd limits were too restrictive, but rather than open Pandora's box by discussing it at WRC-19, they decided to keep a low profile for the time being and to discuss instead a new sharing approach in Q/V bands (33–75 GHz).

9.3. Bringing into use satellite networks and systems

As indicated in RR No 9.1, an administration may initiate the recording in the MIFR of frequency assignments to a satellite network or system project up to 7 years before the planned date of bringing these assignments into use as stipulated in RR No. 11.44, which requires that the notified date of bringing into use of any assignment to a space station of a satellite network shall be no later than 7 years following the receipt of the advance publication information (API). The API is considered the starting regulatory date for the registration and recording, in the MIFR, of frequency assignments to a satellite network or system (RR No. 9.1). To the extent that all required mandatory information and actions are undertaken by the notifying administration, the international rights of the administration in respect of the planned, although yet virtual, system would be maintained for that period. However, a major element of the satellite registration procedure to keep these rights over this period would be to submit the notices for recording in the MIFR under Article 11 of the RR, and to bring the satellite network or system frequency assignments into use (BiU) at the latest by the end of the 7-year time limit, together with administrative due diligence information on the spacecraft manufacturer and launch provider (RR Resolution 49).

9.3.1. *Bringing into use GSO networks*

For GSO networks, the regulations have clearly established the rules in:

> A frequency assignment to a space station in the geostationary-satellite orbit shall be considered as having been brought into use when a space station in the geostationary-satellite orbit with the capability of transmitting or receiving that frequency assignment has been deployed and maintained at the notified orbital position for a continuous period of 90 days. The notifying administration shall so inform the Bureau within 30 days from the end of the 90-day period… (RR No. 11.44B)

9.3.2. *Bringing into use non-GSO systems*

However, for non-GSO systems, no specific rules existed, except some principles similar to the GSO network approach in Rule of Procedure on RR No. 11.44, up to WRC-19 that adopted new provisions, in particular for systems in FSS, MSS and BSS:

> A frequency assignment to a space station in a non-geostationary-satellite orbit network or system in the fixed-satellite service, the mobile-satellite service or the broadcasting-satellite service shall be considered as having been brought into use when a space station with the capability of transmitting or receiving that frequency assignment has been deployed and maintained on one of the notified orbital plane(s) of the non-geostationary satellite network or system for a continuous period of 90 days, irrespective of the notified number of orbital planes and satellites per orbital plane in the network or system. The notifying administration shall so inform the Bureau within 30 days from the end of the 90-day period... (RR No. 11.44C)

The same was also agreed, however, without the 90-day obligation for frequency assignments to a space station in a non-GSO system with "Earth" as the reference body for other services than FSS, MSS and BSS (RR No. 11.44D). For non-GSO systems with a reference body that is not "Earth", the requirement was simplified as the notifying administration only has to inform the ITU Radiocommunication Bureau that a space station with the capability of transmitting or receiving that frequency assignment has been deployed in accordance with the notification information (RR No. 11.44E).

In addition to the definition of BiU, and to ensure that the MIFR reasonably reflects the actual deployment of non-GSO satellite systems specifically radio-frequency bands and services, in particular for the low Earth orbit (LEO) satellites, WRC-19 also reached an agreement on a regulatory procedures for the deployment of non-GSO systems: "A milestone-based approach for the implementation of frequency assignments to space stations in a non-geostationary-satellite system specifically frequency bands and services" to apply specifically frequency bands of FSS, MSS and BSS satellite systems (RR, Resolution 35; ITU 2019).

The milestone-based approach also sought to strike a balance between the prevention of radio-frequency spectrum warehousing, the proper functioning of coordination mechanisms and the operational requirements related to the deployment of non-GSO systems.

Under the newly adopted regulatory regime, these systems will have to deploy 10% of their constellation within 2 years after the end of the current regulatory period for BiU, 50% within 5 years and complete the deployment within 7 years. The consequence of failing to meet a milestone for a satellite system would be a proportional reduction in the total number of space stations in the original ITU filing based on a deployment factor applying to the number of space stations declared as deployed at the date of the milestone.

WRC-19, based on the work of ITU-R Study group (ITU 2020b) and administrations' contributions to WRC-19, recognized the inventive and innovative character of the milestone-based approach, and included the possible waiving off for the first milestone for non-GSO satellite systems filed at the ITU before November 2015, however under strict conditions to be reviewed by the ITU Radio Regulations Board (RRB) and ultimately WRC-23. The Bureau has also been invited to identify and report to WRC-23 on specific frequency bands specifically services for which there may be a problem similar to that which resulted in the creation of RR Resolution 35, as well as any difficulties encountered in the implementation of this resolution.

9.4. Radio Navigation Satellite Systems

The Radionavigation Satellite Service (RNSS) is defined in RR No. 1.43 as a radiodetermination satellite service used for the purpose of radionavigation. The RR and the administrations recognize that the safety aspects (RR No. 1.59) of radionavigation services require special measures to ensure a true harmful interference-free environment and therefore to take this factor into account in the allocation and operation of these frequencies.

Today, the RNSS is globally part of all State critical infrastructures and fully integrated in our daily life as, for example, in all mobile phones.

Historically since the 1980s, the RNSS has been operated by the GPS (United States) and GLONASS (Russia) systems in the frequency bands 1,559–1,610 MHz (L1 signals) and 1,215–1,260 MHz bands (L2 signals). Because of the extensive use of both bands, ITU-R WP 4C (ITU 2020c) started in the 1990s an exhaustive study to find additional spectrum (worldwide allocation) in the bands between 1 and 6 GHz to support RNSS developments in both directions of transmission (space-to-Earth and Earth-to-space). These additional RNSS allocations would enable additional RNSS operations with greatly improved performance characteristics (accuracy, availability and continuity, improved ionosphere corrections and phase tracking capabilities), which were essential for the new, robust and precise satellite-based navigation.

As a result of these studies and enhanced requirements, WRC-2000 adopted a new RNSS allocation in the band 1,164–1,215 MHz (L5 signals). Additionally, based on WRC-12 decisions, the Radiodetermination Satellite Service (RDSS) gained a new shared allocation in the band 2,483.5–2,500 MHz. This new RDSS global allocation, because of the proximity to Mobile service (MS) allocations above 2.5 GHz, would offer attractive synergies with terrestrial MS due to use of shared hardware (embedded in the mobile phone) which was not possible with other RNSS bands.

9.4.1. *RNSS band regulations*

9.4.1.1. *Band 1,164–1,215 MHz (L5 signals)*

The band 1,164–1,215 MHz (L5 signals) is shared between the long-standing Aeronautical Radionavigation service (ARNS) and the newcomer RNSS. The sharing conditions are described in RR Resolution 609 (REV.WRC-07) – Protection of aeronautical radionavigation service systems from the equivalent power flux-density produced by RNSS networks and systems in the 1,164–1,215 MHz frequency band and RR Resolution 610 (REV.WRC-19) – Coordination and bilateral resolution of technical compatibility issues for RNSS networks and systems in the bands 1,164–1,300, 1,559–1,610 MHz and 5,010–5,030 MHz. Administrations are recalled that the implementation of Resolution 609 shall be conducted in such a way as to promote the principle of equality and fairness in ensuring access for RNSS operators and planned RNSS systems in the relevant bands. Forms of coordination RR No. 9.12 (non-GSO against non-GSO systems), RR No. 9.12A (non-GSO systems against GSO networks) and RR No. 9.13 (GSO networks against non-GSO systems) of RR apply to the RNSS.

The basic concept of the RR Resolution 609 to protect the ARNS from harmful interference (HI) (RR No. 1.169) is based on the aggregate epfd calculation (Recommendation ITU-R M.1642-2) of all “actual” RNSS systems only (Milestone criteria) which shall not exceed the level of −121.5 dB (W/m^2) in any 1 MHz band:

– To achieve the objectives of this Resolution, administrations operating or planning to operate RNSS systems will need to agree cooperatively through Consultation meetings to achieve the level of protection for ARNS systems that is stated in this Resolution.

– No single RNSS system shall be permitted to use up the entire interference allowance.

– Based on Recommendation 608 Guidelines for Consultation meetings established in RR Resolution 609, the maximum pfd produced at the surface of Earth by emissions from a single space station in the RNSS, for all angles of arrival, should not exceed −129 dB (W/m^2) in any 1 MHz band under free space propagation conditions.

– Administrations operating or planning to operate RNSS systems or networks in this band shall, in collaboration, take all necessary steps, including, if necessary, by means of appropriate modifications to their systems or networks, to ensure that the aggregate interference into ARNS systems caused by such RNSS systems or networks operating co-frequency in this frequency band is shared equitably among the systems identified and does not exceed the level of the aggregate protection criterion given in this Resolution.

– Currently (November 2021) – based on 18th RES-609 Consultation meeting – the maximum aggregate epfd of satellites associated with the referenced RNSS networks and systems is determined to be no greater than –121.71 dB (W/(m^2·MHz)), that is, 0.21 dB below the Resolution 609 limit of –121.5 dB (W/(m^2·MHz)). It is noted that the result is based on the use of worst-case assumptions in terms of interference from RNSS into ARNS.

9.4.1.2. *Band 1,215–1,300 MHz (L2 signals)*

In the band 1,215–1,300 MHz (L2 signals), the RNSS is not operating independently from the other radiocommunication services in the band, but under the condition that no HI is caused to, and no protection is claimed from, the Radionavigation service (RNS) authorized under RR No. 5.331, and subject to the condition that no HI is caused to the Radiolocation service (RLS). RR Resolution 608 (WRC-03) guarantees no constraint in addition to the operating RNSS systems (GPS and GLONASS) in the frequency band 1,215–1,260 MHz brought into use before June 3, 2000.

9.4.1.3. *Band 1,559–1,610 MHz (L1 signals)*

The band 1,559–1,610 MHz (L1 signals) was used up to January 1, 2015 also by the Fixed (terrestrial) service (FX) on a secondary basis, but today, ITU RR (edition 2020) Article 5 Table of Frequency Allocations, is only allocated to two global co-primary services: RNSS and Aeronautical Radionavigation service (ARNS). The use of this band by the RNSS is subject to the application of the provisions of RR No. 9.12, 9.12A and 9.13. RR Resolution 610 shall also apply.

9.4.1.4. *Band 1,559–1,610 MHz (L1 signals)*

In the band 2,483.5–2,500 MHz, the RDSS is a co-primary allocation with FX, MS, MSS (Mobile-Satellite) services. The use of the band 2,483.5–2,500 MHz by the MSS and RDSS is subject to the coordination under RR No. 9.11A. Administrations are also urged to take all practicable steps to prevent harmful interference to the Radio Astronomy Service (RAS), especially those caused by second-harmonic radiation that would fall into the 4,990–5,000 MHz band allocated to the RAS worldwide. This band is also designated for Industrial, Scientific and Medical (ISM) applications (RR No. 1.15). Radiocommunication services operating within this band must accept harmful interference which may be caused by these applications. ISM equipment operating in this band is subject to the provisions of RR No. 15.13.

9.4.2. *ITU-R Recommendations related to RNSS*

The ITU-R Recommendations (ITU 2020d) constitute a set of international technical and global standards developed by the Radiocommunication Sector of the ITU.

The ITU-R Recommendations are approved by ITU Member States. However, their implementation is not mandatory. As they are developed by experts from administrations, operators, the industry and other organizations dealing with radiocommunication matters from all over the world, they enjoy a high reputation and are implemented worldwide.

The ITU-R Recommendations (M series) provide protection criteria for RNSS systems and RNSS receiver types for particular applications.

9.5. Non-GSO Satellite systems with SDM

WRC-19, following its Agenda Item 7(I) – Modified regulatory procedure for non-GSO satellite systems with SDM – and taking into account results of the studies at ITU-R WP 4A (ITU 2020b) and WP 7C (ITU 2020e), agreed on the regulatory definition of SDM satellite systems, and adopted specific procedures as described in RR Resolution 32 (ITU 2019) – Regulatory procedures for frequency assignments to non-geostationary-satellite networks or systems identified as SDM not subject to the application of Section II of RR Article 9.

WRC-19 also took into account the results of ITU-R studies, as described in Report ITU-R SA.2312, indicating that the SDM satellites typically have a short (1–2 years) development time; are low cost, often using low cost off-the-shelf components; with a short operational lifetime generally from several weeks up to few years; and finally that SDM satellites are being used for a wide variety of applications, including remote sensing, space weather research, upper atmosphere research, astronomy, communications, technology demonstration and education.

WRC-19 also considered the results of ITU-R Report SA.2425 (Studies to accommodate spectrum requirements in the Space Operation service [SOS] for SDM satellite systems), ITU Report SA.2426 (Technical characteristics for telemetry, tracking and command in the SOS below 1 GHz for SDM satellite systems and ITU) and ITU Report SA.2427 (Studies on the suitability of existing allocations to the space operation service below 1 GHz and additional sharing studies on possible new and/or upgraded allocations).

Regarding the registration procedures for such systems, WRC-19 concluded that the application of Articles 9 and 11 of the RR to frequency assignments to SDM satellites should not adversely affect the regulatory treatment of other satellite systems or should not change the sharing status with respect to networks and systems not applying the modified regulatory procedure as described in RR Resolution 32.

9.5.1. *Non-GSO SDM satellite system regulations*

An SDM satellite system, based on RR Resolution 32, is defined as a satellite system with the total number of satellites not exceeding 10 satellites (in case of constellation) and with the maximum period of operation and validity of frequency assignments not exceeding 3 years from the date of bringing into use of the frequency assignments – which is defined as the launch date associated with the first launch (in the case of systems with multiple launches) when the first SDM satellite is placed into its notified orbital plane. There is no possibility of extension of this period of operation and validity, and the Bureau is instructed to cancel the recorded assignment of the SDM satellite after the 3-year operation.

9.5.2. *Space Operation Service frequency bands for non-GSO SDM satellite systems*

WRC-19, in accordance with its Agenda Item 1.7 (Study the spectrum needs for TT&C in the Space Operation Service [SOS] [RR No. 1.23] for SDM satellite systems), added the use of the space operation service (space-to-Earth) for non-GSO SDM systems to the existing SOS allocation in the frequency band 137–138 MHz, subject to RR Resolution 660 (RR No. 5.203C and 5.209A) and allocated the frequency band 148–149.9 MHz for the space operation service (Earth-to-space) by non-GSO SDM systems (RR No. 5.218A).

9.6. Conclusion

Due to the ever-ending progress of technology and the inherent development in the operation of the spectrum and satellite orbits, the management of this natural and limited resources has become a vital issue within national and international decision-making bodies. In line with its commitment to connecting all people around the world – wherever they live and whatever their means – and to protect and support everyone's right to communicate, ITU has developed a legal, technical and regulatory regime, through strategic cooperation and understanding between all stakeholders, to remove, to the extent possible, all obstacles in particular impeding the development and bringing into operation of new satellite networks and systems. As focused on in this document, the last 30 years have paved the way to the emergence of satellite constellations with thousands of satellites, providing broadband Internet access, but also more affordable small and SDM satellites with a myriad of new applications, many still to be invented, and the exponential development of radionavigation satellite system applications… with the ultimate goal to connect the unconnected.

9.7. References

Henri, Y. and Matas, A. (2017). RNSS and the ITU Radio Regulations. *Inside GNSS* [Online]. Available at: https://www.insidegnss.com/wp-content/uploads/2018/04/janfeb18-LAW.pdf [Accessed 20 September 2020].

Henri, Y. and Matas, A. (2019). Space security and frequency management. In *Handbook of Space Security. Policies, Applications and Programmes*, 2 Volumes, 2nd edition, Schrogl, K.-U., Adriaensen, M., Giannopapa, C., Hays, P.L., Robinson, J., Antoni, N. (eds). Springer, New York.

ITU (1992). ITU Constitution. Collection of the basic texts of the International Telecommunication Union adopted by the Plenipotentiary Conference [Online]. Available at: https://www.itu.int/pub/S-CONF-PLEN-2011 [Accessed 20 September 2020].

ITU (2018). Recommendation S.1503 [Online]. Available at: https://www.itu.int/rec/R-REC-S.1503/en [Accessed 22 April 2012].

ITU (2019). Final Acts WRC-19 [Online]. Available at: https://www.itu.int/pub/R-ACT-WRC.14-2019/en [Accessed 20 September 2020].

ITU (2020a). ITU Radio Regulations 2020 [Online]. Available at: https://www.itu.int/pub/R-REG-RR/en [Accessed 20 September 2020].

ITU (2020b). Working Party 4A (WP 4A) – Efficient orbit/spectrum utilization for FSS and BSS [Online]. Available at: https://www.itu.int/en/ITU-R/study-groups/rsg4/rwp4a/ [Accessed 22 April 2012].

ITU (2020c). Working Party 4C (WP 4C) – Efficient orbit/spectrum utilization for MSS and RDSS [Online]. Available at: https://www.itu.int/en/ITU-R/study-groups/rsg4/rwp4c/ [Accessed 22 April 2012].

ITU (2020d). ITU-R recommendations [Online]. Available at: https://www.itu.int/pub/R-REC [Accessed 20 September 2020].

ITU (2020e). Working Party 7B (WP 7B) – Space radiocommunication applications [Online]. Available at: https://www.itu.int/en/ITU-R/study-groups/rsg7/rwp7b/ [Accessed 22 April 2012].

ITU (2023a). Radio conferences [Online]. Available at: https://www.itu.int/en/history/Pages/RadioConferences.aspx?conf=4.89 [Accessed 22 April 2012].

ITU (2023b). ITU-R Preparatory Studies for WRC-23 [Online]. Available at: https://www.itu.int/en/ITU-R/study-groups/rcpm/Pages/wrc-23-studies.aspx [Accessed 22 April 2012].

ITU (2023c). Resolution 609 (Rev. WRC-07) consultation meeting [Online]. Available at: https://www.itu.int/ITU-R/go/space-resolution609/en [Accessed 22 April 2012].

ITU (2023d). RES-609 CM decisions and results [Online]. Available at: https://www.itu.int/en/ITU-R/space/Pages/RES-609-CM-Decisions-and-Results.aspx [Accessed 22 April 2012].

10

Satellite Radio Communications

Philippe ACHILLEAS
Université Paris-Saclay, France

10.1. Introduction

The launch in 1957 of the first artificial satellite, *Sputnik*, placed in a low orbit and emitting its famous "beep-beep", paved the way for space radiocommunications. According to the definition given by the International Telecommunication Union (ITU) in the radio regulations (RR), radiocommunications represent any "telecommunication using radio waves". Contrary to popular belief, telecommunications are not limited to voice communications, but include all means of transmitting information remotely by electronic means, regardless of the technology used and the content transmitted[1]. Today, the vast majority of wireless communications are based on terrestrial networks, in particular the networks of mobile operators, but also on complementary networks such as Wi-Fi. However, in the context of the digital transition of territories and the fight against the digital divide, satellite represents a very effective complementary tool for connecting isolated regions. In addition, satellites remain the best way to obtain earth observation data or to provide location-based services. Since the beginning of the conquest of space, satellites have provided services that have become so familiar to

---

1 According to the ITU, the word telecommunication means "Any transmission, *emission* or reception of signs, signals, writings, images and sounds or intelligence of any nature by wire, *radio*, optical or other electromagnetic systems" (ITU, *Radio Regulations, edition 2020*, ITU, Geneva, 2020, Number 1.3).

us that we often forget their origin in space: telecommunications, media and video, remote sensing, meteorology, navigation, etc. For a long time reserved for the space powers alone, access to space has been democratized with the multiplication of governmental programs and private initiatives encouraged by the development of small, less expensive satellites and by the lowering of launch costs. We are thus witnessing an acceleration in the pace of satellite launches. A record number of satellites were launched into orbit in 2020, according to the Satellite Industry Association's (SIA) State of the Satellite Industry Report. At the end of 2020, there were 3,371 satellites in orbit around Earth, a 37% increase in 2019. Forecasts for the next few years predict an explosion in the number of objects in orbit in the context of the deployment of satellite constellations designed to provide very high-speed services from space. For example, if the major projects are completed, there will be 50,000 new devices, including 42,000 for *SpaceX*'s *Starlink* constellation alone.

Satellite operations are based on the use of frequencies and orbits. The "spectrum-orbit" resource is thus essential to the development of space activities. The exploration and exploitation of outer space depend on its use. According to the ITU, the frequency spectrum encompasses "electromagnetic waves of frequencies arbitrarily lower than 3,000 GHz, propagated in space without artificial guide" (RR, number 1.5). The use of frequencies enables communication with and from the satellite to ensure both its operation and its exploitation. The orbits represent the trajectories that the satellites draw in space around the Earth under the effect of gravity. The geostationary orbit, located at 36,000 km from the Earth, is of particular interest to satellite operators insofar as an object placed in this orbit maintains a fixed position relative to a point on the Earth.

Access to this dual resource therefore has a strategic dimension for States and operators.

The international regime for the use of the "spectrum-orbit" resource derives, on the one hand, from the ITU Constitution of 1992 and the ITU Convention of 1992. On the other hand, it is set out in the RR, which has the legal value of a treaty. The document is revised periodically within the framework of the World or Regional Radiocommunication Conferences (WRC/RRC). The last modification of the RR took place during the WRC organized in Sharm el-Sheikh, Egypt, from October 28, 2019 to November 22, 2019.

It is mainly in order to avoid interference between stations belonging to different administrations and to guarantee equitable access to the "orbital spectrum" resource by all States that its use is the subject of institutionalized international cooperation under the responsibility of the ITU. Within this specialized agency of the United

Nations (UN) system, it is the Radiocommunication Sector (ITU-R), and more particularly the Radiocommunication Bureau, which is responsible for coordinating the needs of States in this area.

The rules defined by the ITU take into account a physical reality: the scarcity of the "spectrum-orbit" resource. Indeed, the spectrum of frequencies allowing the transmission of information is not extensible, while the needs in communication, supported by the development of the digital society, increase rapidly. On its side, the geostationary orbit knows a limited number of positions[2]. The scarce and limited character of the resource is underlined by the Constitution of the ITU:

> Member states shall bear in mind that radio frequencies and any associated orbits, including the geostationary-satellite orbit, are limited natural resources and that they must be used rationally, efficiently and economically, in conformity with the provisions of the Radio Regulations, so that countries or groups of countries may have equitable access to those orbits and frequencies, taking into account the special needs of the developing countries and the geographical situation of particular countries (ITU Constitution, Art. 44, 2).

The ITU texts thus organize the regime for the use of the "spectrum-orbit" resource by respecting the principle of cooperation between States (section 10.2). A large number of the procedures established by the organization relies on the work of national administrations, which are responsible for the implementation of ITU regulations in good faith. The multiplication of the number of satellites launched and their growing importance in the provision of digital services has led the international community to take measures to guarantee the use of the "spectrum-orbit" resource in the interest of all States (section 10.3).

10.2. The regime for the use of the "spectrum-orbit" resource

Despite the nature of the "spectrum-orbit" resource, which is characterized by scarcity and non-appropriation, since it is located in outer space (see above), it is

2 The number of available orbital positions depends on the frequency bands. For example, 120 positions can be used in C-band (fixed satellite service) and 180 in Ku-band (broadcasting satellite service and fixed satellite service). It is possible to place several satellites in the same orbit if they do not use the same frequencies or if they are jointly operated by the same operator. The problem is that operators are targeting strategic positions from a commercial point of view, to cover Europe, North America and Asia-Pacific in particular.

managed directly by States, in accordance with the principle of the sovereign right of each State to regulate its telecommunications, as recognized by the ITU (ITU Constitution, Preamble, para. 1). The organization is therefore neither a regulatory authority nor a spectrum resource management agency. Its mission here is to coordinate the efforts of States to achieve rational and efficient use of the spectrum and orbits. Since its creation in the 19th century, the ITU's role in radiocommunications has changed little and its methods of action have remained traditional, despite the evolution of technology and the ever-increasing importance of the spectrum in our societies (Watine and Gourrierec 1978). Thus, the organization proposes a set of procedures for the use of the resources necessary for space radiocommunications. It sets out the stages of the distribution and use of the "spectrum-orbit" resource (section 10.2.1), the rights and obligations of operators and administrations (section 10.2.2) as well as control and dispute settlement mechanisms (section 10.2.3).

10.2.1. *The distribution stages and use of the "spectrum-orbit" resource*

The first step is to allocate frequencies to the various services by assigning bands dedicated to one or more services (section 10.2.1.1). The national administrations will then authorize the use of the frequencies by means of assignments (section 10.2.1.2) after coordinating to avoid harmful interference (section 10.2.1.3). The recording of assignments in the ITU provides international protection against interference (section 10.2.1.4). Since 2019, a simplified regime has been adopted to govern the launch of small satellites on short missions (section 10.2.1.5).

10.2.1.1. *Allocation of frequency bands*

ITU Member States make frequency band allocations through WRC/RRC. An allocation of a frequency band means an "Entry in the Table of Frequency Allocations of a given frequency band for the purpose of its use by one or more terrestrial or space radiocommunication services, or the radio astronomy service under specified conditions" (RR, No. 1.16). Thus, the usable part of the radio spectrum is divided into bands that are reserved for use by one or more radiocommunication services, that is, a service "involving the transmission, emission and/or reception of radio waves for specific telecommunication purposes" (RR, number 1.19). The RR identifies 39 radiocommunication services (RR, numbers 1.20–1.60) that are divided between terrestrial services (fixed, mobile, broadcasting, radiodetermination, meteorology, radioastronomy, amateur, standard frequencies and time signals) and space services (fixed-satellite, mobile-satellite, radiodetermination-satellite, meteorology-satellite, earth exploration-satellite, space research, amateur-satellite and space operation). The frequency bands are allocated to the various services on an exclusive basis or shared. The allocations are necessarily compatible with each other. A distinction is also made between global and regional allocations. For this purpose, the ITU divides the world into three major regions under the so-called allotment procedure. Region I includes Europe, the former USSR, Turkey and the People's Republic of Mongolia as well as Africa. Region II includes the American continent and Region III Asia and Oceania (RR, numbers 5.2–5.22). In principle, the allocations are different for the three ITU regions, except for the global services.

The complete list of services and frequency bands allocated in the different regions is the Table of Frequency Allocations. It is an integral part of the RR. The national Tables of Frequency Allocations are modeled on the one developed in the ITU. The Table of Frequency Allocations can only be modified by a WRC/WRC, where States negotiate to accommodate additional frequency requests for new services while protecting existing services. Decisions are adopted by consensus or, if necessary, by vote.

10.2.1.2. *Frequency assignment*

The use of the "spectrum-orbit" resource is based, with some exceptions, on an administrative authorization. The assignment of a radio frequency or channel is thus an "authorization given by an administration for a radio station to use a radio frequency or radio frequency channel under specified conditions" (RR, No. 1.18). Therefore, each State identifies or establishes one or more administrations responsible for issuing assignments. In the United States, this mission falls to the Federal Communications Commission (FCC) for commercial services and to the National Telecommunications and Information Administration (NTIA) for government services. In the French national context, the *Agence Nationale des Fréquences* (ANFR) is the sole interlocutor with the ITU for all matters relating to the use of frequencies and issues authorizations for space services. The assignment must comply with the requirements of the International Table of Frequency Allocations (RR, No. 4.4). Furthermore, it can only be issued if it complies with the procedure for the use of the "spectrum-orbit" resource. National administrations may nevertheless make assignments that do not conform to the Table of Frequency Allocations or to other provisions of the RR, provided that they do not cause interference with a foreign service (RR, number 8.4). The assignment issued is accompanied by a set of technical information (frequency, class of station and bandwidth).

10.2.1.3. *International coordination*

For space applications, Member States mainly use the international coordination procedure between interested administrations provided for in Article 9 of the RR[3]. Coordination is required for fixed satellite services, broadcasting satellite and mobile satellite services because of the significant risk of interference. Thus, an entity wishing to use a frequency must apply to its national administration, which must obtain the agreement of all foreign administrations whose services already registered in the ITU or in the process of being registered could be affected by such use. For space services, administrations shall, under the advance publication of information (API) on satellite networks or satellite systems, inform others of their willingness to assign space frequencies without receiving any fee. To this end, they shall send to the Radiocommunication Bureau a general description of the planned network not earlier than 5 years and preferably not later than 2 years before the planned date of bringing it into use (RR, numbers 9.1 et seq.). The Bureau publishes this information in its weekly circular. On the basis of this information, it is up to the administrations concerned to resolve any difficulties through negotiations. In the event of disagreement, the Radiocommunication Bureau will act as mediator and may, if the disagreement persists, provide an unfavorable finding against the project.

10.2.1.4. *Notification and registration of assignments at the IUT*

Any frequency assignment must in principle be notified to the Radiocommunication Bureau. This is particularly the case for the majority of space services (RR, number 11.2). Administrations must therefore send the notices to the Bureau not earlier than 3 years before the date of bringing into use of these assignments (RR, number 11.15)[4] . The Bureau examines the notices in the order in which they are received (RR, number 11.28). This is the famous "first come, first served" principle, according to which the first administration to file its file is the first to obtain international protection of the assigned frequencies. Each notice is examined from the point of view of its conformity with the Table of Frequency Allocations and other provisions of the RR, but also of the probability of harmful interference that may be caused to or by assignments recorded with a favorable

3 For terrestrial services, an alternative procedure is preferred. States shall conclude special agreements in accordance with the provisions of the Radio Regulations. The Secretary General shall be informed.

4 To notify a frequency assignment, the administration shall provide the relevant characteristics listed in Appendix 4 Annex 2 (RR, number 11.15).

finding[5]. Where appropriate, the assignment is examined for conformity with a global or regional plan and its related provisions[6]. When the examination results in a favorable finding, that is, when the assignment is in conformity with ITU rules, the assignment is recorded in the Master International Frequency Register (MIFR). This database contains the frequency assignments and their characteristics as notified to the ITU in accordance with Article 11 of the RR. When the conclusion is unfavorable, the assignment can be recorded in the MIRF for information, if the administration makes a commitment that it will be operated without causing interference to a foreign service. This is called a non-conforming assignment. Otherwise, the notice is returned to the notifying administration with an indication of the action to be taken (RR, number 11.36). The rights and obligations of administrations with respect to their own and other administrations' frequency assignments depend on their entry in the MIFR.

For non-coordinated satellite networks, the procedure can be relatively quick. A national administration can send its notice to the ITU 6 months after the publication of the API by the Radiocommunication Bureau. In practice, some administrations send the API and the notification at the same time. This is the case for space agency satellites, for example, when they are pre-coordinated with each other. The ITU then considers that the notification has been received 6 months after the publication of the API.

10.2.1.5. *Special rules for small satellites*

Since the beginning of the 2000s, there has been an increase in the number of small satellites with a mass of less than 100 kg. This category includes nanosatellites whose mass is generally between 1 and 10 kg, as well as picosatellites, whose mass is usually between 0.1 and 1 kg. These satellites are generally developed in a short time (1–2 years) and carry out a short mission (between several weeks and 3 years at most). Their design is based on inexpensive and commercially available components.

These objects thus provide States with a low-cost means of access to space. They are used by national agencies or universities for scientific or educational purposes, but also by private companies that exploit them for commercial purposes. They are

---

5 The main task is to verify that the assignment corresponds to an allocation in the Table of Frequency Allocations and, if applicable, that it complies with other technical or operating conditions set out in the RR (power limits, authorized classes of emission, minimum elevation angle).

6 Planning is a procedure put in place to ensure equitable access to the "spectrum-orbit" resource. See above.

used in various fields: remote sensing, communication, space weather research, upper atmosphere research, astronomy, etc. The frequencies they use go beyond the framework of telecommunications (fixed and mobile) or broadcasting.

Several small satellites have been launched without having complied with ITU rules, often due to a lack of knowledge about the applicable regime. The organization decided to develop texts and reports on small satellites, providing detailed information that would increase awareness of the procedures applicable to the submission of satellite network notices to ITU[7]. An initial guide published in 2015 by the Office for Outer Space Affairs and the ITU entitled *Guidance on Space Object Registration and Frequency Management for Small and Very Small Satellites* thus represents a good basis for knowing the applicable legal and regulatory framework.

WRC-19 adopted a resolution to simplify the regulatory framework applicable to non-geostationary satellites with short missions[8]. The aim is to set up regulatory procedures that take into account the short development cycle, the limited useful life of these satellites and the typical missions they perform by adapting the application of certain provisions relating to international coordination (RR, Art. 9) and to the notification and recording of frequency assignments (RR, Art. 11).

The resolution applies only to non-geostationary satellites identified by the notifying administration as carrying out short-term missions (maximum 3 years) and operating on assigned frequencies, which are not subject to the coordination provisions under Section II of Article 9 of the RR. In principle, the resolution does not concern fixed, mobile and broadcasting services. Furthermore, the resolution does not apply to non-geostationary networks and systems with a total number of satellites exceeding 10.

10.2.2. *Obligations and rights of administrations and operators*

National administrations must provide the Radiocommunication Bureau with evidence of the existence of the project, failing which the Bureau will delete the entry in the MIFR (section 10.2.2.1). They must also ensure that the satellite is

7 ITU (2015). Improving the dissemination of knowledge concerning the applicable regulatory procedures for small satellites, including nanosatellites and picosatellites, Resolution UIT-R 68.

8 ITU (2019). Resolution 32 (WRC-19) Regulatory procedures for frequency assignments to non-geostationary-satellite networks or systems identified as short-duration mission not subject to the application of Section II of Article 9.

brought into service within 7 years of starting the international coordination procedure (section 10.2.2.2) and in compliance with the characteristics notified to the ITU (section 10.2.2.3). Once in operation, the space object has a right to be operated without causing interference (section 10.2.2.4).

10.2.2.1. *The proof and existence of the project*

One of the contested practices that marked the end of the 20th century involved the misuse of the "first come, first served" rule for speculative purposes. States filed applications for coordination even though they were not sure they would be able to launch a satellite, or even if they knew they would not launch a satellite. These fictitious projects, whose main interest was to obtain frequency rights, are called "paper satellites". The issue of paper satellites came to light following developments in the *Tongasat* case. Between 1988 and 1990, this private company based in the Tonga archipelago presented to the ITU a project concerning 16 orbital positions over the Pacific Ocean. This request, totally disproportionate to the needs of the archipelago, masked the desire to monetize the rights associated with the frequencies and orbital positions to operators who wished to carry out trans-Pacific communications. Other States followed Tonga's example, such as Papua New Guinea, Indonesia, Brazil and Laos, causing disputes with operators wishing to use frequencies already assigned. This unprecedented increase in the number of notices sent to the ITU has had the effect of lengthening the time required by the Bureau to process requests. The main consequence of the development of paper satellites for national administrations remains the impossibility of assigning frequencies recorded in the MIFR or in the process of being recorded, even though they will never be used. This practice blocks access to the spectrum and therefore to space.

To combat the proliferation of paper satellites, Resolution 49, adopted by WRC-97 (REV.WRC-12), requires administrations requesting coordination for a satellite network to prove their good faith by demonstrating the reality of the project. This includes providing the Bureau with information on the identity of the satellite network, the spacecraft manufacturer and the launch service provider. The information must be sent to the Office within 7 years of the API. If the required information is not submitted on time or is incomplete, the networks will no longer be considered and the Bureau will remove the MIFR listing.

10.2.2.2. *Commissioning within 7 years*

The notifying administration must undertake to bring into use the assignments recorded in the MIFR within seven years of the start of the coordination procedure, that is, the receipt by the Bureau of the complete satellite project information (RR, numbers 9.1 and 9.2). Any frequency assignment that is not brought into use within the required time period shall be canceled by the Bureau, which shall notify the administration at least 3 months before the expiry of that period (RR, number 11.44). Finally, the administration shall notify the Radiocommunication Bureau of any change in the characteristics of the assignment concerned so that it can be reflected in the MIFR, if necessary, after further coordination with the administrations of the countries concerned.

The 7-year rule may pose problems for satellite constellation operators. Due to constraints related to the design of the system as well as the availability of launchers, the commissioning of these networks may exceed the regulatory period imposed by the RR. Furthermore, in the absence of a special regime for constellations, a partial or even very partial deployment before the end of the regulatory period for a constellation allows the rights to use the "spectrum-orbit" resource obtained by recording the assignments in the MIFR to be retained. At WRC-19, States decided to adopt a new phased approach for the deployment of non-geostationary satellites, in particular constellations using low Earth orbit[9]. Thus, beyond the basic regulatory deadline of 7 years, 10% of the satellites making up the constellation must be deployed within 2 years, 50% within 5 years and 100% within 7 years. In total, the deployment and commissioning of a constellation can be spread over a period of 14 years.

10.2.2.3. *Regular commissioning*

Administrations shall ensure that the recorded assignments are put into normal use with the notified required characteristics. If it appears, on the basis of reliable information available, that an assignment is not being used regularly, the Bureau may decide to cancel the recording (RR, number 13.6). Apart from the case of registration of fictitious satellites, practice has shown two examples of non-regular use.

Some administrations have found that the condition is met regardless of the duration of the entry into service. The satellite is thus operated for at least 1 day at the notified position, but is then moved to its final location. Thus, the ANFR sent the ITU Bureau a file attesting to the effective bringing into use by Eutelsat of the assignments relating to the orbital position at 29°E recorded in 1989. In fact, Eutelsat had carried out in-orbit tests in 1996 and then 1998 in order not to lose the rights pending the launch of its Europesat direct broadcasting satellite scheduled for 2000. However, in August 1997, the operator SES began occupying the position at

9 ITU (2019). Resolution 35 (WRC-19) A milestone-based approach for the implementation of frequency assignments to space stations in a non-geostationary-satellite system in specific frequency bands and services.

28.2°E. Eutelsat pointed out that the Europesat program had been under development since 1996 and invoked the right to international protection against interference. The Radiocommunication Bureau then ruled that the regulatory deadline had not been met, implicitly considering that simple in-orbit tests could not be taken into account (Malavialle et al. 1999, pp. 183–184).

A second practice involves bringing an assignment into service without respecting the declared technical characteristics. Either all the frequencies recorded are not actually used or the coverage area notified is not identical. For example, a dispute arose between the United Kingdom and Argentina concerning the *P.P.Sat-1* satellite, which the British Administration considered had not been brought into use in accordance with the RR at 81° W. In its initial notification, Argentina provided technical specifications including two beams for the provision of service to the Northern and Southern Hemispheres. Subsequently, Argentina entered into a contract to relocate the Canadian satellite *Anik E2* to 81° W. However, according to the information provided by the United Kingdom, the technical specifications of the relocated satellite could not provide a service in the southern hemisphere[10]. The Administration of the United Kingdom, which notified a satellite network with priority right to the ITU, just after the satellite *P.P.Sat-1*, at the orbital position 81° W, requested the Bureau to return the notice submitted by Argentina and to cancel the frequency assignment to this satellite. Without ruling on the merits of the case, the RR Board was unable to grant the United Kingdom's request for procedural reasons[11].

In view of these converging interpretations, WRC-12 added new provisions in the RR to define the concept of bringing into use of a frequency assignment to a space station in the geostationary-satellite orbit. It is now established that an assignment is brought into use when the satellite has been deployed to the notified orbital position, is maintained at that position and operates on the assigned frequency for a continuous period of ninety days (RR, number 11.44 B). Furthermore, as from January 1, 2013, the maximum period of suspension of the use of a frequency assignment to a space station is 3 years, which corresponds to the duration of in-orbit replacement (RR, number 11.49, para 5.2.10 of Appendices 30 and 30A, Rules of Procedure relating to number 11.49 and paragraph 8.17 of Appendix 30B). Administrations shall apply to the Bureau. The procedure for reinstatement of a suspended frequency assignment is similar to the procedure for initial bringing into use. The Bureau shall cancel the MIFR recording if the deadlines are exceeded.

---

10 RR Committee (2006). Document RRB06-1/6-E.

11 RR Committee (2006). Doc. RRB06-1/9(rev.1)-E, pt 9.

10.2.2.4. *The right not to be interfered with*

Interference is one of the main concerns of any radio operator. It can be unintentional or intentional in nature. For this reason, the ITU has put in place regulations to prevent and deal with interference.

Thus, any frequency assignment recorded in the MIFR with a favorable conclusion is entitled to international recognition against harmful interference[12]. Therefore, other administrations must take this into account when making their own assignments (RR, number 4.2). Member States also undertake to ensure that stations under their jurisdiction are operated in such a way as not to cause harmful interference to the services of other States whose assignments are recorded in the Register (ITU Constitution, Art. 45). If an administration becomes aware of interference from a station under its jurisdiction, it must take the necessary action (RR, number 15.21). Non-conforming assignments recorded for information purposes are not entitled to such protection[13].

National administrations must also carry out emission monitoring to detect and sanction possible infringements of ITU provisions, in particular interference-free operation (RR, numbers 15.19–15.21). When harmful interference occurs despite compliance with ITU rules, States must comply with a specific procedure based on goodwill and cooperation between administrations (RR, No. 15.22 et seq.).

Despite the ITU regulatory framework, there has been a significant increase in interference, particularly between 2009 and 2012. Studies conducted by Eutelsat have shown an increase in interference affecting their networks of 130% between 2011 and 2012. The share of intentional interference increased from 5% in 2009 to 18% in 2012. The duration of the interference also increased. The sources of the interference were mainly from Iran, Syria and Ethiopia (Bonte 2015, p. 55). In 2012, Eutelsat filed a complaint with the ITU for the systematic and massive interference of three of its satellites affecting several international media outlets among them, *BBC Persian*, *Voice of America* in Farsi and *Radio Free Europe/Radio Liberty's Radio Farda* services[14]. These interferences were attributed to Iran. It was clear that

12 Harmful interference is "interference which impairs the operation of a radionavigation or other safety service or which seriously degrades, repeatedly interrupts, or prevents the operation of a radiocommunication service used in accordance with the Radio Regulations" (CS, Sch., No. 1003).

13 In addition, if their use causes interference to a station operating under an ITU notification procedure, the user station must immediately cease all transmissions (RR, number 8.5).

14 Eutelsat (2012). Eutelsat condemns acts of jamming from Iran and calls on international regulators to take decisive action, CP/62/12.

the target of the interference was news programs and documentaries dealing with the political situation in Iran. In 2014, Eutelsat was the victim of another source of interference attributed to Ethiopia targeting the *Oromia Media Network* channel broadcast by the Ethiopian diaspora from Minneapolis. The international organization Arabsat also reportedly suffered jamming targeting the same media (De Selding 2014).

We can see that, despite the establishment of very strict procedural rules, disputes arise between administrations regarding the use of the "spectrum-orbit" resource.

10.2.3. *Control mechanisms and dispute resolution*

The Union functions on the model of a cooperative organization and exercises its competences with respect to the sovereignty of its members. It is therefore understandable that it has a general framework for settling disputes, supplemented by specific procedures, which are very respectful of sovereignty (section 10.2.3.1). In recent years, however, the ITU has sought to strengthen its powers of control and dispute settlement, in particular from the point of view of identifying and designating the author of a violation of the organization's rules and procedures (section 10.2.3.2).

10.2.3.1. *A priority for political settlement methods*

The ITU's dispute resolution mechanisms are based on cooperation between States and administrations. Article 56 of the ITU Constitution states that Member States may settle their disputes by negotiation, diplomatic means, or by procedures established by the parties to the dispute. In the event that none of these means of settlement is adopted, a Member State party to a dispute may have recourse to arbitration (ITU Convention, Art. 41). States may also refer disputes to the ITU bodies. Thus, a dispute may be brought to the attention of the Plenipotentiary Conference, the supreme body of the Union. Composed of delegations representing Member States, the Plenipotentiary Conference is convened every 4 years (CS, Art. 8). It can adopt decisions to settle a dispute or condemn the practices of a member state.

There are three bodies within the ITU Radiocommunication Sector that can deal specifically with disputes relating to the spectrum-orbit resource. First, if harmful interference persists despite the implementation of the cooperation procedure provided for in the RR, the administration responsible for the transmitting station whose emission is interfered with may communicate the details of the case to the

Bureau. It may also request the Bureau to act as mediator. The Bureau may then request the administration allegedly responsible for the origin of the harmful interference to take prompt and appropriate action to terminate the interference (RR, numbers 15.39 et seq.). Second, the RR Board may examine any problem that cannot be solved by the application of the rules of procedure for recording frequency assignments made by Member States. This Board is composed of 12 members elected among qualified personalities by the ITU Plenipotentiary Conference according to a geographical distribution key. They exercise their functions in the service of the Union in an independent manner (CS, Art. 14). Finally, a dispute may be placed on the agenda of a WRC, which may decide by consensus on the complaints of Member States.

Even when Member States refer disputes to these bodies, the political considerations that are very present in the ITU paralyze these mechanisms, so that few States have been formally challenged. For example, the dispute between Cuba and the United States over the jamming of Radio and Television *Marti* signals, which was discussed at several WRCs, was not settled at the ITU. Radio *Marti* and TV *Marti* were US stations aimed at the Cuban population and broadcast from US territory since 1985 and 1990, respectively. The programming included a series of anti-Castro contents, so that the reception of these media was repeatedly jammed by the Cuban authorities, who considered that the conditions of broadcasting were contrary to international law.

The condemnation of Israel in 2010 by the ITU Plenipotentiary Conference remains exceptional. The Conference recognized that Lebanon's telecommunication facilities have been subject to acts of hacking, espionage, jamming and disruption and that Lebanon's fixed and cellular telephone networks have been subject to acts of sedition perpetrated by Israel. Accordingly, the Conference condemned the attacks attributed to Israel. It also instructed the Secretary General and the Directors of the ITU Bureau to monitor the cessation of the violations[15].

10.2.3.2. *Strengthening the ITU's control resources*

The accuracy and transparency of the information recorded in the MIFR is essential, as the recording in the MIFR is the basis for international recognition and the right to be protected against harmful interference. The Radiocommunication Bureau is responsible for maintaining the MIFR and may, on the basis of "reliable information", review an assignment recorded in the MIFR and then retain, suspend or delete it if it has not been brought into use in a regular manner or if it is not used

15 ITU (2010). Actes de piratage et attaques contre les réseaux téléphoniques fixe et cellulaire du Liban. Conférence des plénipotentiaires, Résolution 73.

in accordance with the notified characteristics (RR, number 13.6). This information may be provided by administrations or collected by the Bureau on its own initiative or at the request of an administration. These data and information may be useful in deciding a dispute.

In the absence of autonomous means of control, the Office's task is very tricky from both a technical and a political point of view. Moreover, from a legal point of view, the Board is not competent to determine whether an administration is acting in bad faith, as in the dispute between the United Kingdom and Argentina concerning Position 81 W[16]. In the case concerning interference against Eutelsat attributed by the French operator to Iran, the administration of that country denied for a long time that the interference originated from its territory. However, the Radiocommunication Bureau was convinced by the evidence provided by the ANFR (De Selding 2010).

Article 16 of the RR provides for the use of the international monitoring mechanism to verify the application of ITU rules. The monitoring of emissions is based on cooperation between the organization and administrations, which must designate dedicated stations. Administrations should carry out to the extent that they consider possible, such monitoring as may be requested by other administrations or by the Bureau. This mechanism was only put in place in 2012.

At WRC-12, the ITU-R decided to implement the monitoring mechanism through a series of agreements with Member States. The 2014 Busan Plenipotentiary Conference confirmed the use of satellite system monitoring facilities through the conclusion of cooperative agreements between the ITU and administrations (Resolution 186 on Enhancing the role of the ITU with respect to transparency and confidence-building measures in outer space activities). ITU can now issue an order to an identified monitoring station with a copy to the administration concerned. Requests for harmful interference should be given priority. For the monitoring of compliance with the MIFR, the ITU must transmit information concerning the identity of the satellite, the frequencies and orbital location to be monitored, as well as the duration of the monitoring requested.

Resolution 186 also instructs the Director of the Radiocommunication Bureau to encourage access to information gathered by the monitoring mechanisms and to take steps to maintain a database of reported cases of harmful interference, after consultation with the Member States concerned. The Radiocommunication Bureau has thus developed a public register of interference cases indicating their source. This mechanism makes it possible to directly and publicly designate the country of origin of the interference, following the so-called Naming and Shaming approach,

16 RR Committee (2006). Doc. RRB06-1/9(rev.1)-E, pt. 9.

which, in this very hushed organization, represents the first step toward international condemnation. Iran has thus been officially recognized as being at the origin of the jamming of the Eutelsat satellites.

10.3. The use of the resource in the interest of all States

From the 1970s onwards, a new approach marked the legal regime of the orbital spectrum resource aimed at promoting its use in the interest of all States. Its increasing occupation on a unilateral basis by States, in particular the industrialized States, led developing countries to request the introduction of rules aimed at the recognition of equitable access to the resource by all States (section 10.3.1). Developing countries have also sought to limit the occupation of the resource, while equatorial States have claimed sovereignty over the geostationary orbit, thus opening the way to a debate on the non-appropriation of frequencies and associated orbits (section 10.3.2). More recently, the space race by private operators, in particular for the launch of satellite constellations, is worrying the space community and, more recently, public opinion concerned about the environmental consequences. The projects, sometimes pharaonic, of constellations in low earth orbit, leaves us perplexed as to their profitability and their interest while the more performing terrestrial networks are multiplying. *SpaceX*, for example, intends to launch up to 42,000 satellites to provide broadband service as part of its *Starlink* constellation. Environmental issues are becoming increasingly important in the regulation of satellite radio communications (section 10.3.3).

10.3.1. *Equitable access by all countries regardless of their stage of development*

The debates on the rights of developing countries were crystallized at WRC-79. On the eve of the conference, the ITU had 154 member states, but 90% of the frequency spectrum was allocated to industrialized countries. At the center of the debate was the revision of the "first come, first served" rule (RR, number 11.28), which ensured a right of priority to the States that notified the assignment of a frequency to the ITU and which essentially benefited the industrialized countries for a long time the only ones capable of launching satellites. The developing countries and the non-space powers, which feared being excluded from any possibility of using space, wanted the ITU to guarantee each State the right to have frequencies and associated orbital positions, even if these could not be used immediately, given the technical backwardness of the States or the absence of a space project. The industrialized countries wanted to guarantee freedom of access to these resources and their exploitation according to the current needs of the States. Furthermore, any freezing of frequencies for the benefit of developing countries was considered a waste of the resource (Courteix 1980, pp. 635–636). These discussions were also linked to the broader debates around the New World Information and Communication Order and the New International Economic Order. After affirming the principle of equitable access to the "spectrum-orbit" resource (section 10.3.1.1), the organization implemented it within the framework of planning procedures (section 10.3.1.2). Today, the principle of equitable use is in danger of being called into question by the deployment of mega-constellations generating the risk of establishing in-orbit monopolies (section 10.3.1.3).

10.3.1.1. *The progressive affirmation of the principle of equity*

In 1971, the World Administrative Space Telecommunications Conference adopted Resolution SPA 2.1 on the use by all countries, with equal rights, of the frequency bands allocated to the space radiocommunication services[17]. The Conference decided in paragraph 1 of the text "that the registration with the ITU of frequency assignments for space radiocommunication services and their use should not provide any permanent priority for any country or groups of countries and should not create an obstacle to the establishment of space systems by other countries". Therefore, according to paragraph 2 of the text, "a country or a group of countries having registered with the ITU frequencies for their space radiocommunication services should take all practicable measures to realize the possibility of the use of new space systems by other countries or groups of countries so desiring". The ITU Constitution now incorporates these principles by confirming the rule that frequencies and associated orbits "must be used rationally, efficiently and economically [...] so that countries or groups of countries may have equitable access to those orbits and frequencies, taking into account the special needs of the developing countries and the geographical situation of particular countries" (ITU Constitution, Art. 44, para. 2).

10.3.1.2. *Implementing equitable access through planning*

The first concrete departure from the "first come, first served" rule came with the planning of satellite broadcasting services. Established in 1977 for Region I (Europe) and Region III (Asia), and in 1983 for Region II (America), this planning allowed each State to obtain the right to use five broadcasting channels for Regions I and III and a minimum of 16 channels for Region II for the provision of a national service. The plan also allocated frequencies for broadcasting services to all States, giving them the right to use specific frequencies that they could not change. In 2000, at the Istanbul Conference, the number of channels allocated per State was increased

17 ITU (1971). Actes finals de la Conférence administrative mondiale des Télécommunications Spatiales (Geneva 1971), pp. 311–312.

to the equivalent of 10 or 12 analog channels, depending on the State, to take account of digital broadcasting capacities. In addition, with the 1985 and 1988 Conferences, planning was applied to part of the frequencies intended for the fixed satellite service.

10.3.1.3. *Monopolistic risks*

The satellite constellation projects are likely to reopen the debate on equitable access to the "spectrum-orbit" resource. Based on the principle that MIFR systems have a right to be free from interference, national administrations may oppose any new project that might cause interference. However, the first constellations to occupy the low earth orbit to provide broadband services will benefit from a right of priority that may prevent the deployment of competing networks. Indeed, they will have to demonstrate that they will not cause interference to the registered systems. *SpaceX* is ahead of the game with the *Starlink* constellation, and if the FCC finalizes the 42,000 satellite process at the ITU, any future coordination with the *Starlink* system may become so complex that few competing systems will be allowed to provide satellite broadband services. The creation of private monopolies in orbit would undoubtedly limit access to the "spectrum-orbit" resource, not only by companies, but also by states, thereby restricting their freedom of space.

10.3.2. *From non-appropriation to permanent non-occupation*

The question of the appropriation of space frequencies and associated orbits has been dealt with only incompletely. However, international telecommunication law reflects a desire to limit the occupation by States of this resource (section 10.3.2.1), whereas space law, which logically applies at least to orbits, enshrines the principle of non-appropriation (section 10.3.2.2).

10.3.2.1. *The non-appropriation of orbits by space law*

The issue of non-appropriation has been specifically raised for the geostationary orbit. Eight of the 13 equatorial States[18] have claimed sovereignty over the portion of the geostationary orbit located "above" their territories. In a Declaration signed in Bogotá on December 3, 1976, after recalling the absence of delimitation of space, they emphasized that the geostationary orbit, located on the plane of the Equator, is a natural resource physically dependent on the Earth and therefore on their respective territories (Mateesco Matte 1982, pp. 456–460). Taking up the principle of permanent sovereignty of States over their natural resources as proclaimed in

18 Brazil, Colombia, Democratic Republic of Congo, Ecuador, Indonesia, Kenya, Uganda and Republic of Congo.

UNGA Resolution 1803 [XVII] of December 14, 1962, they intend to subject the use of this resource to a national authorization procedure issued by the authorities of the equatorial State concerned. The segments of the geostationary orbit located "above" international maritime zones would be declared "common heritage of humanity". The space powers immediately denounced this position, which was contrary to the principle of non-appropriation of space: "outer space, including the Moon and other celestial bodies, is not subject to national appropriation by claim of sovereignty, by means of use or occupation, or by any other means" (1967 Treaty, Art. II). Article II applies independently of the definition of space, including the geostationary orbit (Courteix 1978, pp. 895–902). The debate, which raises questions of space law, is outside the competence of the ITU. The States expressed themselves within the framework of the Committee on the Peaceful Uses of Outer Space of the United Nations (COPUOS). Each year the debate was thus reopened in COPUOS. Finally, at the 38th session of the Scientific and Technical Subcommittee held in February 2001, some delegations pointed out that it was possible to reach a consensus on the following statement: "the geostationary orbit, characterized by its particular properties, is part of outer space"[19]. The Legal Subcommittee has not yet reached a decision on this point. No official solution has been adopted at the international level. Colombia, on the other hand, still holds to the position set out in Article 101 of its Constitution, which deals with territory. The text emphasizes that the segment of the geostationary orbit located above Colombia and the associated electromagnetic frequencies are part of the national territory. It seems reasonable to reject the claims of the Equatorial States. In fact, the principle of non-appropriation proclaimed by the 1967 Treaty applies to the geostationary orbit since it is located in the outer space zone, even if the delimitation is not clearly defined. Practice and the majority of the doctrine are largely in favor of applying the 1967 Treaty to the geostationary orbit.

10.3.2.2. *The limitation of occupation by telecommunications law*

The ITU has sought to avoid recognizing a perpetual right of use for the first occupant. Since WRC-79, it is thus expressly established that in order to favor the access of all States, whatever their degree of development to the "spectrum-orbit" resource, the registration of frequency assignments and the use of these assignments do not confer any permanent priority to any country or group of countries and do not in any way hinder the creation of space systems by other countries[20].

19 AGNU (2001). Rapport du Sous-Comité scientifique et technique sur les travaux de sa trente-huitième session tenue à Vienne du 12 au 23 février 2001. Doc. A/AC.105/761, p. 25.

20 ITU (2003). Resolution 2 (REV.WRC-03). Equitable use, by all countries, with equal rights of the geostationary-satellite and other satellite orbits and frequency bands for space radiocommunication services.

Since the ITU has recognized the principle of rational and efficient use of the frequency spectrum and the geostationary orbit, particularly with a view to ensuring that all countries have access to these resources, the question of limiting the duration of the validity of assignments within the framework of a harmonized procedure naturally arose for the bands not subject to planning. WRC-79 opened this debate between developing and industrialized countries. Today, it pits the countries with at least one satellite against the others. To date, the question remains a national matter. The administrations themselves propose this duration according to their operational service needs, while taking into account the amortization of the considerable investments made for the development of space radiocommunications, which must be spread over a predetermined and realistic period. While the ITU does not wish to impose on administrations a fixed regulatory period that is identical in all cases, WRC-03 nevertheless clarified the obligations of States[21]. A frequency assignment to a space station is deemed to be permanently abandoned beyond the operating period indicated on the notice, counted from the date of bringing the assignment into use. This period shall be limited to that for which the satellite network was designed. The Bureau shall then invite the notifying administration to proceed to the cancellation of this assignment. An administration may, however, extend the period of operation by informing the Bureau more than 3 years before the end of the period in question.

10.3.3. *Environmental protection*

The environmental protection of the "spectrum-orbit" resource first manifested itself in the pollution caused by space debris (section 10.3.3.1). Recently, the environmental issues are aimed at fighting against light and radio pollution (section 10.3.3.2).

10.3.3.1. *Space debris*

Following the publication in 1999 of the COPUOS technical report on space debris[22], the question of preventing space pollution by debris has gradually become a priority. In the absence of an international legal definition, we will retain here the technical definition given by the Inter-Agency Debris Committee (IADC) and taken up by COPUOS, according to which debris represents: "all man made objects including fragments and elements thereof, in Earth orbit or re-entering the

21 ITU (2003). Resolution 4 (REV.WRC-03). Period of validity of frequency assignments to space stations using the geostationary-satellite and other satellite orbits.

22 United Nations publication, Sales No: F.99.I.17.

atmosphere, that are non functional"[23]. There are various sources of debris[24]. Most of it comes from the accidental or intentional disintegration of space objects and especially from explosions in orbit[25]. Moreover, even if we stopped sending objects into space, the amount of debris would continue to grow automatically by a phenomenon of self-feeding caused by the collision of debris between them. The debris represents a real risk for space operations that can go as far as the destruction of an operational satellite in orbit. For example, on February 10, 2009, the non-functional Russian Cosmos-2251 satellite collided with the American Iridium-33 communications satellite, causing its total destruction, as well as 600 additional pieces of debris[26]. On the other hand, the risk of damage on Earth is less likely, as the re-entry into the atmosphere causes the destruction of the debris. Two space zones are mainly polluted: the low orbits and the geostationary orbit. The problem is becoming urgent because if the international community does not react quickly, the number of debris created by the phenomenon of self-feeding will become so significant that it would make access to space impossible. Humanity would then become a prisoner of its own planet. This catastrophic scenario was envisaged, as early as 1978, by NASA astrophysicist Donald J. Kessler. It is the famous Kessler syndrome (Kessler and Cour-Palais 1978).

We note a beginning of technical standardization to prevent the production of debris and their harmful consequences on space activities first by the ITU, then by the UN. Thus, in 2010, the ITU-R adopted a Recommendation on the Protection of the geostationary satellite orbit environment[27]. The text mentions four measures: reducing as much as possible the amount of debris dropped into the geostationary satellite orbit region during the orbiting of a satellite; reducing the lifetime of debris in elliptical transfer orbits whose apogees are at or near the orbit; the elimination of the satellite from the geostationary region before the complete exhaustion of its fuel by transferring it to a so-called "graveyard" orbit; and the adoption of necessary measures to ensure that the said transfer is carried out with the utmost care in order to avoid interference with the radio frequency links of active satellites. From a space law perspective, the IADC adopted the Space Debris Prevention Guidelines on

23 AGNU (2007). Rapport du Comité des utilisations pacifiques de l'espace extra-atmosphérique, Doc. A/62/20, Annexe § 1, p. 52.

24 We count the devices at the end of their life or at the end of their mission (satellites, upper stages of launchers), as well as the objects left by man during extravehicular activity.

25 Thus, in 2007, China proceeded to a destruction test of one of its satellites by a missile causing a cloud of 3200 catalogued debris.

26 For an analysis of the case from a liability law perspective, see Kerrest (2009).

27 IUT-R (2010). Environmental protection of the geostationary-satellite orbit, ITU-R S.1003-2.

October 15, 2002[28]. This technical document, revised in 2007, served as the basis for the adoption in the same year by COPUOS of its Space Debris Mitigation Guidelines[29], which were endorsed by the UNGA in paragraph 26 of its Resolution 62/217 of 2007. The text, while emphasizing the voluntary nature of the Guidelines, states that they are consistent with existing practices developed by various national and international organizations and invites Member States to implement them through national mechanisms. In particular, Guideline 6 deals with the reduction of the prolonged presence of spacecraft in the low-Earth orbit region after the end of their mission, which includes deorbiting in a controlled manner. Guideline 7 addresses limiting the prolonged disturbance of spacecraft in the geosynchronous orbit region after the end of their mission, including the use of a higher graveyard orbit.

10.3.3.2. *Light and radio pollution*

The May 23, 2019, cluster launch of some 60 communications satellites by the U.S. company *SpaceX* as part of the *Starlink* project deployment has raised immense concern in the astronomical community. The International Astronomical Union (IAU), in a June 3, 2019 appeal, echoed a twofold concern and the need to protect the "dark sky" and the "silent sky" (IAU 2019). On the one hand, the surfaces of satellites are often made of highly reflective metal, which causes white streaks as they move through the field of telescope detectors. The light trails caused by the satellites have disrupted the observation of the sky. Even as the company *SpaceX* was considering non-reflective coatings, scientific observation of the sky, as well as amateur astrophotography, will be equally affected, this time by black trails. On the other hand, signals emitted by satellite constellations can always threaten radio-astronomical observations. The *Starlink* project, with its 42,000 satellites, is the focus of concern, but if all the satellite constellations are launched, astronomy, whose first traces go back to prehistoric times, would be profoundly and probably permanently affected. In 2020, the IAU published the results of a study to simulate the impact of future constellations based on a large sample of over 25,000 satellites from various projects (IAU 2020). It appears that the number of satellites above the horizon at any given time is between 1,500 and a few thousand, depending on the latitude. It confirms the serious disturbances caused to the activities of astronomers. Of course, the IAU stresses that there are ways to counteract this. For example, the effects of new satellites could be mitigated by accurately predicting their orbits and by interrupting observations, if necessary, during their passage. Similarly, it would be possible to process the observation data to eliminate the traces left by these

28 IADC (2002). Space Debris Mitigation Guidelines, IADC- 02-01.

29 Official Records of the General Assembly, Sixty-second Session, Supplement No. 20 (A/62/20), paras. 117 and 118 and annex.

objects. However, these practices generate costs and complicate the programming and operation of astronomical observations. Even worse, according to the IAU study, while the vast majority of satellites will not be visible to the naked eye, the appearance of the pristine night sky, especially when observed from dark sites, will nevertheless be altered, especially during the deployment phases of the satellites before they reach their final orbit. We could then witness a parade of artificial and moving lights throughout the night. The sky, as it has existed since the beginning of humanity, would disappear. A new appeal by astronomers posted online in 2020 calls on states to take concrete legal steps to protect ground-based astronomical facilities and to limit communications satellite fleets to what is strictly necessary (Astronomers' Appeal n.d.).

Some astronomers are considering a referral to the International Court of Justice (ICJ) to ask whether the sky cannot be protected under the Convention concerning the Protection of the World Cultural and Natural Heritage of 1972 (UNESCO Convention). Application of the UNESCO Convention is not an option since, even if it is accepted that the starry sky can be considered a natural or cultural heritage element (associated with astronomical heritage), the mechanism provided by Article 2 of the Convention aims to have properties and sites within the territories of the States Parties included on the World Heritage List (Charlier and Bourgeois 2013)[30]. Other scientists are considering filing suit against the Federal Communications Commission (FCC) for allegedly licensing the *Starlink* project in violation of the National Environmental Policy Act (NEPA). This law, passed in 1970, requires federal agencies to assess the environmental impact of their decisions. The legal war between astronomers and satellite operators is now open. In France, environmental protection associations have mobilized against *Starlink*. In 2022, they succeeded in having the *Conseil d'Etat* annul the authorizations for the use of frequencies issued by the *Autorité de régulation des communications électroniques, des postes et de la distribution de la presse* (Arcep) for the coverage of the entire territory from space, on the grounds that the regulator had not carried out a prior public consultation[31]. Following a new request from the company *Starlink* for these frequencies, Arcep opened a public consultation in accordance with the decision of the Council of State. In view of the contributions received demonstrating the interest of the *Starlink* service in developing connectivity in mobile "white zones" or areas poorly served by fiber networks, Arcep granted a new authorization to the company[32].

30 On the relationship between the starry sky and UNESCO World Heritage, see Charlier and Bourgois 2013.

31 Conseil d'État, April 5, 2022, No. 455321, ECLI:FR:CECHR:2022:455321.20220405.

32 Arcep, May 25, 2022, Decision No. 2022-1102.

Beyond environmental issues, the launch of mega-constellations of satellites may raise questions from the point of view of space or telecommunications law. First of all, while the commercial exploitation of outer space is protected under the freedom of use of space, this cannot be carried out without taking into account the general interest. Thus, Article 1 of the 1967 Treaty specifies that the exploration and use of outer space "shall be the province of all mankind" and "shall be carried out for the benefit and in the interests of all countries" (1967 Treaty, Art. I). A state could not invoke the freedom of space if a project had harmful consequences for one or more countries. While satellite constellations provide services in the general interest, the disproportionate nature of certain projects poses a problem at a time when very high-speed terrestrial networks are multiplying. From the point of view of telecommunications law, these same projects run up against the principles of rational and economic use of frequencies and equitable access to the "spectrum-orbit" resource studied earlier. How can we not question the interest, even commercial, of excessive projects when the needs of the populations would be covered by the terrestrial networks or could be covered by more restricted satellite systems? Are we not witnessing the waste of a scarce resource?

The race for satellite broadband connectivity illustrates the evolution of commercial space exploitation as part of the New Space movement. This movement, initiated by American entrepreneurs and billionaires who made their fortune in the digital industry, intends to support a new space economy. They rely on the promises of an Eldorado offering infinite development prospects. Their strength of conviction has attracted investors to support the craziest projects ranging from the exploitation of natural resources of celestial bodies to the establishment of space colonies. Many of these projects are simply unfeasible in the state of scientific and technical knowledge. Others have no commercial interest. Satellite constellations are no exception to the rule. Many projects will be abandoned for lack of funding or demand. In the context of the COVID-19 health crisis, the American *Oneweb* project was one of the first victims of the withdrawal of investors. After filing for Chapter 11 bankruptcy protection in March 2020, the company was finally taken over by an Anglo-Indian consortium with which Eutelsat was associated. The regime set up by the ITU, which seeks to promote the use of a rare and common resource by all States, represents a little-known model of cooperation. It could, however, serve as a basis for the establishment of an international mechanism for consultation in the very critical area of the commercial exploitation of the natural resources of celestial bodies. However, with the multiplication of satellites in orbit, the procedures put in place by the ITU at the beginning of the space age to prevent harmful interference could prove to be inoperative.

10.4. References

Astronomers' Appeal (n.d.). Appeal by Astronomers [Online]. Available at: https://astronomersappeal.wordpress.com/ [Accessed April 29, 2022].

Bonte, A. (2014). Eutelsat: Freedom of information, "regardless of frontiers"... In *Governing the Geostationary Orbit: Orbital Slots and Spectrum Use in an Era of Interference*, Penent, G. (ed.). Ifri, Paris.

Charlier, B. and Bourgeois, N. (2013). "Half the park is after dark". Les parcs et réserves de ciel étoilé : nouveaux concepts et outils de patrimonialisation de la nature. *L'Espace géographique*, 42(3), 200–212.

Courteix, S. (1978). Questions d'actualité en matière de Droit de l'espace. *Annuaire français de droit international*, 24, 890–919.

Courteix, S. (1980). La Conférence administrative mondiale des télécommunications de 1979 et le nouvel ordre juridique de l'éther. *Annuaire français de droit international*, 26, 625–647.

IAU (2019). Statement on satellite constellations [Online]. Available at: www.iau.org [Accessed April 29, 2022].

IAU (2020). Understanding the impact of satellite constellations on astronomy [Online]. Available at: www.iau.org [Accessed April 29, 2022].

Kerrest, A. (2009). Actualités du droit de l'espace : la responsabilité des États du fait de la destruction de satellites dans l'espace. *Annuaire français de droit international*. 55, 615–628.

Kessler, D.J. and Cour-Palais, B.G. (1978). Collision frequency of artificial satellites: The creation of a debris belt. *Journal of Geophysical Research*, 83(A6), 2637–2646.

Mateesco Matte, N. (1982). *Droit aérospatial – Les télécommunications par satellite*. Pedone, Paris.

Malavialle, A.-M., Pasco, X., Sourbès-Verger, I. (1999). *Espace et puissance*. Ellipses, Paris.

de Selding, P.B. (2010). ITU implores Iran to help stop jamming. *Space News* [Online]. Available at: https://spacenews.com/itu-implores-iran-help-stop-jamming/ [Accessed July 13, 2022].

de Selding, P.B. (2014). Eutelsat blames Ethiopia as jamming incidents triple. *Space News* [Online]. Available at: https://spacenews.com/40818eutelsat-blames-ethiopia-as-jamming-incidents-triple/ [Accessed July 13, 2022].

Watine, M.F. and Gourrierec, A. (1978). L'Union internationale des télécommunications face aux nouveautés techniques. *Annuaire français de droit international*, 24, 589–604.

11

Navigation Satellite Systems

Magda COCCO[1] **and Helena CORREIA MENDONÇA**[2]

[1] *Vieira de Almeida & Associados, Lisbon, Portugal*

[2] *Vieira de Almeida & Associados, Lisbon, Portugal*

11.1. Introduction

Navigation satellite systems are systems of satellite networks that provide positioning, navigation and timing (PNT) information through the transmission of signals. PNT data are critical in countless areas and sectors of society and are bound to become ever more important to achieve the United Nation's Sustainable Development Goals (SDGs) and tackle the green and digital transition.

This chapter provides a brief analysis of PNT systems from a legal point of view. An initial section covers basic concepts and a brief description of the main PNT systems to frame the analysis made afterward.

11.2. Initial concepts and overview

11.2.1. *Concepts and definitions*

PNT networks comprise three main segments: the space segment, the ground segment and the user segment. The space segment includes satellites, which continuously transmit PNT data. The ground segment includes the monitoring, command, and control stations, as well as the receiving stations. In addition to these, there are communication links, that is, the downlink and uplink signals that are

Space Law,
coordinated by Thomas LECLERC. © ISTE Ltd 2023.

transmitted from and to the satellites (Chatzipanagiotis and Liperi 2017). The user segment is composed of the entities and individuals resorting to PNT data.

PNT networks can be classified as core systems or augmentation systems. Core systems are the basic PNT networks that provide PNT data. Core systems can be global (covering the whole planet) or regional. Global systems (GNSS) include the United States (US) GPS, the European Union (EU) Galileo, the Russian GLONASS and the Chinese BEIDOU Navigation Satellite Systems (BDS). Regional systems include the Japanese Quasi-Zenith Satellite System (QZSS) and the Indian Regional Navigation Satellite System (IRNSS/NavIC), as well as the BEIDOU regional component (Plattard and Sitruk 2017a).

Augmentation systems improve PNT data provided by core systems and aim at providing better accuracy, integrity, continuity and availability (Kahveci 2017), especially needed for traffic monitoring and management, such as in aviation and maritime transport. They include satellite-based augmentation systems (SBAS) and ground-based augmentation systems (GBAS). Both provide more accurate data through fixed ground stations that deliver a corrected PNT signal to the receiver. However, while SBAS use satellites as relay stations, and thus cover vaster areas (Smith 2015), GBAS use ground stations only.

SBAS systems include the US Wide Area Augmentation System (WAAS), the European Geostationary Navigation Overlay Service (EGNOS), the Japan Satellite Based Augmentation System (MSAS) and the Indian GPS Aided GEO Augmented Navigation system (GAGAN). Other SBAS being developed or operational include the Russian System for Differential Corrections and Monitoring (SDCM), the BEIDOU satellite-based augmentation system (BDSBAS), the Southern Positioning Augmentation Network (SPAN) from Australia and New Zealand, the Korea Augmentation Satellite System (KASS) and the A-SBAS by the Agency for Air Navigation Safety in Africa and Madagascar (ASECNA).

11.2.2. *PNT Systems*

The US Global Positioning Service (GPS) was developed by the US Department of Defence (DoD) and is operated by the US Space Force. The National Executive Committee for Space-Based Positioning, Navigation and Timing, which is a joint civil and military body chaired jointly by the Deputy Secretaries of Defence and Transportation, coordinates GPS-related matters across several federal agencies[1]. The space segment is composed of a set of satellites with the United States

1 Available at: https://www.gps.gov/governance/excom/.

committing to maintaining the availability of at least 24 satellites 95% of the time[2]. The GPS system provides two main services: the civilian GPS service or Standard Positioning Service (SPS) (which is open and free to all users on a continuous, worldwide basis) and the military GPS service or Precise Positioning Service (PPS) available to the United States and allied armed forces as well as approved Government agencies.

Galileo was developed within the framework of the EU and is currently operated by the EU Agency for the Space Programme (EUSPA). Unlike the other GNSS, Galileo is civilian in nature (Plattard and Sitruk 2017b). The network comprises 24 satellites along with spares. In accordance with the current legal framework[3], the services provided by Galileo shall comprise a Galileo open service (GOS), a high-accuracy service (HAS), a signal authentication service (SAS), a public regulated service (PRS)[4], an emergency service (ES) and a timing service (TS). In addition, Galileo shall also contribute to (i) the search and rescue support service (SAR) of the COSPAS-SARSAT, which is the International Satellite System for Search and Rescue, (ii) integrity-monitoring services standardized at the Union or international level for use by safety-of-life services and (iii) space weather information via GNSS Service Centre[5] and early warning services via the Galileo ground-based infrastructure.

GLONASS was developed by and is the property of Russia (and fully financed by the federal budget) and is under the supervision of the Russian Federal Space Agency (ROSCOSMOS) together with the Ministry of Defence (Smith 2015)[6]. It

2 As of April 22, 2022, there were a total of 30 operational satellites in the constellation, excluding the decommissioned, on-orbit spares. Available at: https://www.gps.gov/systems/gps/space/.

3 European Parliament and EU Council (2021). Regulation (EU) 2021/696 establishing the Union Space Programme and the European Union Agency for the Space Programme and repealing Regulations (EU) No. 912/2010, (EU) No. 1285/2013 and (EU) No. 377/2014 and Decision No. 541/2014/EU (Space Programme Regulation).

4 See, in this respect, European Parliament and EU Council (2011). Decision No. 1104/2011/EU on the rules for access to the public regulated service provided by the global navigation satellite system established under the Galileo programme.

5 As referred to in Commission Implementing Decision (EU) 2016/413 of 18 March 2016 determining the location of the ground-based infrastructure of the system established under the Galileo programme and setting out the necessary measures to ensure that it functions smoothly.

6 General coordination of works on maintenance, development and use of GLONASS are also assumed together with the Ministry of Transportation, the Ministry of Foreign Affairs, the Ministry of Communications, and other federal agencies and ministries within their powers.

includes 25 satellites (22 of them operational and three in maintenance)[7]. GLONASS provides a standard precision (SP) signal, which is publicly accessible, and a high precision (HP) signal, restricted to military and government purposes.

BEIDOU Navigation Satellite System (BDS) was developed and is owned by China and is operated by the China National Space Administration (CNSA), with the third generation (BEIDOU-3) comprising 30 operational satellites, three in geostationary orbit (GEO), three in inclined geostationary orbit (IGSO), and the remaining 24 in the medium Earth orbit (MEO). Along with positioning, navigation and timing services, the BDS-3 system can provide a variety of value-added services such as global search and rescue assistance, short message communication, ground- and satellite-based augmentation, as well as precise point positioning[8].

In addition to the Global PNT systems briefly described above, two regional systems are operational. The Japanese Quasi-Zenith Satellite System (QZSS) was commissioned by the Japanese Government, with Quasi-Zenith Satellite System Services Inc. (QSS) operating the system[9]. The Cabinet Office – CAO – is in charge of planning, designing and coordinating policies concerning space exploitation, and implementing policies regarding development, maintenance and operation of the QZSS[10]. The network comprises four satellites. Its services were officially deployed on November 1, 2018[11], and a future seven-satellite constellation independent of GPS is being planned. The Indian Regional Navigation Satellite System (IRNSS/NavIC) was developed and is operated by the Indian Space Research Organization (ISRO), which is under India's Department of Space. It comprises seven satellites (three GEO and four IGSO) that cover India and a region extending 1500 km around India[12], and with additional satellites being planned. It aims at providing a standard and a restricted/precision service[13].

With relation to SBAS systems, the US WAAS was developed by the US Federal Aviation Administration and is deployed in continental United States, Mexico and Canada. Its purpose is to augment the GPS signal for aviation. WAAS comprises three GEO satellites. The European EGNOS was developed by the European Space Agency (ESA) in cooperation with the European Commission and

---

7 Available at: https://www.glonass-iac.ru/en/GLONASS/index.php.

8 Available at: http://en.beidou.gov.cn/SYSTEMS/System/.

9 Available at: https://qzss.go.jp/en/overview/services/sv02_why.html.

10 Available at: https://www8.cao.go.jp/space/english/index-e.html.

11 Available at: http://qzss.go.jp/en/overview/notices/qzss_181101.html.

12 Available at: https://www.isac.gov.in/navigation/irnss.jsp.

13 Available at: https://www.isro.gov.in/irnss-programme.

the European Organization for the Safety of Air Navigation (Eurocontrol). EGNOS is owned by the European Commission, and its exploitation has been delegated to EUSPA, with the services being delivered by the European Satellite Services Provider (ESSP), to whom EUSPA awarded the EGNOS operations and service provision. It comprises three GEO satellites. The Japanese MSAS is operated by Japan's Ministry of Land, Infrastructure and Transport and the Japan Civil Aviation Bureau and aims to augment GPS signals and provide coverage for the hemisphere centered on 140°, which includes Japan and Australia. It comprises two GEO satellites. The Indian GAGAN has been jointly developed by the Indian Space Research Organization (ISRO) and the Airports Authority of India (AAI). It comprises three GEO satellites.

11.2.3. *Governance and cooperation*

Governance of each PNT system is determined by the legal and/or policy frameworks of each State or entity (such as in the case of Galileo and EGNOS) within which the network is developed. But because of the global or regional reach of PNT systems, mechanisms for governance and international cooperation have been developed.

The International Committee on GNSS (ICG) was established by the UNGA resolution 61/111 of 2006. The ICG is an informal body aimed at enhancing cooperation in PNT services and promoting interoperability, compatibility and transparency between the different system providers. ICG is open to States members of the UN, international organizations and entities that are responsible for GNSS and their augmentations operating under governmental authority or that are involved in implementing or promoting GNSS services and applications.

In addition to the ICG, international cooperation on PNT matters has also been pursued through bilateral agreements, which aim to increase interoperability and compatibility of the different systems and thus allow their seamless use. For instance, the United States has concluded Joint Statements with several countries and the EU (and its Member States) have also concluded several international agreements in this respect (Chatzipanagiotis and Liperi 2017).

Other international entities are worth referring given the nature of PNT networks (which include a space segment and links) and the vital relevance of PNT data in certain sectors. UN Committee on the Peaceful Uses of Outer Space (COPUOS) and ITU play a significant role within PNT systems, as COPUOS is one of the major international forums addressing the governance of exploration and (peaceful) uses of outer space and the International Telecommunications Union (ITU) is the UN agency responsible for assigning orbital slots and managing frequencies. At the sectoral level and given the initial role of PNT data in aviation and the maritime sectors, the International Civil Aviation Organisation (ICAO) and the International Maritime Organisation (IMO) continue to play a relevant role in addressing and discussing issues arising from the use of PNT signals in these sectors.

11.3. Legal framework applicable to PNT

11.3.1. *Background*

There is no dedicated international legal framework for PNT, but only a patchwork of several applicable legal regimes. What exists are (i) provisions regulating each PNT system and (ii) provisions on PNT for specific sectors – such as in aviation and maritime, which were the first cases of PNT used at a broader level.

Nevertheless, provisions applying to the network (the space segment, the ground segment and the links), as well as applying to the availability of the signal and the provision of data and/or services based on PNT, are also relevant. The first ones include the UN Space Treaties (see *supra*, Chapter 1) and National Space Laws for the space segment, national laws for the establishment of ground stations (especially radio stations) and provisions for the assignment of orbital slots and spectrum. The second ones include provisions on contracts, protection of intellectual property (especially with relation to signal and/or data) and insurance.

Naturally, other provisions may apply and do also certainly come into play: tax, trade law, environmental law, property law, among others. But analysis of the PNT legal framework will be limited to the above points. Liability, on its turn, merits a separate analysis due to its relevance for PNT systems.

11.3.2. *Legal framework specific to PNT*

11.3.2.1. *Frameworks dedicated to each PNT system*

The legal and policy frameworks regulating PNT are the ones that are applied to each specific PNT system. These rules and provisions are aimed essentially at regulating the implementation and exploitation of the PNT systems, including their governance.

For instance, the US GPS is addressed in Title 10 of the US Code (US Code, Title 10, Subtitle A, Part IV, Chapter 136, section 2281), where GPS[14] is acknowledged as a dual-use system with "essential civil, commercial, and scientific uses" (US Code, (a) (3) Findings) in addition to the military ones (Harrington 2017). The Federal Radionavigation Plan (FRP)[15], on its turn, contains the PNT policy and planning for the Federal Government. Among other points, it details agencies' roles and responsibilities (notably of the DoD, Homeland Security and Transportation). In this scope, and as indicated above, the National Executive Committee for Space-Based PNT is the organization which advises and coordinates federal departments and agencies on GPS matters. In accordance with the US Code, it is however the Secretary of Defence who is responsible for both the sustainment and operation of GPS capabilities for military purposes and for commercial, civil and scientific purposes and without direct user fees (US Code, Title 10, Subtitle A, Part IV, Chapter 136, section 2281 (a)).

With relation to the EU Galileo and EGNOS, it is the Space Programme Regulation that contains the main provisions on the European satellite navigation systems. The Regulation expressly indicates that both are civilian systems under civilian control and requires them to be compatible and interoperable. The financial contribution of the EU to the space program and the governance structure are also addressed. With relation to this last point, the Commission has the overall responsibility for the program and supervises its implementation. EUSPA ensures the security accreditation of the program's components (being further responsible for the operational security of Galileo and EGNOS and for ensuring the operation of the Galileo Security Monitoring Centre) and undertakes communication, market development and promotion activities concerning services offered by Galileo and EGNOS, particularly activities relating to the market uptake and coordination of user needs. It is also entrusted by the Commission with the management and exploitation of Galileo and EGNOS. ESA is responsible for and shall be entrusted with systems evolution and design and development of parts of the ground segment, and of the satellites, including testing and validation. In accordance with the Regulation, the Commission shall conclude with EUSPA and ESA a Financial Framework Partnership Agreement (FFPA) defining the roles, responsibilities and

14 Which is defined as "consisting of a constellation of satellites and associated facilities capable of providing users on earth with a highly precise statement of their location on earth" (US Code, (a) (1) Findings).

15 The FRP is a regular report jointly prepared by the Secretaries of Defence, Transportation, and Homeland Security. See DoD (2021). Federal Radionavigation Plan. https://www.navcen.uscg.gov/sites/default/files/pdf/2021_Federal_Rdionavigation_Plan.pdf.

obligations of each of them and necessary coordination and control mechanisms. On June 2021, the FFPA was concluded.

As with the US GPS and the EU PNT systems, the national legal framework for the other systems is also very much focused on their exploitation and governance.

For instance, regarding GLONASS, several acts regulating various aspects of the system exist, including Federal Laws on navigation activities, calculation of time and State Automated Information System ERA-GLONASS[16] (Volynskaya 2017). In relation to BEIDOU, there are also official documents relating to the system[17]; however, detailed information on its organization framework and financial features is not widely available (at least in English). In Japan, a Cabinet Decision on September 30, 2011 (Japan 2011) establishes the basic policy on the implementation of the operational QZSS project (Kosuka 2017). In India, there does not seem to exist publicly available legal or regulatory acts on the exploitation and governance of the IRNSS system[18], but there are policy documents on the IRNSS program[19].

The national PNT-specific legal and regulatory regimes are therefore very diverse and with varied levels of detail. While some of them consist of structured frameworks with comprehensive provisions, others are more limited in scope, more fragmented or even composed mainly of reports or policy documents of the agencies responsible for the PNT programs. Regardless of the approach that is taken when it comes to the national PNT, the legal rules, policy provisions or guidelines share in general one common trait: they address the scope, purpose and/or services of the system, its deployment and exploitation, its governance and, often, its financing.

---

16 This is the first-tier of laws. The second tier comprises over 30 more detailed regulatory acts by the President and the Government. The third tier is about 20 Acts adopted by ministries and agencies. There are also over 100 intergovernmental and national standards concerning the use of GLONASS and related issues (Volynskaya 2017).

17 There are policy documents including on technical aspects (interface and development) and on the development and promotion of the system (Wang 2017).

18 "A policy on distribution of IRNSS signals has not yet been announced. A responsive law and procedural regime would best serve to realize the full potential of IRNSS and GAGAN non-civil aviation end users" (Kaul 2017).

19 Given its operation by ISRO, it is the reports and policy documents of this agency that are most relevant here. For instance, it is the Report of ISRO Working Group on "Space" that contains the Indian satellite navigation space program's goals. ISRO. See ISRO. Report of Working Group on "Space" on the Eleventh Five Year Plan Proposals 2007–2012 for Indian Space Programme [Online]. Available at: https://dst.gov.in/sites/default/files/rep-space.pdf. See, also, Kaul (2017).

11.3.2.2. *Frameworks dedicated to the use of PNT in specific sectors*

11.3.2.2.1. Aviation

The regulatory framework of the aviation sector addresses the use of PNT systems. Indeed, the Convention on International Civil Aviation of 1944 (Chicago Convention)[20] dedicates its Annex 10 to GNSS air navigation Standards and Recommended Practices (SARPs), and further relevant provisions are to be found in the Chicago Convention's Annexes 2, 4, 6 Parts I and II, 11, 14 Vol. I and 15, as well as in the respective guidance in ICAO's GNSS Manual and its Circulars 257 and 278 (Chatzipanagiotis and Liperi 2017).

A "Charter on the rights and obligations of States relating to the GNSS services" was also issued in 1998 by ICAO under its Assembly Resolution A32-19 of 1998, containing fundamental principles on this matter. This Resolution establishes that GNSS services should primarily be used for the maintenance of safety in the international framework of civil aviation. Authority and control are granted to every State over operations developed by aircrafts over their territory, imposing provider States the duty of assuring the services' availability, continuity, integrity, reliability and accuracy. It further establishes that access to the use of GNSS services should be granted to all States and their aircrafts, without any discrimination and on a uniform basis (Abeyratne 2011).

Furthermore, ICAO also issued its Assembly Resolution A32-20 of 1998 named "Development and Elaboration of an Appropriate Long-term Legal Framework to Govern the Implementation of GNSS". This Resolution recognized the complexity of the legal aspects of the implementation of CNS/ATM[21], including the use of GNSS in this scope, and the necessity of an appropriate long-term legal framework to govern the implementation of GNSS in the aviation sector. However, the difficulty in amending the Chicago Convention dictated that efforts in this regard would prove unfeasible (Abeyratne 2011).

The principles established in the ICAO Assembly Resolutions were further reiterated in its Assembly Resolution A33-15 of 2001, particularly regarding their application to CNS/ATM.

---

20 The use of GNSS aims at meeting ICAO requirements prescribed in Article 44 of the Chicago Convention, which encourages "the contracting states to implement a satellite based navigational system as a means of raising the safety levels" (Smith 2015).

21 CNS stands for Communications, Navigation and Surveillance; ATM stands for Air Traffic Management. For more detail on CNS/ATM and GNSS, see Abeyratne 2011.

11.3.2.2.2. Maritime

Maritime-specific regulations also address the use of PNT. IMO approved its Assembly Resolution A.915 of 2022 which establishes, among other points, that GNSS should serve mainly general navigation purposes and recommends that augmentation provision should be harmonized worldwide to enable ships to carry only one shipborne receiver.

In addition, the International Convention for the Safety of Life at Sea of 1974 (SOLAS) establishes that merchant and passenger ships must have a receiver for a GNSS or a terrestrial radionavigation system, or other means suitable, for use at any time along the voyage, in order to determine the ship's position by automatic means (SOLAS, Reg. 19, para 2.1.6) (Yamada 2017). There are also performance standards for such equipment regarding mainly accuracy, continuity and integrity monitoring (Chatzipanagiotis and Liperi 2017).

11.3.2.2.3. Road vehicles and railway

There is no international framework for the use of PNT systems in the road sector, despite its relevance especially for autonomous driving[22]. In this respect, it is important to note that the Vienna Convention on Road Traffic of 1968 established a set of requirements with relation to road traffic which prevented, up until March 2016, autonomous vehicles. Amendments taking place on March 23, 2016 allowed autonomous driving technologies provided they are in compliance with the applicable UN Regulations[23] or provided they can be deactivated by the driver – meaning that, in the end, autonomous driving is permitted but the systems shall be able to be controlled or deactivated by the driver[24]. A new amendment proposed in

22 See the levels of autonomy as determined by SAE International. SAE International (2018). SAE J3016™: Taxonomy and Definitions for Terms Related to On-Road Motor Vehicle Automated Driving Systems. Available at: https://www.sae.org/standards/content/j3016_201806/.

23 As per the 1958 UNECE Agreement, which contains a set of regulations with technical requirements and protocols for homologation of vehicles. See Agreement concerning the Adoption of Harmonized Technical United Nations Regulations for Wheeled Vehicles, Equipment and Parts which can be Fitted and/or be Used on Wheeled Vehicles and the Conditions for Reciprocal Recognition of Approvals Granted on the Basis of these United Nations Regulations, E/ECE/TRANS/505/Rev.3. Note also the 1998 UNECE Agreement on UN Global Technical Regulations (UN GTRs), which contains a set of technical global regulations.

24 See Report of the Global Forum for Road Traffic Safety on its eighty-first session, ECE/TRANS/WP.1/173/Add.1. A Group of Experts on drafting a new legal instrument on the use of automated vehicles in traffic was also set up. Its purpose is to draft a new legal instrument for the safe deployment of automated vehicles in international traffic. The Group of Experts will have a two-year duration (with a possibility of extension), having started on July 1, 2021.

2020 further provides that the driver requirement is deemed to be satisfied while the vehicle is using an automated driving system which complies with (a) domestic technical regulations, and any applicable international legal instrument, concerning wheeled vehicles, equipment and parts which can be fitted and/or be used on wheeled vehicles and (b) domestic legislation governing operation.

At the EU level, it is worth mentioning, for instance, the EU eCall Regulation[25], which requires manufacturers to ensure that "receivers in the 112-based eCall in-vehicle systems are compatible with the positioning services provided by the Galileo and the EGNOS systems. Manufacturers may also choose, in addition, compatibility with other satellite navigation systems" (Regulation (EU) 2015/758, Art. 5, 4).

In railway systems, the importance of GNSS is also increasingly acknowledged, despite once again the lack of an international or continental legal framework. At the EU level, the use of satellite-based train localization in railway signaling is however being called upon (EUSPA 2021), including through regulatory frameworks with a focus on digitalization and space technologies.

11.3.3. *Other legal frameworks relevant for PNT*

11.3.3.1. *Legal framework applicable to the PNT network*

11.3.3.1.1. Space segment

The space segment of a PNT framework is first and foremost regulated by Space Law, especially by International Space Law and the UN Space Treaties (see *supra*, Chapter 1). We will make a brief overview on the topics of responsibility and registration.

Regarding responsibility, Article VI of the Treaty on Principles Governing the Activities of States in the Exploration and Use of Outer Space, including the Moon and Other Celestial Bodies of 1966 (OST) establishes that States Parties to the Treaty shall bear international responsibility for national activities in outer space whether such activities are carried on by governmental agencies or by non-governmental entities. Operation of PNT systems as above described is carried out by States and hence international responsibility would fall upon them for their own space activity. The exceptions are the EU systems, owned by and exploited by

25 European Parliament and EU Council (2015). Regulation (EU) 2015/758 concerning type-approval requirements for the deployment of the eCall in-vehicle system based on the 112 service and amending Directive 2007/46/EC.

the EU. Article XIII provides that the OST applies to activities carried out jointly by States, including cases where they are carried on within the framework of international intergovernmental organizations. It is argued whether the EU can be considered an "international intergovernmental organization" for the purposes of this provision, as one needs to bear in mind that the EU sees itself more as a community of states (Smith 2015) with some supranational features rather than an international intergovernmental organization. In addition, Article XIII establishes that practical questions arising in connection with activities carried on by international intergovernmental organizations shall be resolved by the States Parties either with the appropriate international organization or with one or more States members of that international organization, which are Parties to the OST. Responsibility seems therefore to fall upon the Member States of the EU, especially the ones directly involved in the PNT systems, but complaints seem to be able to be addressed against the EU or the Member States (Chatzipanagiotis and Liperi 2017) (if and to the extent that the EU can be seen purely as an intergovernmental organization). The role of ESA, as the entity responsible for systems' development, testing and validation, may come also into play especially to the extent it may be considered to perform "activities in outer space" under the OST, and as a result questions arising from the EU systems could potentially also be brought against ESA or its Member States[26]. Article VI also indicates that States shall authorize and supervise the activities of non-governmental entities in outer space. It is important to note that most National Space Laws apply to activities carried out not only by their nationals and/or entities therein established, but also carried out in their territories. The extent to which this requirement would entail that Member States of the EU would have to authorize and supervise the launch and operation of the PNT space segment in their territories is a question that would require an analysis of the applicable National Space Laws, especially of its scope of application to the EU or to international organizations.

The same issues arise with respect to registration. In accordance with the Convention on Registration of Objects Launched into Outer Space of 1974 (Registration Convention)[27], space objects, such as the space segment of PNT

26 To what extent the seat of EUSPA, the placement of the ground segment or even the investments into the systems might cause specific responsibility of specific Member States may also be evaluated (von der Dunk 2017).

27 See Recommendations on enhancing the practice of States and International Intergovernmental Organizations in Registering Space Objects, UNGA Resolution 61/111 of 2007, which further describes practices and elements of registry and notification. The Group of Governmental Experts on Transparency and Confidence-Building Measures in Outer Space Activities also issued a report which highlights the importance of provision of registration information. See GGE (2013). Report of the Group of Governmental Experts on Transparency and Confidence-Building Measures in Outer Space Activities. UN Doc A/68/189.

systems, are required to be registered (Registration Convention, Art. II). Registration shall be made by a launching State which, under the Registration Convention, is determined by four alternative criteria: (i) a State which launches or procures the launching of a space object or (ii) a State from whose territory or facility a space object is launched (Registration Convention, Art. I, a). The registration is twofold: first, there is a national registration in launching States (Registration Convention, Art. II, 1). Second, there is an international registration at the UN (Registration Convention, Arts. III and IV) to be carried out by the launching State. Article VII provides that references to States shall be deemed to apply to any international intergovernmental organization which conducts space activities if the organization declares its acceptance of the rights and obligations – but the EU has not, unlike ESA, declared its acceptance to the Registration Convention (von der Dunk 2015a) (once again, the EU sees itself more as a community of States with supranational features rather than an intergovernmental organization[28]). Taking this into consideration, and that the EU is the entity procuring the launch, the establishment of the launching State would have to be made with regards to one of the remaining alternative criteria apart from the criteria of the State that procures the launching of a space object. Hence, the obligation to register seems to fall upon the State from whose territory or facilities the PNT satellite is launched and/or upon ESA as the entity launching if applicable[29]. This has relevant implications when it comes to jurisdiction and control under the OST, because it is the State on whose registry an object launched into outer space is carried that retains jurisdiction and control over such object under the OST (OST, Art. VIII).

In the light of the above, and in summary, regulation of the space segment requires looking both at International Space Law (especially the UN Space Treaties)

---

28 Considering that the EU has not qualified as an intergovernmental organization and that, as a result, registration by the EU is not possible – including because the Registration Convention precludes the possibility to register satellites in two or more states at the same time (Registration Convention, Art. II, 2) (von der Dunk 2017).

29 Differently, arguing that "the European Union remains a group of states and, if anything, a sui generis 'supranational' state-like entity – where following Article II of the Registration Convention it is a 'launching State' that should register its satellites. The European Union operates out of Brussels, so that Belgium as principal state would, in the absence of registration by ESA, be required to undertake this step" (Smith 2015). Arguing that "although the Act remains silent about this, Belgium seems to have taken the practical standpoint that certain space navigational activities conducted for non-commercial or (inter)governmental purposes by international organizations that have an administrative or technically dedicated establishment on the Belgian territory do not fall under the Belgian Space Act nor under jurisdictional control of the Belgian authorization regime" (Michielsen 2016).

and the applicable National Space Law in order to determine the provisions that apply to the launching and operation of PNT satellites.

11.3.3.1.2. Ground segment

The legal framework applying to the ground segment is national and it usually covers the provisions on licensing of ground stations, especially of radio-electric stations. Hence, the deployment of ground stations is usually dependent upon the national laws where the ground stations are placed. Once again, determination of their scope of application (especially of to whom such laws apply – if only to private entities or also the State and/or international organizations) requires a case-by-case analysis. This, on its turn, requires also, naturally, a joint analysis with the PNT-specific legal frameworks and to what extent they may provide for rules that may waive the requirements of laws on licensing of stations, especially radio-electric stations.

For instance, the EU legal framework (Space Programme Regulation, Arts. 33(1), 34(6) and 47(c))[30] addresses ground stations but mainly when it comes to the protection of ground infrastructure (which is a responsibility of the Commission with Member States also having obligations relating to the protection of ground infrastructure located on their territory) and the location of the centers belonging to the ground-based infrastructure of Galileo and EGNOS (which shall be determined by the Commission in accordance with security requirements, following an open and transparent process). Hence, national laws relating to their installation do continue to be relevant.

11.3.3.1.3. Links

The links (i.e. the signals) require the use of the electromagnetic spectrum. Assignment of spectrum (as well as of orbital slots[31]) follows an international

30 See, also, EU Commission (2016). Commission Implementing Decision (EU) 2016/413 determining the location of the ground-based infrastructure of the system established under the Galileo programme and setting out the necessary measures to ensure that it functions smoothly (as amended) and EU Commission (2017). Commission Implementing Decision (EU) 2017/1406 determining the location of the ground-based infrastructure of the EGNOS system.

31 The assignment of orbital slots under ITU provisions is connected not with the assignment of a specific physical position for satellites in orbit but with the use of the spectrum by the satellites. Indeed, the principle of freedom of use and exploration established in the OST precludes limits with relation to placement of satellites except those expressly provided therein (von der Dunk 2015b).

regime under ITU[32] and a national regime (von der Dunk 2015b). International and national rules on the use of spectrum aim at guaranteeing their coordinated use and avoid harmful interference[33] on the basis of two fundamental principles: efficient use and equitable access (ITU Constitution, Art. 44, 2)[34].

At the international level, and in general, spectrum is allocated to different categories of radiocommunication services taking into consideration, among other points, the three regions of the world established by ITU (RR, Art. 5). Specifically, when it comes to PNT, ITU has been working on the allocation of frequency bands for Radio Navigation Satellite Services (RNSS). Spectrum is then allotted to States (RR, Art. 1, 17). Each State has its own national frequency plan determining the allocation of spectrum in line with ITU provisions. In the EU, and without prejudice to Member States' frequency plans, Decision No. 243/2012/EU[35] establishes a multiannual radio spectrum policy program aimed at guaranteeing the planning and harmonization on the use of spectrum in the EU. The EU has also approved a European Radio Navigation Plan (ERNP) in 2018[36] in order to facilitate the introduction of GNSS applications in different sectors. The EU framework on PNT already addresses the importance of guaranteeing that the radio spectrum necessary for the EU space program is available and protected (Space Programme Regulation, Rec. 43 and Art. 27, 4). In this scope, it is the EU that has the power to negotiate frequency issues for Galileo and EGNOS, as well as to negotiate and conclude international compatibility and interoperability agreements (Chatzipanagiotis and Liperi 2017)[37].

---

32 This legal regime, agreed upon by the ITU Member States at World Administrative Radio Conferences is codified through the ITU Constitution of the International Telecommunications Union of 1992 (ITU Constitution), including the ITU Radio Regulations of 2020 (RR).

33 According to the ITU, "harmful interference" is defined as the interference which endangers the functioning of a radionavigation service or of other safety services or seriously degrades, obstructs, or repeatedly interrupts a radiocommunication service operating in accordance with ITU Radio Regulations (RR, Number 1.169). Note that the ITU does not differentiate intentional from unintentional harmful interference.

34 For a detailed analysis in this respect, see Henri et al. (2017).

35 European Parliament and EU Council (2012). Decision No. 243/2012/EU establishing a multiannual radio spectrum policy program (as amended).

36 European Radio Navigation Plan (2018). [Online]. Available at: https://ec.europa.eu/docsroom/documents/33024 [Accessed 22 June 2022]. The Space Strategy for Europe already addressed this issue, referring to the release of the radio navigation plan in order to facilitate the introduction of GNSS applications in different sectors. See EU Commission (2016). Space Strategy for Europe. COM (2016) 705 final.

37 See, also, Space Programme Regulation, Rec. 12.

With relation to orbital and spectrum resources for satellite networks, these are either pre-allotted to countries (the so-called a priori planning procedures which aim at guaranteeing equitable access to orbit/spectrum resources for future use) or may be required through a mechanism of first-come first-served before ITU (the so-called coordination procedures aimed at guaranteeing the efficiency of orbit/spectrum use and interference-free operation) (Henri et al. 2017)[38]. Requiring frequencies and orbital slots before ITU shall be done by national administrations and, once they are granted, the national administration assigns them to interested entities (RR, Art. 1, 18). Terrestrial spectrum, on its turn, is assigned on a national basis by the competent national authorities in accordance with their regulatory framework. When it comes to the EU space program, the EU cannot, as a result, submit fillings before ITU, as it is not a State. Hence, the EU has to submit filings through one or more national administrations of its Member States[39]. The use of terrestrial spectrum has also to be dealt with the States concerned, including mainly the ones where radio electric reception stations are installed.

11.3.3.2. *Legal framework applicable to the availability of signal/provision of services*

The legal framework on the availability of the signal and provision of services relates essentially with determining what is the legal framework that applies to making available the signals to the user segment or among operators, and to the provision of services (value-added services) resorting to PNT data.

The first legal framework that seems applicable – to the signal in space – is National and International Space Law, both when the signal is made available directly by the space segment to other PNT systems or to the downstream user segment. Differently, when the signal is made available through ground stations (resorting to terrestrial spectrum) or when what is at stake are downstream value-added services, then it is our understanding that Space Law does not apply.

Making available the signal or value-added services may be in addition subject to national laws. Because the use of PNT data and services may require the conclusion of a contract with the data/service provider, as well as the acquisition and

38 Pre-allotted orbital slots are for the broadcasting satellite service (BSS) (Appendices 30/30A) and fixed satellite service (Appendix 30B), and the coordination procedure is addressed in the Radio Regulations (RR, Arts. 9 to 11).

39 For Galileo, a group of Member States makes available their rights of use to the EU under a Memorandum of Understanding (Baumann 2018).

use of receivers[40], national legal provisions on contracts, product warranties and consumer protection may all come into play.

Another relevant legal regime relates with intellectual property (IP). In addition to the IP on the PNT networks themselves[41], value-added providers usually process PNT signals and provide a service resorting to applications (software[42]) and data organized in a structured manner (database[43]). The frameworks on IP (including at the international level, such as the Paris Convention on Industrial Property of 1883, the Berne Convention on Copyright of 1886 and the Agreement on Trade-Related Aspects of Intellectual Property Rights of 1994 [TRIPS], and at the national level) come all to play with relation to the allocation of IP rights among all stakeholders in the PNT value chain. Specifically, ownership of the data (raw and processed data, background and foreground data) and their use needs to be determined and addressed taking into consideration applicable regimes (for instance, on open data, data governance and data sharing) and contractual agreements.

Another relevant legal framework for PNT systems is insurance. Liability insurance is a compulsory requirement in most National Space Laws, and insurance may cover launch and in-orbit operation, with commercial operators usually ensuring for signal loss in orbit. But the extent to which insurance also covers third-party damages (in the user segment) is also largely dependent upon whether operators would be liable in such cases (see below on this subject). In any case, there does not seem to exist alignment in either national laws or practice in contracting insurance, and standard liability insurance for PNT systems does not exist yet (Gaubert 2015). In the end, the fact is that most operators do not take insurance because they are public entities (the State) (Smith 2015). With relation to the user segment, especially value-added providers, insurance is usually taken by commercial operators capped to the level of liability they have under contract and

40 See, for instance, European Parliament EU Council (2014). Directive 2014/53/EU on the harmonization of the laws of the Member States relating to the making available on the market of radio equipment and repealing Directive 1999/5/EC.

41 Ownership of the PNT systems is determined by PNT specific provisions, as indicated above. See, for instance, in the EU, Article 9 of the Space Programme Regulation.

42 See, for instance, European Parliament EU Council (2009). Directive 2009/24/EC on the legal protection of computer programs.

43 The EU has a specific regime for the protection of databases. See European Parliament EU Council (1996). Directive 96/9/EC on the legal protection of databases (as amended).

taking into consideration potential rights of redress they may have against the stakeholders up above the value chain[44].

11.4. Liability

11.4.1. *International liability of PNT operators as States*

International liability of States for their role as PNT operators emerges first from International Space Law. In this scope, Article VII of the OST and the Convention on International Liability for Damage Caused by Space Objects of 1971 (Liability Convention) are the ones to be looked at.

In order to be within the scope of the Liability Convention, the damage must result from a "space object". Space object is defined (in a somewhat circular way) as including "component parts of a space object as well as its launch vehicle and parts thereof". Damage, on its turn, is defined as "loss of life, personal injury or other impairment of health; or loss of or damage to property of States or of persons, natural or juridical, or property of international intergovernmental organizations" (Liability Convention, Art. I, a and d). Only launching States[45] (Liability Convention, Art. I, c) are liable for damages and liability can be absolute (or strict) liability, for damage on the surface of the Earth or to aircraft in flight; and fault liability, for damage elsewhere than on the surface of the Earth to a space object (Liability Convention, Arts. II and III).

The determination of whether a State, as a PNT owner/operator, is internationally liable under Space Law, depends upon assessing the following:

a) *The source of the damage*: if it is not a space object, then there is no liability. But even if it is a space object, the question arises whether the space object causing the damage includes only its physical elements or also non-tangible elements, such as when the source of the damage is the software, data or signals of a space object.

b) *The damage*: liability for damages in space covers only damages to space objects and thus the same question shall be asked: whether a space object comprises only its physical elements or also includes the software, data and signals. Liability for damages on Earth covers damages to property and the question is whether "property" is a concept requiring a physical element or if non-tangible items (such

44 Note that in the space sector, contracts rely greatly on waivers of redress and hold-harmless agreements (Smith 2015).

45 It is the same definition of the Registration Convention.

as software and data of the ground stations of the PNT system) can also be considered "property".

c) *The causation link*: the main issue relates to assessing whether damages caused by a space object to the PNT user segment trigger the applicability of the Liability Convention, that is, whether the damages on the user segment are proximate enough to the source of the damage.

With relation to the concept of "space object" and "property", most scholars are of the opinion that only physical elements (and especially only physical damage) are addressed by the Liability Convention (e.g. Kerrest and Thro 2017). Another view argues, however, that an "object" or "property" cannot be seen only as comprising physical elements. In accordance with this view, an object includes both physical and logical elements (including data residing in the object), and therefore all should be considered part of the object (Mendonça et al. 2015). As a result, for instance, cyberattacks resorting, for example, to a virus from space could be considered caused by a space object and damages to the software or data could be considered damage to a space object. Likewise, it is argued that the intangible elements of a physical object could also be considered a part of "property". For instance, "intellectual property" is considered "property" in copyright legal regimes, despite the fact that this type of property does not fall upon physical items but upon immaterial items[46]. Hence, it could be argued that damage to software and data, including disruption, interruption or loss that cause loss of functionality, availability or performance, could give rise to absolute liability in the ground segment and to fault liability in the space segment.

With relation to signals, it seems that as fully unincorporated elements, they would be more remote from the concept of "object", even in the digital world[47]. When it comes to "property", however, it has been argued that "property" can be interpreted as including fully unincorporated elements such as signals in an approach similar to the one of the copyright and neighboring rights regime, where

46 See, for instance, European Parliament EU Council (2001). European Directive 2001/29/EC on the harmonization of certain aspects of copyright and related rights in the information society (as amended), which mentions in Recital 9: "Intellectual property has therefore been recognised as an integral part of property". Likewise, Directive (EU) 2019/790 of the European Parliament and of the Council of 17 April 2019 on copyright and related rights in the Digital Single Market indicates, in Recital 70, "the right to property, including intellectual property".

47 "The object causing the damage has to have material and physical properties, and thereby does not include non-material masses, such as signals or electronic interferences" (Kerrest and Thro 2017).

similar transmissions are viewed as property (Mountin 2014)[48]. An approach such as the one above would seem to lead to the conclusion that disruption of signals in space would not trigger liability (as damage shall be caused to a space object in space[49]) but could lead to liability in case of disruptions of terrestrial spectrum, all the more serious given that liability would be absolute in this case. This surprising conclusion makes clear that there is a lack of convergence or dispersion of the several rules applicable to space activities and that satellite operations should be viewed as a whole, in turn requiring a convergence of the applicable normative frameworks. A more traditional approach, however, by which both the signal in space and the terrestrial signal would be outside the Liability Convention would address this issue, and this is indeed the opinion of the majority of scholars (Smith and Kerrest 2009; Kerrest and Thro 2017)[50]. Another approach, whereby signals would also be considered part of the notion of "space object" would also lead to coherence in this matter. In this scope, it has been highlighted that the purpose of the OST and the Liability Convention was to create a binding system of effective compensation for victims of accidents resulting from space activities, and although at the time of the drafting physical damage by the physical elements of a space object were of the most concern, this may be interpreted as not necessarily leading to the exclusion of damage when it comes to signals. Hence, it is also argued that signals are an integral part of space operations, without which space objects cannot operate (Smith 2015; Chatzipanagiotis and Liperi 2017).

With relation to damages to the user segment, the issue of the causation link (i.e. whether damages on the user segment are proximate enough to the source of the damage) is central. The issue is complex because the user segment comprises different levels of stakeholders: from the value-added service providers (such as air-traffic services in aviation or providers of mapping services), users (such as

48 See Convention Establishing the World Intellectual Property Organization of 1967, as amended 28 September 1979, pursuant to which broadcasts are "intellectual property" (Art. 2, viii).

49 Or to "persons or property on board such a space object". Reference to "property" could be interpreted as covering signals, though the expression "on board" may be interpreted as requiring physical property or at least that the unincorporated elements are within the space object and not being emitted by it.

50 Note also the 2010 UNIDROIT's study on Third Party Liability for GNSS Services, where it is indicated that the OST and the Liability Convention do not apply to third party liability for accidents caused by GNSS failure or malfunction in transmitting the signal. See UNIDROIT (2010). Study LXXIX – Preliminary Study "An instrument on third party liability for Global Navigation Satellite System (GNSS) services: a preliminary study". International Institute for the Unification of Private Law, Rome.

aircraft operators or fleet management providers, and even consumers using PNT data to obtain directions and routes) to beneficiaries of the value-added services (such as passengers). The further we go in the value chain, the more distant we may be from the initial cause of the damages. Despite the fact that the Liability Convention does not address this issue expressly, it has been considered that only direct damages were covered by it (Larsen 2015; Soucek 2015) or that, in the case of indirect damages, the causation link must be such that the damage would not otherwise have occurred had the space object not caused the initial damage (Smith and Kerrest 2009; Kerrest and Thro 2017). Hence, it could be argued that both the primary effects of the action as well as the secondary and even tertiary effects could be covered and thus the damages suffered by the user segment could be taken into consideration under the Liability Convention. However, we must recall that the Convention aims to compensate third parties' victims to a space activity, and a user of PNT data or service may not be considered a third party (Kerrest and Thro 2017). These would only be, it seems, the beneficiaries of the PNT data/service (such as aviation passengers) and bystanders that may also suffer damages as a result of failures of the PNT system (e.g. victims of an aviation disaster who are not passengers).

In the case of damage caused by a terrestrial physical or non-physical source to the space segment, the ground segment or the signals, it seems clear that the Liability Convention would not apply, as the source of the liability would clearly not be a space object[51].

In all cases where the Liability Convention does not apply, and without prejudice to other provisions of the OST[52], States may also, for instance, trigger State responsibility under the Draft Articles on Responsibility of States for Internationally Wrongful Acts, UNGA Resolution 56/83 of 2002 (ARSIWA), require the unlawful conduct to cease (ARSIWA, Art. 30), and obtain restitution, compensation or satisfaction (ARSIWA, Arts. 34 to 37).

Hence, at the international level, PNT operators, as States, may be internationally liable for damages caused by the PNT system under Space Law or, at least, be

51 It should be noted that if the source of the damage is nevertheless in space, it could still be considered a "space activity". Therefore, Articles III and VI of the OST and the Draft Articles on Responsibility of States for Internationally Wrongful Acts could also be used for repairing damages arising therefrom (von der Dunk 2015c).

52 For instance, and as above referred, victims might obtain reparation for breach of Article III and VI of the OST.

internationally responsible under the OST and the Draft Articles on State Responsibility[53].

When the PNT system is not owned or operated by a State – which is the case of Galileo and EGNOS, owned and exploited by the EU – a further analysis should be done.

It is true that some authors argue that the provisions on liability contained in the OST are customary international law, which would mean that the EU itself would be bound to the rules of liability of International Space Law (Lyall and Larsen 2009; Smith 2015). In this respect, we note that the EU has legal personality distinct from its Member States (Treaty on European Union [TEU] of 2007, Art. 47) and that it further has competences in outer space which are shared with its Member States (Treaty on the Functioning of the European Union [TFEU] of 2007, Art. 89)[54]. However, it is not entirely clear that even if adopting a position in which the EU would be bound to the OST, that would lead to a conclusion that the EU would be liable under the Liability Convention, in light of the concept of "launching State". Only launching States can be liable and the EU is not a State, even if it is procuring a launch. It is true that Article XXII of the OST establishes that reference to States shall be deemed to apply to any international intergovernmental organization which conducts space activities. However, not only would it be necessary to conclude that the EU is an international intergovernmental organization as seen above, but also it would require the organization to declare its acceptance to the Convention and the EU has not, unlike ESA, made a declaration of acceptance of the provisions of the

53 Note, in this respect, that the International Law Committee (ILC) also adopted the 2006 Draft Principles on the Allocation of Loss in the Case of Transboundary Harm Arising Out of Hazardous Activities, A/61/10. These principles aim to ensure prompt and adequate compensation to victims of transboundary damage and to preserve and protect the environment (Principle 3). This regime is without prejudice to ARSIWA, and, although at the time they were drafted, space activities might not have been at the forefront of ILC's thought, the Commission stated that "transboundary damages" could involve other possibilities that may not be readily contemplated. In addition, the ILC states that "transboundary damage" refers to damage "caused in the territory or in other places outside the territory but under the jurisdiction or control of which the hazardous activities are carried out". See ILC (2006). Report of the International Law Commission on the work of the fifty-eight session, Yearbook of the International Law Commission. Vol. II, Part Two, A/61/10, International Law Commission, New York.

54 It is a shared or parallel competence.

Liability Convention. Hence, liability of the EU by direct application of the Liability Convention to the EU faces some obstacles[55].

11.4.2. *Liability of PNT operators as signal or service providers*

PNT operators may also be liable as signal or service providers under internal provisions[56].

The only PNT system that expressly addresses the question of liability is the EU system (Smith 2015)[57]: Article 97 of the Space Programme Regulation[58] expressly addresses EUSPA's liability by indicating that (i) "The contractual liability of the Agency shall be governed by the law applicable to the contract in question" and (ii) "In the event of non-contractual liability, the Agency shall, in accordance with the general principles common to the laws of the Member States, make good any damage caused by its departments or by its servants in the performance of their duties".

As the agency is responsible for exploiting Galileo and EGNOS and for the provision of their services (Space Programme Regulation, Art. 29, 2 a, and 44), it is thus the agency that is liable for damages arising therefrom, including for damages suffered by users.

In this respect, and when it comes to contractual liability, the first point to be analyzed is whether a contract for the use of PNT data or services exists. It has been widely considered that the availability of open signals within PNT systems does not

55 Some authors are also of the position that when a State is bound by the OST, the Liability Convention is also applicable, even if it is not party to the convention itself, due to the fact that it is a development of the principle set out in its Article VII (von der Dunk 1998). Regardless, the issue discussed above of the application of the notion of "launching State" to the EU could still be an obstacle in this case.

56 Note that Article XI (2) of the Liability Convention also establishes that "nothing in this Convention shall prevent a State, or natural or juridical persons it might represent, from pursuing a claim in the courts or administrative tribunals or agencies of a launching State" (Liability Convention, Art. XI, 2). Hence, for instance, even if one were to consider that disruption of signals was included in the Liability Convention, a victim could nevertheless choose to bring a claim under national law.

57 It is possible that the greater flexibility shown in this case results from the civilian nature of Galileo, whereas other GNSS are (also) military in nature (Larsen 2015).

58 These provisions reflect Articles 340(1) and 340(2) of the TFEU which govern the contractual and extra-contractual liability of the EU institutions under EU law.

constitute a contract (as they are open access and it is not possible for the PNT operator to monitor who receives the signal) (von der Dunk 2004; Chatzipanagiotis and Liperi 2017), while the availability of the signal only to certain stakeholders may configure a contractual relationship. In our opinion, the determination of whether there is a contract or not relies on the existence of provisions determining the conditions for access and use of the signal.

When there is a contract, liability would first be assessed on the basis of the contract, which could lawfully exclude or limit liability (Smith 2014). The extent to which liability can be excluded contractually is a matter for national law to determine. In EU countries, for instance, it is fairly common for legal provisions to allow exclusion of liability for simple negligence only (and not for gross negligence or intentional misconduct). Limits on liability exclusions are especially clear in case of subscription contracts that cannot be negotiated: the Directive on unfair terms in consumer contracts[59] contains provisions prohibiting the seller or supplier from excluding (and even limiting) its liability in certain cases, such as in case of death, personal injury or total or partial non-performance or inadequate performance.

To the extent that there is not a contract, liability can only be extra-contractual[60]. In this respect, and among other points, it would have to be determined whether there is a legal obligation or an otherwise duty of care in the provision of the signal that is breached (Smith 2015)[61].

We note that Article 10 of the Space Programme Regulation indicates that the services, data and information provided by the EU space program – including thus Galileo and EGNOS – shall be provided without any express or implied warranty concerning quality, accuracy, availability, reliability, speed and suitability for any purpose, and that the users of such services, data and information shall be duly

59 EU Council (1993), Directive 93/13/EEC on unfair terms in consumer contracts.

60 In this regard, principles relating to liability for risk or for hazardous activities may come to play, which exist regardless of fault (strict liability). Nevertheless, the Court of Justice of the European Union has not dealt so far with cases of damages caused by dangerous activities such as satellite operations (Smith 2015).

61 Note, however, that "the group of persons most likely to be negatively affected by GNSS malfunction – generally known as third parties – does not fall within the category of those towards whom the law always imposes a legal duty. Consequently, there is no general duty in law or under current GNSS operations to provide compensation for third-party loss" (Smith 2015). The author further highlights that currently third parties are only covered in specific regimes mainly those of transport conventions (see below).

informed of this[62]. This provision seems to result in an exclusion of EUSPA's liability for Galileo and EGNOS services.

Unlike Galileo and EGNOS, other PNT systems do not expressly address liability issues. It has been considered, as a result, that unless national provisions allow otherwise, no liability would exist due to the fact that PNT systems are owned and explored by States and thus enjoy sovereign immunity – which precludes a litigant from asserting a case against a State unless the State so consents (von der Dunk 2004)[63].

It is important to note that sovereign immunity does not apply to the relationship between a State and its nationals. In this case, and lacking specific PNT provisions, the general legal provisions of national law would have to be looked at in order to determine if the State would be liable for the PNT signals/data it makes available to its citizens.

But, even here, there may be difficulties: for instance, unless there is a strict liability regime/a legal liability for risk or for dangerous activities, liability would require fault, which may be difficult to assess for two main reasons: first, because outer space is congested and (still) lacks a system for space traffic management (Smith 2015), making it complex to assess who caused which damage and whether there is fault. Second, because the PNT value-chain includes many stakeholders and it may not be clear where the damage originated.

---

62 In accordance with Recital 23 of the Space Programme Regulation, "to encourage the widest possible use of the services offered by the Programme, it would be useful to stress that data, information and services are provided without warranty, *without prejudice to obligations imposed by legally binding provisions*" (Space Programme Regulation, Rec. 23) (our emphasis).

63 The Convention on Jurisdictional Immunities of States and their Property of 2004, not yet into force, makes an interesting differentiation between contractual and extra-contractual liability under Article 10, under which, to the extent there is not a contract (which would be most likely the case of open signals made available without limitation) then the principle of sovereign immunity would apply. Differently, if there is a contract, then the State would be contractually liable (though it could exclude its liability in the contract). The European Convention on State Immunity of 1972 – which was however ratified or accepted only by eight states, none of which is the owner of a PNT system – also contains a provision (Article 4) under which a State cannot claim immunity from the jurisdiction of the courts of another State if the proceedings relate to an obligation of the State, which, by virtue of a contract, falls to be discharged in the territory of the State of the forum, unless, among other cases, the contract is concluded between states or the parties have agreed otherwise. Indeed, sovereign immunity can only be invoked if the State is acting as a state and not like a private person. See Roma et al. (2019).

One final note shall be made relating to National Space Laws: indeed, national laws may exclude liability for signal disruption to the extent that liability is connected with space activities and operations and these are limited to launching, operation and return of space objects[64]. To the extent such laws also apply to the State and not only to private operators, they limit the rights of victims under national law, even if under international law the right to reparation of damage could exist.

11.4.3. *Liability of the downstream PNT stakeholders*

The determination of the liability of downstream PNT stakeholders, such as those providing value-added services, is also relevant and shall be done taking into consideration national provisions on contractual and extra-contractual liability.

In the EU, for instance, provisions applying to consumer contracts establish a set of conditions such contracts shall comply with. For example, the Directive on consumer rights[65] contains, among other points, a set of obligations with relation to information that shall be given in distance contracts and rights of withdrawal of the consumer. These provisions apply to services contracts as well as to the provision of digital content, such as PNT data. In addition, in accordance with the Directive on certain aspects concerning contracts for the supply of digital content and digital services[66], the digital content or digital service shall, in particular, where applicable, be of the description, quantity and quality, and possess the functionality, compatibility, interoperability and other features, as required by the contract, as well as be fit for any particular purpose for which the consumer requires it and which the consumer made known (Directive (EU) 2019/770, Art. 7).

It is further important to note that when it is national laws that shall apply to the determination of liability, one must first determine which law applies. The applicable law depends usually on the type of claim and is determined by rules of national private international law[67]: contractual claims are often governed by the law

64 See, for example, the French Law No. 2008-518 of 3 July 2008 concerning space operations (as amended), which explicitly excludes from its scope, in Article 1 (1°), any damage that is the consequence of the use of the signal emitted by a space object.

65 European Parliament EU Council (2011). Directive 2011/83/EU on consumer rights.

66 European Parliament EU Council (2019). Directive (EU) 2019/770 on certain aspects concerning contracts for the supply of digital content and digital services.

67 Conventions have been approved to unify the conflicts of laws rules arising in cross-border disputes such as the Convention on the Law Applicable to Products Liability of 1973.

chosen by the parties or by the law of the State closest to the dispute[68]. Extra-contractual claims are usually governed by the law where the harmful event occurred[69].

11.4.4. *Other heads of liability: product supplier and certification entity*

The manufacturer of the hardware that integrates the PNT system and the manufacturer of the receiver may also be liable under product liability provisions[70]. Under EU law, the Product Liability Directive[71] introduced a regime of strict liability for all the producers involved in the production process of defective products, their representatives, as well as importers. The Directive on contracts for the sale of goods[72] further applies to tangible movable items and requires, among other points, conformity of the good with the contract and that it shall be fit for any particular purpose for which the consumer requires it and which the consumer made known to the seller.

In addition, the certification entity may also be brought to play. In this respect, we note, for instance, that the Space Programme Regulation determines that EUSPA is responsible for ensuring, through its Security Accreditation Board, the security accreditation of all of the space program's components (Space Programme Regulation, Arts. 29, 1 and 36)[73]. The EU Cybersecurity Act[74], on its turn, entrusts

68 For example, European Parliament EU Council (2018). Regulation (EC) No. 593/2008 on the law applicable to contractual obligations (Rome I), Arts. 3 and 4.

69 For example, European Parliament EU Council (2007). Regulation (EC) No. 864/2007 on the law applicable to non-contractual obligations (Rome II), Art. 4.

70 "Claims under these headings in the space sector will be limited in practice. While it is conceivable that claims be brought in relation to navigational receiving equipment or 'hardware' that relies on GNSS software, they will not provide a reliable basis for claims based on error or malfunction of SiS [Signals in Space]" (Smith 2015).

71 EU Council (1985). Directive 85/374/EEC on the approximation of the laws, regulations and administrative provisions of the Member States concerning liability for defective products.

72 European Parliament EU Council (2019). Directive (EU) 2019/771 on certain aspects concerning contracts for the sale of goods, amending Regulation (EU) 2017/2394 and Directive 2009/22/EC, and repealing Directive 1999/44/EC.

73 Article 34.3 further provides that EUSPA is responsible, in Galileo and EGNOS, for their operational security and shall, to that end, carry out risk and threat analysis and all the necessary activities to ensure and monitor their security, in particular setting of technical specifications and operational procedures. In addition, in accordance with Article 34.4, Galileo Security Monitoring Centre (the operation of which is ensured by EUSPA) is responsible for monitoring the security of Galileo. The operation of the Centre shall be ensured by EUSPA also in accordance with the instructions developed under the scope of

upon the European Union Agency for Cybersecurity (ENISA) a set of duties on certification of ICT products, services and processes, including preparing candidate European cybersecurity certification schemes – under which products, services and processes may be certified as compliant by a national cybersecurity certification authority or a conformity assessment body (Cybersecurity Act, Arts. 56, 58 and 69)[75]. Without prejudice to issues of sovereign immunity if applicable, it is not clear if a certification entity can be made liable for damages in the context of its certification functions (Smith 2015). If, however, the damage is also caused because of reliance upon the certification (without which, for instance, certain signals, data, services or products would not be used, especially at the public level), it could be argued that certification entities would be liable in case of a negligent or wrongful act or omission (Chatzipanagiotis and Liperi 2017).

11.4.5. *Liability under sector-specific provisions*

Sector-specific regimes in aviation and the maritime sector contain provisions on liability for damages which also apply in case of malfunctioning of the PNT system, as they provide for strict liability of the carrier in benefit of passengers and third parties (and hence even if the flaw is on the PNT system which is not under the control of the carrier), in addition to fault-liability (which can for instance apply in case of malfunctioning receivers and without prejudice to the right of redress against the manufacturer).

In the aviation sector, and for passengers of an aircraft, the Convention for the Unification of Certain Rules Relating to International Transportation by Air, Warsaw of 1929 and the Convention for the Unification of Certain Rules for International Carriage by Air of 1999 provide for compensation based on fault-liability and on strict liability (up to a limit) (respectively) by the carrier. Hence, a problem with the PNT system affecting the passengers would allow these to act against the carrier. For third-party damage on the ground, the Rome Convention on Damage Caused by Foreign Aircraft to Third Parties on the Surface of 1952 provides also for strict liability by the carrier. The 2009 Montreal

---

Council Decision (CFSP) 2021/698 of 30 April 2021 on the security of systems and services deployed, operated and used under the Union Space Programme which may affect the security of the Union.

74 European Parliament EU Council (2019). Regulation (EU) 2019/881 on ENISA (the European Union Agency for Cybersecurity) and on information and communications technology cybersecurity certification and repealing Regulation (EU) No 526/2013 (Cybersecurity Act).

75 Note, however, that conformity self-assessment by the manufacturer or provider of ICT products, services or processes may also be permitted (Cybersecurity Act, Art. 53).

Conventions (Convention on Compensation for Damage to Third Parties, Resulting from Acts of Unlawful Interference Involving Aircraft of 2009, not yet into force, and Convention on Compensation for Damage to Third Parties, Resulting from Acts of Unlawful Interference Involving Aircraft of 2009, not yet into force) also provide for strict liability up to certain specified insurance limits.

In the maritime sector, the Athens Convention Relating to the Carriage of Passengers and Their Luggage by Sea of 1974 (Athens Convention) establishes the liability of the carrier in case of damages suffered by passengers (including death, injury and loss or damage to property). This liability is however subject to limits, which have been increased by the Protocol to the Athens Convention relating to the Carriage of Passengers and their Luggage by Sea of 2002. This Protocol also introduces strict liability (and compulsory insurance) in connection with passenger claims.

In the road sector, and as indicated above, there are no international provisions addressing liability for the use of PNT data, a point which is most relevant due to the advent of autonomous driving. Hence, it will be national laws that, for the time being, will have to be looked at to determine how liability is addressed in case of accident (Smith 2015)[76].

Other sector-specific conventions may also be relevant on a case-by-case basis in the light of the increasing use of PNT systems for diverse applications[77].

11.5. Conclusion

PNT systems play an increasingly central role in today's society, contributing to the operation and sustainability of several applications, infrastructures and services. Despite their international nature, their role, the increasing interoperability/ interaction among systems and the impacts of defects on or disruptions of PNT signals, there are no uniform provisions on PNT systems, including on liability. This creates a complex legal situation that may lead to legal uncertainty and even to different results in potential litigations on the basis of the laws that are applied, the stakeholders that may be liable and the users who may be affected. Most relevant is the situation of third parties or mere beneficiaries/bystanders, who may be unable to see their damage repaired due to the lack of a contractual relationship with PNT

76 Note that national laws that address autonomous driving, even if only for tests, may already address liability, even if not with a focus on the use of PNT systems.

77 For example, in nuclear facilities, which bring forward the application of the relevant nuclear conventions (Roma et al. 2009).

suppliers and the remoteness of their damage under extra-contractual frameworks. Furthermore, the legal status of the signal in space is not fully defined, giving rise to further difficulties, only made more complex due to the chain of stakeholders that intervenes and the complexity in determining who caused the damage.

This situation has led to some calls for a convention or at least an international legal framework or guidelines (Roma et al. 2009) regulating rights and duties in PNT systems. However, there is no broad support for this approach (Smith 2015) not the least because of the military uses of PNT systems. Thus, we are left only with sector-specific conventions to deal with liability that may arise from PNT systems in their areas of competence and with general legal frameworks not fully adapted to PNT.

Nevertheless, at least at a regional level (such as in the EU), the approval of rules coupled with non-binding guidelines and best practices, including on liability, could be evaluated. This would go a long way to both protect victims and PNT operators and providers, who would have their rights and responsibilities clearly defined.

11.6. References

Abeyratne, R. (2011). *Space Security Law*. Springer, Berlin.

Baumann, I. (2018). 2018 – A decisive year for European Space Policy [Online]. Available at: https://www.bho-legal.com/wp-content/uploads/2018/03/2018-A-decisive-year-for-European-Space-Policy_BHO-Legal_Baumann.pdf [Accessed 22 June 2022].

Chatzipanagiotis, M. and Liperi, K. (2017). Regulation of global navigation satellite systems. In *Routledge Handbook of Space Law*, Jakhu, R.S. and Dempsey, P.S. (eds). Routledge, Abingdon, Oxon.

von der Dunk, F.G. (1998). The 1972 Liability Convention – Enhancing adherence and effective application. *Space, Cyber, and Telecommunications Law Program Faculty Publications*, 50.

von der Dunk, F.G. (2004). Liability for global navigation satellite services: A comparative analysis of GPS and Galileo. *Space, Cyber, and Telecommunications Law Program Faculty Publications*, 3.

von der Dunk, F.G. (2015a). European space law. In *Handbook of Space Law*, von der Dunk, F.G. (ed.). Edward Elgar Publishing, Cheltenham.

von der Dunk, F.G. (2015b). Legal aspects of satellite communications. In *Handbook of Space Law*, von der Dunk, F.G. and Tronchetti, F. (eds). Edward Elgar Publishing, Cheltenham.

von der Dunk, F.G. (2015c). International space law. In *Handbook of Space Law*, von der Dunk, F.G. and Tronchetti, F. (eds). Edward Elgar Publishing, Cheltenham.

von der Dunk, F.G. (2017). Space law and GNSS – A look at the legal frameworks for "outer space". *Space, Cyber, and Telecommunications Law Program Faculty Publications*, 93.

EUSPA (2021). EU Parliament calls for fast adoption of satellite-based train localization in railway signalling [Online]. Available at: https://www.euspa.europa.eu/newsroom/news/eu-parliament-calls-fast-adoption-satellite-based-train-localization-railway [Accessed 22 June 2022].

Gaubert, C. (2015). Insurance in the context of space activities. In *Handbook of Space Law*, von der Dunk, F.G. and Tronchetti, F. (eds). Edward Elgar Publishing, Cheltenham.

Harrington, A.J. (2017). Regulation of navigational satellites in the United States. In *Routledge Handbook of Space Law*, Jakhu, R.S. and Dempsey, P.S. (eds). Routledge, Abingdon, Oxon.

Henri, Y., Matas, A., Santos, J. (2017). Regulation of telecommunications by satellites: ITU and space services. In *Routledge Handbook of Space Law*, Jakhu, R.S. (ed.). Routledge, Abingdon, Oxon.

Japan (2011). Basic policy on the implementation of the operational Quasi-Zenith Satellite System (QZSS) Project. *Cabinet Office* [Online]. Available at: www8.cao.go.jp/space/english/basicpolicy.html [Accessed 22 June 2022].

Kahveci, M. (2017). Recent advances in GNSS: Technical and legal aspects. In *IEEE Conference Proceedings, 8th International Conference on Recent Advances in Space Technologies (RAST)*, Istanbul.

Kaul, R. (2017). Regulation of navigational satellites in India. In *Routledge Handbook of Space Law*, Jakhu, R.S. and Dempsey, P.S. (eds). Routledge, Abingdon, Oxon.

Kerrest, A. and Thro, C. (2017). Liability for damage caused by space activities. In *Routledge Handbook of Space Law*, Jakhu, R.S. and Dempsey, P.S. (eds). Routledge, Abingdon, Oxon.

Larsen, P.B. (2015). International regulation of global navigation satellite systems. *J. Air L. & Com.*, 80(2), 365–422.

Lyall, F. and Larsen, P. (2009). *Space Law: A Treatise*. Ashgate Publishing, Garnham.

Mendonça, H.C., Cocco, M., Santos, J. (2015). International laws regulating satellite communications and their intentional disruption in times of peace and conflict. *Annals of Air and Space Law*, XL.

Michielsen, P. (2016). The Belgian Space Act: An innovative legal safeguard to boost the space industry. *Air & Space Law*, 41(2), 89–118.

Mountin, S. (2014). The legality and implications of intentional interference with commercial communication satellite signals. *Int'l. L. Stud.*, 90(101), 101–197.

Plattard, S. and Sitruk, A. (2017a). The governance of Galileo. Report 62, European Space Policy Institute.

Plattard, S. and Sitruk, A. (2017b). Looking at the future governance of Galileo. Report, 68th International Astronautical Congress, Adelaide.

Roma, A., Schrogl, K.-U., Sánchez, M. (eds) (2009). Policy aspects of third party liability in satellite navigation, preparing a roadmap for Europe. ESPI report, Vienna.

Smith, L.J. (2014). The current challenges of liability for loss of satellite-based services. In *Proceedings of the International Institute of Space Law 2013*, Jorgenson, C.M. (ed.). Eleven International Publishing, The Hague.

Smith, L.J. (2015). Legal aspects of satellite navigation. In *Handbook of Space Law*, von der Dunk, F.G. and Tronchetti, F. (eds). Edward Elgar Publishing, Cheltenham.

Smith, L.J. and Kerrest, A. (2009). The 1972 Convention on International Liability for damages caused by space objects (Liable). In *Cologne Commentary on Space Law: In Three Volumes*, Volume 1, Hobe, S., Schmidt-Tedd, B., Schrogl, K.-U. (eds.). Carl Heymanns Verlag, Cologne.

Soucek, A. (2015). *Space Law Essentials*, Volume 1. Linde Praktikerskripten, Vienna.

Volynskaya, O.A. (2017). Regulation of navigational satellites in the Russian Federation. In *Routledge Handbook of Space Law*, Jakhu, R.S. and Dempsey, P.S. (eds). Routledge, Abingdon, Oxon.

Wang, G. (2017). Regulation of navigational satellites in China. In *Routledge Handbook of Space Law*, Jakhu, R.S. and Dempsey, P.S. (eds). Routledge, Abingdon, Oxon.

Yamada, H. (2017). IMO and the GNSS – Navigating the seas. *Inside GNSS* [Online]. Available at: https://insidegnss.com/wp-content/uploads/2018/01/sepoct17-LAW.pdf [Accessed 22 June 2022].

12

The Legal Framework of Remote Sensing by Satellites: The Challenge of the New Space

Thierry LEMAIRE
General Counsel, Marseille, France

12.1. Introduction

In recent years, the space sector has undergone a profound transformation marked by the increased intervention of commercial companies in a field that was previously reserved for States. This phenomenon, called New Space, began in the United States in the early 2000s with the appearance of start-ups whose objective was to develop new applications based on space technologies to meet new societal needs. The emergence of FAAAM (Facebook, Alphabet, Apple, Amazon, Microsoft) and their Chinese counterparts BATX (Baidu, Alibaba, Tencent and Xiaomi) in the space economy has accelerated innovation and fostered the introduction of technologies from the computer and digital sectors into the space sector, thus providing new opportunities for services available from space infrastructures.

In the field of Earth observation, the intervention of private companies has, in a few years, favored the emergence of a real commercial market[1]. For example, the development of the software named Earthview, which became GoogleEarth following the purchase of the US company Keyhole by Google in 2004, marked a milestone in the use of space images. Based on a virtual globe, this software offers a visualization of the entire planet with an assembly of aerial and satellite photographs of a resolution on the ground previously reserved for military satellites[2]. This assembly enables, depending on the location, buildings, cars or even some large wild animals to be clearly distinguished. Now coupled with ground shots, the software allows users to virtually move around certain cities and, with a click of the mouse, to visualize the facade of a private house or of a competing business. This software, which is free for most users, has democratized access to spatial imagery and has opened the door to new investors, who are aware of the multitude of applications resulting from the use of spatial images, particularly in the field of geographic information.

The emergence of new commercial applications has led to new needs for space-based data and has ultimately contributed to the development of new, more efficient, mass-produced satellites at very competitive prices. The continuous decrease in the cost of launching satellites (currently about $2,000 per kilogram in low earth orbit) has enabled the deployment of satellite constellations. For example, the US-based company Planet, founded in 2010, has the world's largest operating constellation of optical satellites to date, capable of taking pictures of any location on Earth in near

1 According to the latest report from Euroconsult published on October 6, 2021, the commercial Earth observation market generated about $1.6 billion in revenue in 2020 and could reach $2.5 billion by 2030. The defense sector accounts for about 45% of the total Earth observation market and generates about $1.8 billion in revenue, followed by infrastructure development and natural resource management. The value-added services market could be worth about $5 billion by 2030, driven by the development of satellite constellations and BigData. Another report on the geospatial imagery analysis market study predicts growth from $9.1 billion in 2021 to $37.5 billion in revenue by 2026. See MarketsandMarkets (2022). Geospatial Imagery Analytics Market by Type, Collection Medium, Application, Deployment Mode, Organization Size, Vertical and Region – Global Forecast to 2026 [Online]. Available at: https://www.reportlinker.com/p05123121/?utm_source=GNW [Accessed 22 May 2022].

2 On November 27, 2019, the Indian space agency successfully launched the Cartosat-3 satellite with a panchromatic ground resolution of 27 cm into orbit. The ninth satellite in the Cartosat series offers the highest resolution commercially available to date. It now surpasses the performance of the US commercial satellite *Worldview-3*, launched in 2014, with an optical resolution of 31 cm and *Worldview-4* (formerly *GeoEye-2*) launched in 2016 with an optical resolution of 34 cm (the latter suffered a failure of its gyroscopic actuator in 2019 which rendered it inoperable since that date).

real time[3]. This company, which was listed on the New York Stock Exchange at the end of 2021, has considerable space capabilities, enabling it to archive changes in urban and natural areas, to know the traffic in a given port or to offer images of military installations or conflict zones. The US company Planet has recently provided space images of the Russian-Ukrainian conflict, showing very precisely and in near-real time the movement of troops or the destruction of habitats, industries and means of communication. Other constellation projects have been launched in recent years with varying degrees of success, such as the 22-satellite constellation in operation by the US company Orbital Insight, the 14-satellite constellation in orbit by the US company BlackSky or the project by the Canadian company UrtheCast, which, despite serious financial difficulties, still has the ambition of operating a constellation of 16 radar satellites. In contrast, the US company EarthNow had to abandon in 2021 its project to put into orbit a constellation of satellites that would have had on board superpowerful processing capabilities capable of analyzing the images captured in real time.

In addition, the use of artificial intelligence and machine learning has considerably improved data analysis because of powerful algorithms that allow the automatic identification of observed elements and are progressively replacing the intervention of specialists in the interpretation of spatial imagery that was previously indispensable. These technological advances contribute to the marketing of new products such as the deployment of monitoring and alert software, programmed to automatically detect certain predetermined elements, thus facilitating the supply of pre-selected spatial images adapted to the needs of customers for the management of their activities or the monitoring of those of their competitors, for example.

From the above, it is clear that the New Space economy has turned data, and more specifically space data, into a high value-added product by responding to new user needs, thanks in particular to a greater capacity for revisiting observation satellites, more frequent data updates, and greater information storage, all at lower cost. In this context, the States with space capabilities very early on sought to assert their leadership in this increasingly lucrative sector by creating regulatory conditions favorable to the intervention of private companies in this sector and by transferring to them the responsibility for the development and operation of remote sensing systems. This transfer of responsibility has been accompanied by a transfer of technologies previously reserved for military observation satellites, which has led to continuous improvement in the performance of commercial satellites.

---

3 *Planet* has 200 satellites in orbit in 2022 providing images with a resolution of 30 cm with a revisit frequency of 10 times per day. *Planet* plans to launch a constellation of 32 satellites in 2023 with a resolution of 30 cm [Online]. Available at: https://www.planet.com/ [Accessed 22 May 2022].

The multiplication of private actors in the remote sensing sector raises the question of the suitability of the national and international legal framework governing the activities of remote sensing by satellites with the liberalization of the market. Moreover, the development of new commercial applications derived from space images requires questioning the adequacy of the existing legal regime of protection of the remote sensing data. Finally, given the precision of the indications now provided by remote sensing activities, the analysis of the impacts of the commercialization of space images with regard to the protection of individual rights and fundamental freedoms shall also be addressed.

12.2. The international legal framework applicable to remote sensing by satellites

The general principles of United Nations General Assembly (UNGA) Resolution 41/65 of December 3, 1986 (Principles on remote sensing) is the only international text to date governing the activities of remote sensing by satellite for civilian purposes. Resolution 41/65 defines in its first principle remote sensing by satellite as the sensing of the Earth's surface from space for the purpose of improving natural resources management, land-use and the protection of the environment. This Resolution was adopted by consensus after more than 20 years of debate between the developed countries, which had satellite capabilities, and the developing countries, which claimed the right to control not only the observation of their territory but also the dissemination of the data collected by remote sensing satellites. The issue at stake was essentially for the observed countries to retain sovereignty over their natural resources. Conversely, the space powers of the time demanded total freedom of observation, dissemination and access to the data collected. The decision of the developing countries to renounce the principle of prior authorization for the observation of their territories allowed the international community to adopt Resolution 41/65 and thus to consecrate the fundamental principle of freedom of view and data collection on the observed territories as well as that of non-discriminatory dissemination of data.

12.2.1. *The principle of freedom of collection and non-discriminatory distribution of data*

The freedom to take photographs from space and collect data derives from the principle of freedom of exploration and use of outer space, set forth in Article 1 of the Treaty on Principles Governing the Activities of States in the Exploration and Use of Outer Space, including the Moon and Other Celestial Bodies, opened for signature on January 27, 1967 (1967 Treaty), which was reiterated in Principle IV of

Resolution 41/65. It should be noted that this freedom has not been questioned by States since the adoption of Resolution 41/65 and constitutes the legal basis for the regulations adopted by many space agencies or international organizations in the field of Earth observation. This continuous and uncontested practice thus gives the principle of freedom of view and data collection a customary scope and a more binding legal value.

With regard to the principle of non-discriminatory dissemination of data, Resolution 41/65 defines in Principle XII the rules applicable to the distribution of data, depending on whether it is primary data, processed data or analyzed information.

First, this principle states that observer States must provide the observed State with all primary or processed data concerning the latter's territory as soon as these data are produced, that is, prior to any commercialization of the data. This distribution must be non-discriminatory and reasonably priced, which prevents observer States from restricting the access of observed States to primary and processed data for political, social or economic reasons. Observer States cannot in principle oppose the existence of exclusive contracts for the sale of images to observed States which would have the effect of excluding the latter from access to the data collected. It is necessary to specify that Resolution 41/65 does not require observer States to disseminate data free of charge or at uniform prices and it is therefore, in practice, the play of free competition that will determine the price and reasonable conditions applicable to the dissemination of primary and processed data (Ravillon 2002). However, it seems to be accepted that if the price leads de facto to the exclusion of the observed State from accessing these data, the latter may invoke the violation of the obligation of non-discriminatory dissemination of data laid down by the Resolution.

Furthermore, Principle XII defines the rights of observed States with respect to access to the analyzed information, which must be made available to observer States as long as it is in the possession of any State participating in remote sensing activities, without discrimination and under the same conditions. This obligation of non-discriminatory dissemination is incumbent only on States and does not refer to private persons under their jurisdiction. Thus, it is recognized that private companies do not have an obligation to provide the observed State with analyzed information in their possession. Unlike primary and processed data, Resolution 41/65 does not require that the dissemination of analyzed information be made at reasonable prices. This flexibility thus allows private companies to invest in primary and processed data interpretation and analysis capabilities and to recoup these investments through market prices. However, nothing prevents the observer States from imposing this obligation on private companies operating remote sensing satellites and compelling them to disseminate the analyzed information relating to their territories to the observed States that request it, without discrimination and under the same conditions.

In addition to the obligation to disseminate data in a non-discriminatory manner, States participating in remote sensing activities also have a general obligation to inform the observed States about the remote sensing activities they carry out. This obligation to inform results from Principles X and XI of Resolution 41/65, concerning respectively the protection of the Earth's natural environment and the protection of mankind against natural disasters. The purpose of these two principles is to encourage cooperation between the States participating in remote sensing activities and the States observed.

One example is the Copernicus program, the European component of the Global Earth Observation System of Systems (GEOSS), initiated by the European Union (EU), the United States, Japan and South Africa. Since its launch in 2012, the objective of Copernicus is to bring together all the data obtained from existing environmental satellites or those developed specifically for this program (such as the Sentinel satellites, the first of which was launched in 2014) and on-site measurement instruments in order to produce a global and comprehensive view of the state of our planet. This information is shared free of charge with all users without discrimination in order to improve understanding of the Earth's natural events and prevent their impacts that could be detrimental to humanity. The use of the images collected within the framework of this program can also bring solutions to prevent or fight against industrial risks, marine pollution, deforestation or the optimization of water management and control more generally the impacts of human action on the environment. Finally, this program also has a broader objective to contribute to the respect of international treaties, to support peacekeeping operations, or even to provide assistance in the surveillance of European borders.

The International Charter "Space and Major Disasters"[4], created in 1999 on the initiative of the European Space Agency (ESA) and the Centre National d'Études Spatiales (CNES), is also a practical example of the implementation of the obligation to provide information for protection against natural disasters, as defined in Principle XI of the Resolution. The objective of this Charter is to develop international cooperation for the coordinated use of space assets in the event of a natural or technological disaster. This cooperation has been progressively extended to many

4 Charter on cooperation for the coordinated use of space assets in the event of natural or technological disasters, signed on October 20, 2000 [Online]. Available at: https://disasterscharter.org/fr/web/guest/home [Accessed 22 May 2022].

space agencies that have committed themselves under the Charter to provide users with their satellite data for the management of disasters around the world, whether natural or man-made. In addition to space agencies, Charter members also include private companies and public organizations that operate and/or use space-based systems. Authorized users of space imagery include civil protection, rescue, defense and security agencies of the Charter countries. The objective of this cooperation is to provide relief organizations with so-called "value-added" maps that give them an accurate view of the situation on the ground, without requiring special knowledge of space imagery, so that they can provide assistance to disaster victims as quickly as possible. In this context, the use of high-resolution remote sensing satellites is invaluable for providing images at much finer scales and better organizing relief efforts, assessing damage and managing the crisis. However, this need for precision is also accompanied by a requirement for rapid transmission of processed information and all-weather access capability (Chappez 2007), which requires an ever-increasing number of satellites and data acquisition ground stations that can be called upon for effective implementation of the Charter.

Finally, it should be added to these examples of cooperation that high-resolution remote sensing has also become an essential element for achieving the objectives of sustainable development that are defined at the international level. Indeed, space imagery systems provide valuable information on key aspects of the economic life of countries, not only on the state of their natural resources but also on the level of their economic activity, as evidenced by the variety of uses of space imagery used for planning and monitoring of arable land management, land use planning and urban development, protection of natural and historical sites, etc.

12.2.2. *The application of the principle of freedom of collection and non-discriminatory distribution of data by private companies*

The involvement of private companies in the civilian sector of remote sensing by satellites is now unavoidable and is the main driving force for the development of space imagery in the world.

While Resolution 41/65 of December 3, 1986 does not explicitly refer to remote sensing activities carried out by private companies, it is clear from reading Principle XIV that States operating remote sensing satellites are responsible not only for activities carried out by governmental bodies or international organizations to which these States are parties, but also for activities carried out by non-governmental entities. This principle recognizes that the activities of private companies fall within the scope of Resolution 41/65, thus placing an obligation on States to ensure that the principles of the Resolution are respected by these private companies.

Indeed, the placing on the commercial market of high-resolution space images could constitute a violation of the principles of the Resolution or of a rule of international law and thus infringe the interests of the observed or observer States. In such a case, the international responsibility of the State under whose jurisdiction the private company intervenes could be engaged, considering that the precision of space images may now infringe people's right of ownership on the property observed or the right to privacy of individuals, which are protected by international conventions. Moreover, private companies can intervene, at the request of observer States, to collect information in more sensitive areas such as the verification of compliance with disarmament treaties or the surveillance of certain sites deemed dangerous, or within the framework of the settlement of border disputes, or even for the observation of troop movements, as was recently the case within the framework of the conflict between Ukraine and Russia. In this last case, if one can consider that the surveillance activity by commercial remote sensing satellites remains to be carried out for civilian purposes, the collection of such data could be contrary to the interests of the observed State and could be used by certain States in relation to their public opinion in order to justify, for example, a military intervention against the observed State. Finally, the use of space images by "dubious" clients could endanger the national security of certain countries, encourage the preparation of terrorist acts against sensitive installations or justify attempts at political destabilization by disseminating certain space images to public opinion for propaganda purposes. These various examples show that the freedom to collect space images can pose problems when the observation is carried out by private companies because of the risks of engaging the international responsibility of the State under whose jurisdiction they operate.

Moreover, the risk of violation of the principles of the Resolution could also exist in the event that a private company fails to inform the observed State of the identification of indications that could prevent a phenomenon harmful to the environment, or to transmit processed data or analyzed information that could be useful to States that are victims of natural disasters or that could be imminently victimized by them, in disregard of the international rules set forth in Principles X and XI of the Resolution.

In addition, the implementation of commercial agreements for the distribution of data that have the effect of favoring one buyer or category of buyers over another may undermine the principle of non-discriminatory dissemination of data and analyzed information. This is the case, for example, with exclusive distribution contracts, which consist, for private operators of remote sensing satellites, of granting a distributor located abroad the exclusive right to distribute data over a given geographical area, in exchange for the purchase of a minimum quantity of images over a given period. This commercial practice could have the effect of restricting or delaying the distribution of space images for the benefit of the observed State and thus cause it harm.

Finally, infringements of the freedom of distribution could arise when the State under whose jurisdiction the distribution company operates imposes, for reasons of national security, restrictions on the dissemination of data at the request of the observed State, whereas the State or the private company operating the satellite has not imposed such restrictions. This situation could thus generate commercial disputes between the satellite operating company and the distribution company when they are not under the same jurisdiction if the distribution contract does not include a clause allowing one of the parties to suspend its contractual obligations toward the other party, in order to honor the international commitments of the State under whose jurisdiction it operates.

For these reasons, in order to guarantee the compliance with their international obligations and the possible consequences of their liability, the main space powers have defined a national legal framework to regulate Earth observation activities and set the rules and procedures for the distribution of data from remote sensing systems under their jurisdiction.

12.3. State control over private remote sensing activities

Principle XIV of Resolution 41/65 contains no provision requiring States to establish a system of prior authorization for the conduct of space observation activities by private enterprises. However, the reference in this Principle to Article VI of the 1967 Treaty has led States with space observation capabilities to regulate these activities not only with respect to the use of the space segment, which is expressly referred to in Article VI, but also with respect to the commercialization of space images, following the practice followed by the United States and France, in order to control the entire sector of Earth observation by satellite.

12.3.1. *The legal framework for remote sensing activities in the United States*

The development of the remote sensing sector in the United States is the result of cooperation between the private sector and the US government, which opened the remote sensing market to private companies in 1984[5].

After the debacles of the Landsat program in the 1980s, which led to the bankruptcy of the private company Eosat, due to a poorly calibrated commercial and pricing policy of the space images it marketed, the US government took over the operation and development of space infrastructures for Earth observation. The Land Remote Sensing Policy Act, adopted in 1992[6], and still in force today, defined the legal principles for opening the remote sensing market in the United States to private operators and the general conditions for the development and operation of private remote sensing systems. The implementation of this law was placed under the responsibility of the US Department of Commerce, which entrusts the National Oceanic and Atmospheric Administration (NOAA) with the examination of authorization requests for new private remote sensing systems.

This regulatory system thus allows the US government to control the arrival of new players on the remote sensing by satellites market and the collection and dissemination of data by private operators. In order to encourage the intervention of private players, the President of the United States proceeded as early as 1994 to declassify the technologies of certain military intelligence satellites, for the benefit of private US companies, thus encouraging the emergence of private high-resolution remote sensing systems for civilian purposes. However, it was not until the early 2000s that the first high-resolution civilian satellites, developed entirely with private funds, appeared, capable of providing images with a resolution of 1 m, such as the Ikonos-2 satellite of the US company Space Imaging.

In addition, the US government has encouraged federal agencies to use products and services from US commercial satellites, particularly to meet space-based imagery requirements for intelligence, defense, foreign policy and homeland security. It also prohibited the development of new remote sensing capabilities with public funds when the needs could be met by the commercial sector. The goal is to eliminate competition between government and commercial systems that could undermine the competitiveness of US industry.

5 Land Remote Sensing Commercialization Act, Public Law No. 98-365, 98th Congress, H.R 5155, 17 July 1984; 98 Stat. 451; 15 U.S.C 4201.

6 Land Remote Sensing Policy Act, Public Law 102-555, 102nd Congress, H.R 6133, 28 October 1992; 51 U.S.C 601; 106 Stat. 4163.

On the regulatory front, the US government has clearly set a goal of easing regulations and phasing out all restrictions on the development of future high and very high resolution imagery systems by, among other things, allowing the licensing of private companies to operate remote sensing systems with a resolution of less than 50 cm. This new approach has revealed a real paradigm shift for the US government, which now considers that US foreign policy interests are best served by a strong and competitive US industry in the international remote sensing space market and that the implementation of more flexible regulations should help attract private investors in this sector.

For this reason, NOAA's new Licensing on Private Remote Sensing Space Systems regulations, effective from July 20, 2020 (NOOA 2020), substantially changed NOAA's previous regulations issued in 2006. The in-depth analysis of applications for licensing new remote sensing systems is now reserved only for systems using remote sensing technologies or instruments that are not yet available in the US market or in other countries, whether used by private companies or public agencies. Unlike the 2006 regulations, the constraints on the operation of space-based remote sensing systems are only applicable for a temporary period, not to exceed 2 years, and are re-evaluated during the system's operational phase.

Without calling into question the US government's desire to promote the commercialization of space images by private companies, the possibility of using these new high-resolution capabilities for military purposes has forced the government to limit the liberalization of the sector and to maintain its right to intervene in satellite operations in order to limit or control the dissemination of space images because of the risk of undermining the national security of the United States or that of an allied state with which it is bound by international cooperation agreements, or on whose territory it has national economic interests or military installations, for example.

This control, known as "shutter control", is not unique to the United States and has been implemented by many space powers such as the USSR in the 1990s. In 1996, the United States adopted a text stipulating that the US government could not grant a license for the collection and dissemination by a non-governmental entity of satellite images of the territory of Israel or of another state designated by the President of the United States with a ground resolution of less than 2 m, unless images of this precision were already available from other commercial sources[7]. This was supplemented in 2000 by the provisions of an interdepartmental

7 Prohibition on Collection and Release of Detailed Satellite Imagery Relating to Israel, Rep. No. 104-278, 104th Congress 2nd Sess. S.1745 (1996), sec. 1064 - Public Law No. 104-201 (The 1997 Defense Authorization Act).

agreement, appended to NOAA's July 2020 rule, which states that any restrictions must be limited to the smallest area and for the shortest period necessary to protect US national security, international obligations, or foreign policy commitments. The US government now allows private US companies to sell data of the same quality and accuracy as that available on the international market. Restrictions on raw and pre-processed data (unenhanced data) are only applicable when using innovative technologies that produce data not marketed by foreign competitors. The new US regulations are therefore now structured in three categories, depending on whether the data produced are available on the international market, on the domestic market or not available on the market.

It should be noted that alternative measures to the prohibition of collection or distribution exist, such as delaying the transmission and distribution of data, restricting the system's field of view or encrypting the data, which make it possible to control the use of the data. These measures are applicable not only to the licensee of the system but also to its co-contractors such as ground receiving station operators and data distributors located outside the US territory. In addition, licensees must notify the US Administration of their intention to enter into any substantial or significant agreement with a foreign entity, so that the US Administration may make modifications or adjustments to such agreements, based on national security constraints, international commitments, or US foreign policy.

Many commentators considered that these restrictive measures were detrimental to the competitiveness of the US remote sensing industry, considering that the conditions for the use of such restrictions were not precise enough and that the Administration had significant discretionary power to interrupt or significantly affect the economic activity of private companies at any time, which could impact their potential profitability and competitiveness in a market that has become highly competitive. Other commentators have raised the risk of infringement of public liberties due to the control of information by the State, which, under the pretext of protecting the country's national security, could in reality infringe on the freedom of information of citizens and thus contravene the First Amendment of the American Constitution and the Freedom of Information Act (FOIA). Indeed, space images are a source of information for the media, which can use them to illustrate or prove certain situations or sometimes reveal to the public sensitive information on a developing crisis situation. Thus, by implementing restrictions, the government could prevent the public from knowing information that could be detrimental to national security or to the foreign policy of the country, including during foreign military operations.

For this reason, the US government has chosen to limit the use of direct restrictions on the collection or dissemination of space imagery by companies operating commercial remote sensing systems, preferring to apply a policy of bulk purchase of space imagery. Thus, the two main government agencies that use space imagery, the National Geospatial Intelligence Agency (NGA) – formerly the National Imagery and Mapping Agency (NIMA) – and NOAA have implemented a policy of multi-year contracts for the purchase of space images from US commercial companies, which in return grant them exclusive reproduction and distribution rights to the images, particularly those concerning conflict zones or high-risk areas[8]. In return, these contracts stipulate that the commercial company undertakes not to sell, distribute or share the space images relating to this zone with any client other than the US government. Thus, the generalization of these contracts could constitute an infringement of the principle of disseminating space images on a non-discriminatory basis, given the lack of availability of these images on the market for the benefit of other private or public non-US users. However, the multiplication of remote sensing satellites operated by non-US companies tends to limit the effectiveness of such a practice in controlling the dissemination of space imagery.

12.3.2. *The legal framework applicable to remote sensing activities in France*

The French system for disseminating remote sensing satellite data has long been based on a contractual framework agreement signed in the 1980s between the French government, represented by CNES, the satellite owner, and the commercial company Spot Image (now Airbus DS Geo), which is responsible for marketing SPOT satellite data. This family of satellites was initially considered by the French authorities as a strategic tool to guarantee the country's sovereignty, rather than a means to develop a commercial activity, even if the commercialization of the system quickly appeared to be an excellent way to make the public investment profitable. The agreement with the French government, initially conceived as a public service delegation, gradually evolved into a public–private partnership with a progressive transfer to Spot Image of an increasing share of the operating costs of the SPOT family of satellites and the financing of the development of the ground segment. The French government's ultimate goal was to encourage the financing of the development and operation of new observation satellites mainly from private funds.

Like the United States, the guarantee of non-discriminatory access to data for the benefit of users was recognized by the French government and enshrined in this

8 By way of illustration, in 2016, the private US company DigitalGlobe generated approximately 65% of its revenue with the US administration, 55% of which was with the NGA, with which it has signed a multi-year framework contract.

Convention. However, this principle was not absolute and the French government could decide, when the situation required it, to implement certain permanent or temporary restrictions on the dissemination of space images for reasons of national security or compliance with France's international obligations.

Marketing contracts were also signed with the direct reception stations, defining the terms and conditions for distributing and marketing the data. Each distributor received a license from Spot Image to reproduce and market SPOT satellite data in a defined geographical area in return for payment of a fee for each copy of the images sold. On this basis, Spot Image also expanded its offer by obtaining authorization to distribute space images from other operators, such as ESA's Earth Remote Sensing (ERS-2) and Envisat radar satellites, some Indian Remote Sensing (IRS) satellites and the US satellites Quickbird and Orbview-3.

However, toward the end of the 1990s, France questioned the need to review the legal framework applicable to its space activities and, in particular, that relating to remote sensing by satellites because of the arrival of new civilian observation satellites whose resolution was close to that of military systems and whose use could potentially constitute a risk of infringement of national security imperatives and international defense agreements concluded by the French government. However, it was not until the adoption of Law No. 2008-518 of June 3, 2008 on space operations (LOS) that the French authorities established an authorization regime for space activities carried out from French territory or by French nationals, which could engage the responsibility of the State. In particular, the law regulates the activities of primary space data operators, which are defined as any person who programs an Earth observation satellite system or receives Earth observation data. The law now provides that the activities of these operators are subject to a prior declaration issued by the competent administrative authority, the Secretary General of National Defense (SGND). Assisted by an interministerial commission on space data, SGND is responsible for determining and coordinating the monitoring of national policy on private remote sensing activity and for defining the technical characteristics of activities subject to declaration as well as the restrictive measures that may be prescribed for reasons related to national defense, foreign policy and France's commitments[9]. In particular, the objective of this commission is to advise the SGND as to whether the fundamental interests of the Nation or France's international

9 Decree No. 2009-640 of June 9, 2009 implementing the provisions of Title VII of Law No. 2008-518 of June 3, 2008 on space operations (2009 Decree). This decree was amended by Decree No. 2013-654 of July 19, 2013 relating to the supervision of the activity of primary operators of data of space origin, itself amended and repealed by Article 10 of Decree No. 2019-1379 of December 18, 2019 abolishing administrative commissions of an administrative nature (Decree of 2019).

commitments could be impacted when executing an agreement or contract for programming, for direct reception or for production of space-derived data from satellites operated by the primary operators of space derived data, for which the approval of SGND is required.

Under this law, the French government may also take temporary or permanent restrictive measures, in particular on programming, on direct reception for the production of new data or on the dissemination of primary data archived under the responsibility of the primary operators of space-derived data concerning an area or territory (LOS, Art. 24 and Decree of 2009, Art. 5). These data dissemination restrictions may include (i) the immediate suspension, in whole or in part, of the issuance of data concerning specific geographic areas, for a temporary renewable period; (ii) the obligation, in whole or in part, to defer the issuance of data concerning specific geographic areas, for a temporary renewable period; (iii) the total or partial limitation of the technical quality of the data concerning certain geographical areas that can be delivered, for a temporary renewable period; (iv) the permanent limitation of the technical quality of the data concerning certain areas of the national territory that can be delivered. The French system has recently been reinforced with new potential restrictions such as a permanent ban on programming or on reception and the geographical limitation of shooting areas (Decree of 2019).

This legislative framework replaces a governmental practice that had no real legal basis and thus provides a legal basis for private operators who would be obliged, at the request of the French State, to refuse to sell space images and who could be held liable in contract or in tort for the implementation of such restrictions.

12.3.3. *The legal framework applicable to remote sensing activities in the rest of the world*

Remote sensing activities have also been regulated in other countries that have developed space-based Earth observation infrastructures. The implementation of the principles of Resolution 41/65 and the provisions of the 1967 Treaty takes different forms depending on whether the State concerned decides to regulate the development and operation of remote sensing satellites, the collection and dissemination of data, or both.

In this general overview, it is worth mentioning India, which adopted a prior authorization regime in 2001 to control the acquisition and distribution of data from Indian and foreign remote sensing satellites for civilian use in India. Under this law, permission from the Indian government was required to operate a remote sensing satellite from India and to distribute images within the country. The legislation also included measures that allowed authorities to ensure that sensitive areas were masked on the images before they were distributed, in order to protect the country's national security interests. This law has been replaced by the Remote Sensing Data Policy (RSDP) Act of 2011 (Indian Remote Sensing Data Policy 2011) which now allows distribution on a non-discriminatory basis of images up to 1 m resolution on the ground. While not opening up the Indian remote sensing market to the private sector, the Act removes all restrictions of the 2001 Act on the distribution of space-based imagery unless the resolution is less than 1 m. In this case, the images must be checked before distribution by the National Remote Sensing Centre, which has been designated as the only authority that can purchase and distribute in India all images produced by Indian remote sensing satellites or foreign satellites with sub-meter resolution.

In 2005, Canada also implemented a specific licensing system for the operation of the RADARSAT-2 radar satellite and the distribution of remote sensing data, taking into account the dual uses of radar technology[10].

Russia, which has had long experience in the development of military and civilian remote sensing satellites since 1965[11], belatedly adopted legislation on space activities in 1993 covering satellite remote sensing activities, the provisions of which have not been published. This law, amended in 2015, entrusts the Russian space agency, Roscosmos, with the implementation of the national policy for the collection and dissemination of space images, which includes providing free access to all users to images with a resolution greater than 30 m and, on a commercial basis, to images produced by Russian satellites with a resolution of less than 30 m.

In addition, other States have put in place rules to regulate satellite observation activities on the basis of their general space legislation, such as the United Kingdom, Australia, Japan, South Africa, South Korea, Spain, Ukraine and the United Arab Emirates.

Finally, while the European Union does not yet have legislation governing the space activities of European companies that would be binding on European states, ESA has adopted a policy for the dissemination of remote sensing data from space that aims to provide free access to all data collected by all its satellites (including the Sentinel satellites of the Copernicus program) (ESA xxx). This new policy, amended in 2010, now facilitates the widest possible use and reuse of this information and

10 Remote Sensing Space Systems Act and Regulations, 2007. S.C 2005 Chap. 45.

11 Several high-resolution remote sensing satellites are in operation, such as the *Resurs-P* and *Resurs-DK* family of satellites capable of providing sub-meter resolution images, the *Kanopus-V* and *Kanopus V-IK* family of satellites and the *AIST-2D* satellite.

data by the general public, the media, the education sector, ESA's partners and anyone else wishing to exploit this content.

From the above, it should be noted that the emergence of new commercial space imagery providers points to the need to develop a more effective international and/or regional framework that would make it possible to coordinate the various national policies for the collection and dissemination of space images and the possible restrictive measures that are essential to guarantee the national security of States and compliance with their international commitments. Indeed, national measures aimed at restricting the marketing of space imagery have limited effects with respect to foreign suppliers that are not subject to the restrictions taken by States that do not have jurisdiction over them and prove to be counterproductive, on the other hand, for national companies whose commercial positions are weakened on the international market. However, a generalized practice of restrictions on the collection and dissemination of data would not only undermine the principles of freedom of view and non-discriminatory dissemination of data resulting from Resolution 41/65, but would even risk calling into question the international compromise that allowed the adoption of this Resolution.

12.4. Legal protection of remote sensing data

The legal protection of remote sensing data appears to be an indispensable condition for the marketing of space images and for the success of the liberalization of this sector of activity. Indeed, no private operator would have any interest in investing in the setting up of space and terrestrial means of Earth observation, the costs of which are very high, if the product resulting from this investment could not be protected against unauthorized use by third parties.

However, the intangible and volatile nature of remote sensing data poses difficulties for observer States in implementing effective legal protection against the unauthorized reproduction and dissemination of remote sensing data that may have significant economic and strategic value. This protection should not hinder the free flow of data but, on the contrary, should facilitate their transmission in order to stimulate demand and foster the development of new digital solutions and services derived from space data.

Neither international space law nor Resolution 41/65 has defined a legal mechanism to protect remote sensing data. Therefore, data protection measures are, in accordance with Article VI of the 1967 Treaty, a matter for the national provisions of each State. Thus, the major remote sensing States, including the United States and France, have based their space imagery protection regimes on concepts derived from intellectual property law, such as copyright, which are usually used to protect intellectual creations. In both cases, the use of copyright as a means of legal protection raises questions about its application to remote sensing data from space.

12.4.1. *The protection of remote sensing data by author's right*

The protection by author's right, recognized by many countries that are signatories to the Berne Convention for the Protection of Literary and Artistic Works, adopted in 1886, the most recent amendment of which came from the Paris Act of July 24, 1971, amended on September 28, 1979, and by certain instruments regulating international trade[12], confers on its holder, for a specific period of time, an exclusive right of exploitation, allowing them to control the reproduction and dissemination of their work, as well as protection of their moral rights. This protection allows the author to authorize the commercial exploitation of their work by taking advantage of the reproduction and representation rights that may be granted to third parties, while protecting the originality of the work. Author's right therefore applies to works of the mind, which raises questions about its application to remote sensing data from space.

The work of the mind can be defined as the product of a free intellectual work, expressing the personality of the creator and embodied in an original form (Edelman 1999, p. 15). From this definition, it follows that a work of the mind must be able to be concretized in a form perceptible to the human senses and be communicable. It is the communicable character of the work that makes it possible to appreciate its originality, that is, the expression of the personality of its author.

With regard to remote sensing by satellites, it is appropriate to consider the application of the originality criterion to satellite data that are expressed in different forms depending on the stage of their production and processing, ranging from raw data, in the form of computer data (made up of 0s and 1s), to a derived product, in the form of a paper medium or a digital product. As far as the raw data are concerned, the data must undergo a series of treatments and corrections so that the information the data contain becomes communicable in a form perceptible to the human senses. It is only after their realization, that the raw data can be used by a user. Therefore, the criteria of author's rights do not seem to be fulfilled with regard to raw data. On the other hand, one could consider that pre-processed or corrected data and, a fortiori, derived products are true intellectual creations, given the

12 In particular, within the framework of the World Trade Organization (WTO), the Agreement on Trade-Related Aspects of Intellectual Property Rights (TRIPS).

necessity, to obtain these types of data, to carry out a certain amount of processing, which obligatorily requires human intervention and that of self-programmed computer tools. As far as derived products are concerned, such as satellite images or geographical maps, the conditions of originality could be considered as fulfilled, provided an added value is brought to the product as a result of human intervention or, in the case of product obtained through an automatic process, that such product bears the imprint of the personality of its author. Now, this originality could be challenged when the satellite image or the map backgrounds, for example, use elements considered to be part of the public domain (coastal lines, roads, etc.).

In addition to originality, the protection by author's right requires that the author of the work be identifiable. However, given the way in which data are produced by space-based means requiring the intervention of multiple actors, the difficulty arose in determining which entity could acquire the status of author, since this status can only be attributed to a natural or legal person, but not to a technical system or a machine. In the case of remote sensing data, it is possible to consider that it is the fruit of a collective intellectual work carried out between the company that owns the satellite, the satellite operator, the ground stations and the client, each of whom intervenes to make proposals concerning the time of the shooting, the angle of view, the framing, the shutter speed or the method of processing the image. However, it is commonly accepted that satellite data are obtained under the impulse of the satellite operator who acts as a coordinator between the various natural persons participating in the design and operation of the satellite, which confers on him/her the quality of the author of the work.

From the above, it is clear that protection by author's right is imperfect and does not cover all categories of remote sensing data and that it is appropriate to turn to other forms of legal protection such as that of databases[13]. Indeed, the application of this legal regime allows the database to be protected not only by author's right when it results from a genuine intellectual creation of its author, with regard to the choice and organization of the elements constituting the database, but also to protect the database by a sui generis right, distinct from author's right, when it is the product of a substantial investment.

If the protection by author's right of remote sensing databases seems equally questionable and uncertain, whether the data are considered individually or as a grouping in the form of a database, the introduction of a new criterion of protection based on the finding of the existence of a substantial investment made by the producer of the database may allow the application of this legal regime to remote

13 Directive 96/9/EC of the European Parliament and of the Council of March 11, 1996 on the legal protection of databases, L 77/20.

sensing data. Indeed, it is easily demonstrable that the production of remote sensing data results from a significant financial and professional investment, taking into account the implementation of considerable material and human means on the part of the operator of the observation system to produce the data that constitute the database. The interest of this sui generis right is to essentially prohibit any form of extraction or reuse of a qualitatively or quantitatively substantial part of the information sets contained in the database, whether they are original or not.

As with author's right, the question arises as to who owns the protection of the remote sensing database. This question becomes all the more important if one considers that the existence of a database does not appear as soon as the data are produced in space, but during their processing. If one considers that the operator is the one who had the initiative to create the database and the catalog, this would allow the satellite operating company to claim the application of this sui generis right to protect the space images and the digital derived products, while the owner of the satellite could be the owner of the right on the data, depending on the nature of the data. Indeed, the application of the sui generis right for the protection of databases is without prejudice to the existing rights on the content protectable by author's right. Thus, this legal construction would protect all the stages of the process, without interruption, from the transmission by satellite of the raw data to the supply of the most sophisticated digital product. This double mode of protection is moreover the one retained by certain space imagery companies in their sales contracts which expressly aim at author's right and also at the legislation applicable to the protection of databases (Airbus Defence and Space 2017).

12.4.2. *The protection of remote sensing data by copyright*

The criteria for the application of copyright are quite similar to those used for author's right, with regard to the existence of a work of the mind or the criterion of originality. However, in the copyright system applicable in the United States for example, the author appears as the merchant of their work and they maintain a relationship of ownership with it, which leads the doctrine to qualify copyright as an exclusively economic right (Edelman 1999, p. 26). Thus, copyright is a more restrictive right than authors right since it consists mainly of protecting the copyright holder against any reproduction of a material element without their authorization. The material element is thus at the center of this protection, which is more focused on the protection of economic interests than on cultural or artistic interests, as in the case of author's right, for which the personality of the author and the moral element are essential.

The main difference between copyright and author's right is the requirement in the case of copyright that the work be fixed on a material medium in order to be protectable, a material medium that allows it to be received, reproduced and communicated[14]. Thus, copyright could apply to remote sensing data as soon as they are fixed on a material support, upon their reception on the ground by ground stations. However, as with author's right, it seems that raw satellite data cannot be considered a copyrightable work, due to the automatic nature of the data acquisition and their lack of originality, even though the text governing copyright in the United States does not refer to the notion of intellectual creation to grant its legal protection.

Contrary to author's right, the question of identifying the beneficiary of the protection of space data does not seem to be posed in the same terms because of the absence of search for the author of the creation in terms of protection by copyright. Thus, there should be no difficulty in recognizing that the private company, owner of the satellite and having made the investment to develop it, is considered the beneficiary of the protection of the data produced. As such, it would have an exclusive right to reproduce the protected works, create derivative works and distribute copies (Copyright Act of 1976, Art. 106).

To conclude, it appears that the legal protection of remote sensing data by author's right or copyright is not perfect, which could create some legal insecurity and limit the initiatives of private companies in the development of digital solutions and services derived from space images.

12.5. Impacts of remote sensing on individual rights and freedoms

The improved performance of space observation satellite sensors, which can now reach a resolution of 25 cm on the ground, and the increased data collection and processing capacities make it possible to market new digital products and to obtain real-time images of the Earth's surface because of satellite constellations. Space images are therefore no longer only a means of identifying objects on the ground but also a means of analyzing human activity in a more dynamic mode.

Also, private operators could see their ability to commercialize space images hampered by the existence of private rights on the observed elements or by the infringement of individual rights such as the right to privacy or to the protection of personal data, in the case of the use of space images as a complement to personal information files, for example.

---

14 Copyright Act of 1976, Pub. L. No. 94-553, 90 Stat. 2541 (codified as 17 U.S.C. §§ 101-810 (1978), Art 102(a).

12.5.1. *Risks to privacy and the protection of private property*

Faced with the marketing of increasingly precise space images, the question arises as to the extent to which, for example, the owner of a building or a movable asset, observed from space, could oppose the taking of a picture of it or the diffusion of a photograph representing it by asserting the existence of patrimonial rights on the observed assets or by opposing the existence of extra-patrimonial rights such as the right to privacy, for example.

With regard to the right to respect for private property, it seems that the absence of sovereignty rights over outer space that overhangs the territory of a State implies that the latter cannot grant a private individual more rights than it possesses and therefore cannot attribute to the individual the right to authorize or refuse the taking of a picture of their property from space, on the basis of their property rights over the property observed.

Concerning the application of the extra-patrimonial rights of individuals with respect to observation, the question seems to be different because of the consecration of some of these rights on the international and national levels. Indeed, even if space imagery techniques do not allow to visualize a person from space, the identification of an individual is now possible by cross-checking information available elsewhere. In this context, it is for example possible to invoke the protection of the Covenant on Civil and Political Rights, adopted on December 16, 1966, which protects, in Article 17, individuals against any violation of their home or intrusion in their private life without prior authorization. The infringement of other attributes of the personality could be raised on the basis of international texts of regional scope, such as the European Convention for the Protection of Human Rights and Fundamental Freedoms, opened for signature on November 4, 1950, which provides, in Article 8, that everyone has the right to respect for their private and family life, home and correspondence.

At the national level, as in France for example, respect for privacy is protected by an increasingly abundant case law on the question of the infringement of personality rights due to the taking of a photograph of a property without prior authorization, in particular when such a photograph makes it possible to identify an individual or to locate their home and thus to infringe upon the intimacy of their private life, reputation or honor[15]. Thus, case law has ruled on the illegality of an aerial photograph, on the basis of the infringement of the right to privacy, allowing individuals to object to the disclosure of information concerning their resources, such as the ownership of a swimming pool or a tennis court, when the area observed

15 Cour de cassation, Chambre civile 2, June 5, 2003, 02-12.853.

is inaccessible to the public[16]. Similarly, the use of a drone to ascertain the failure of a property owner to comply with a reporting obligation has been recognized as a form of interference in the private life of an individual[17]. Thus, it appears from the jurisprudence of the French courts that the use of space images by the Administration in the context of the investigation of criminal offences could be rejected, in particular when the evidence is obtained in circumstances constituting an offence or an excessive interference in the private life of an individual. On the other hand, this does not seem to be the case for the use of spatial images published on websites such as Google Maps, which has been recognized by the French tax authorities as a lawful means of constituting evidence of an offence, in particular with regard to the identification of swimming pools that have not been declared by their owners to the tax authorities[18].

At the European level, it appears that the administrative authorities are authorized to use photo-interpretation of satellite images or aerial photographs to carry out control operations, in particular for the allocation of Community aid in the agricultural sector[19].

In other legal systems, such as the United States, the open skies doctrine has also been used to justify the use of aerial observation to prove an offence without infringing on private property[20]. However, the Supreme Court considered that the observation techniques used must be publicly known and that the use of certain observation systems could constitute an invasion of privacy and protection of the home if these systems made it possible to learn about an individual's activities inside the home by means of techniques not available on the market[21]. Conversely, a federal court of appeal more recently recognized that crop fields were considered areas not covered by the protection of the private home and that space images could thus be used by the US Administration to track insurance fraud[22].

A balance must therefore be found between the development of techniques to monitor human activities and the protection of the privacy of individuals, whose infringements must be justified and authorized in advance.

---

16 Cour de cassation, Chambre civile 1, May 29, 1984, 82-12.232.

17 Cour de cassation, Chambre criminelle, March 21, 2007, 06-89444.

18 JO Sénat 11/01/2018, 94; JO Sénat 05/03/2020, 1159.

19 For example, see the provisions of Regulation (EC) No. 73/2009 of January 19, 2009 establishing common rules for direct support schemes for farmers under the common agricultural policy.

20 Dow Chemical v. United States, 476 U.S 227, 106 SCt 1819 (1986).

21 Kyllo v. United States, 533 US 27 (2001).

22 United States v. Fullwood, 101342 F.3d 409 (5th Cir. 2003).

12.5.2. *Risks of personal data protection being breached*

Today, the command of information is a major stake for the development of many commercial companies and the interest of the FAAAM for the development of new space technologies and constellations of observation satellites allowing information on human activities to be obtained in quasi-real time, tends to confirm it. The collection of data, including spatial geolocation data of their customers, allows these New Space companies to obtain useful and valuable information on the consumption and life habits of their customers, for example, in order to be able to offer them adapted services[23]. Thus, even if an image is not information as such, it could be considered as a medium for information and become, after processing, nominative data that would necessarily become subject to the rules of personal data protection.

In Europe, the protection of personal data is governed by the General Data Protection Regulation (GDPR)[24], the provisions of which are applicable in all EU countries since May 25, 2018. GDPR applies to any entity, European and non-European, that collects or processes personal data relating to a European resident[25] by obliging them, in particular, to inform the data subject of the collection of personal data and to limit the length of time the data is kept. While today space images very clearly allow real estate properties to be identified as well as their owners by cross-checking with other information, it is clear that images containing this information have been widely distributed for a few years on geolocation platforms such as Google Maps, Mappy or the French government's Géoportail website and are not subject to any time limit or distribution limit.

---

23 *Google Maps*, *Apple Maps*, *Bing Maps* (*Microsoft*), *OpenStreet Map*, *MapQuest* and many other applications are associated with features that connect with restaurants, shopping malls, parking lots or other points of interest in order to facilitate consumption or use of transportation. All of this information can be used to track trips, places frequented and to define a typical consumer profile using personal information.

24 Regulation (EU) 2016/679 of the European Parliament and of the Council of 27 April 2016 on the protection of individuals with regard to the processing of personal data and on the free movement of such data, L 119/1.

25 Article 4(1) of the GDPR defines personal data as "any information relating to an identified or identifiable natural person (hereinafter referred to as 'data subject'); an 'identifiable natural person is one who can be identified, directly or indirectly, in particular by reference to an identifier, such as a name, an identification number, location data, an online identifier, or to one or more factors specific to physical, physiological, genetic, mental, economic, cultural or social identity".

In addition, the Regulation states that the data controller must guarantee the rights of the data subject with respect to the processing of their personal data, including the right to object to the collection and/or dissemination of their personal data (GDPR, Art. 21). This right of opposition could be invoked in the event of an infringement of an individual's privacy and in the event of an abnormal disturbance caused to the latter by the dissemination of the image of a property they own, for example. The implementation of the right to object remains nevertheless a delicate exercise and few geographic data platforms provide for it. A fortiori, when the collection or dissemination of images is based on a legal basis, such as the execution of a public service mission for example, it is unlikely that the person concerned will be able to benefit from his or her right to object, which could limit the Administration's right to communicate and inform the public.

In any case, the application of the European Regulation could be limited by the international nature of the dissemination of remote sensing data and the intervention of countries outside the EU that are not subject to this legislation. However, the Regulation provides that the transfer to a third country of personal data, processed or intended to be processed after their transfer, can only take place if this third country ensures an adequate level of protection of the privacy, freedoms and fundamental rights of the persons concerned by these data.

12.6. Conclusion

Satellite earth observation is a tool for inventory, analysis, forecasting and simulation, which is essential to meet the scientific, economic and strategic challenges of today's world. In addition to the real-time availability of collected information, space imagery has undeniable advantages such as the ability to cover large areas, to provide access to remote regions, with the possibility of multitemporal monitoring, which is not possible with other observation tools such as drones. Space imagery has thus become essential to improve the daily life of humans and to address global issues such as climate change and environmental protection or a major strategic tool in monitoring armed conflicts or population displacement.

The involvement of commercial enterprises has accelerated the development of new technologies and value-added products and services from space-based remote sensing data that are available at no or low cost. The explosion in demand for space-derived data now requires a robust and efficient satellite data processing and distribution system to enable companies and commercial organizations to continue to develop new applications and generate new needs. In addition, as seen previously, the emergence of low-cost small satellites and the improved technological performance of observation instruments have enabled New Space companies to revolutionize their business model, now based on high-resolution images broadcast in near-real time. However, in order to offer new solutions and digital products to customers, it is necessary to transform satellite data into useful information. The enormous amount of data generated by observation satellites represent both a challenge for operators in terms of data dissemination and storage – with the issue of cloud computing in particular – and an opportunity for major players in the information and communication technologies sector.

While the freedom to collect remote sensing data from space contributes to the progress of all humankind, thus meeting the objectives set by Resolution 41/65 and the 1967 Treaty, it is necessary for States to continue to monitor the activities of private companies under their jurisdiction to avoid any use of space images for aggressive purposes or that could infringe on individual rights and freedoms, while at the same time limiting provisions restricting the dissemination of the data that could be a hindrance to technological development and the massive use of the images for all users.

For this reason, it seems desirable that any limitations on the collection and dissemination of space images be adopted within a multilateral framework, which would thus protect the commercial activity of private companies and make it possible to better adapt and coordinate surveillance measures or restrictions on commercial activity for reasons of national security, for example. However, this coordination is less and less easy to implement in the field of space observation because of the increase in the number of countries with remote sensing technologies and the need to find an agreement of all the States concerned on a precise framework of intervention. This mission could be entrusted to a new international agency for earth observation, resulting from an international treaty or a conventional framework for cooperation decided between States, which would govern, more particularly, the consequences of the intervention of private companies in this sector, notably in terms of the dual uses of high-resolution space images and restrictions on the collection and dissemination of images. This agency could also have the role of defining international standards that would make it possible to harmonize the various national policies in this field, whether in terms of the protection of the content of space images and any authorized restrictions, the protection of individual rights and freedoms, or the use of space images as a means of proof in the settlement of disputes between States. Finally, the definition of international standards in contractual matters for the distribution and marketing of space images or the regulation of the storage of data and digital products could be entrusted to it.

12.7. References

Airbus Defence and Space (2017). License to use satellite products [Online]. Available at: https://www.intelligence-airbusds.com/files/pmedia/public/r13117_9_eula-pleiades-vuk-october2017.pdf [Accessed 22 May 2022].

Chappez, J. (2007). Les technologies spatiales au service de la lutte contre les catastrophes naturelles. In *L'adaptation du droit de l'espace à ses nouveaux défis. Mélanges en l'honneur de Simone Courteix*, Kerrest, A. (ed.). Pedone, Paris.

Edelman, B. (1999). *La propriété littéraire et artistique*, 3rd edition. Que sais-je ?, PUF, Paris.

ESA (2010). Data Policy for ERS, Envisat and Earth Explorer missions [Online]. Available at: https://earth.esa.int/eogateway/documents/20142/1564626/ESA-Data-Policy-ESA-PB-EO-2010-54.pdf [Accessed 22 May 2022].

Indian Remote Sensing Data Policy (2011). Indian Remote Sensing Data Policy 2011 [Online]. Available at: https://lib.icimod.org/record/8866#:~:text=The%20new%20policy%202011%20replaced,resolution%20data%20for%20developmental%20activities [Accessed 14 June 2022].

NOOA (2020). Licensing on Private Remote Sensing Space Systems. Report, NOOA, 15 CFR Part 960.

Ravillon, L. (2015). L'exploitation commerciale de l'espace extra-atmosphérique. *Jurisclasseur droit international*, 2, Fasc. 141–20.

13

Small Satellites and the Regulation of Outer Space Activities

Steven FREELAND[1,2]
[1] Western Sydney University, Australia
*[2] Bond University, Robina, Australia**

13.1. The changing nature of space technology

In March 2019, reports emerged that a United States (US) private company, Swarm Technologies, had arranged for several of its tiny micro-satellites to be launched into orbit in January by an Indian launch vehicle, notwithstanding that the US regulator, the Federal Communications Commission (FCC), had expressly declined to authorize the launch. These "Swarm Bees" were very small – about one-fourth of the size of a standard CubeSat – and the FCC therefore had formed the view that the difficulty in tracking these micro-satellites once they were in orbit meant that they posed what was regarded as an unacceptable risk[1].

Even though it appears that there have been no negative consequences in terms of safety concerns following on from this unauthorized launch, the Swarm Technologies incident has highlighted some of the complexities arising from the changing nature of space technology – much of which has involved the development

* This chapter was completed in September 2020.

1 The FCC has subsequently fined Swarm Technologies US$900,000 for its unauthorized launch. See Coldeney (2018).

of small satellite technology. This follows a pattern of technological advancement for space since we were first able to find ways to enter into this new (final?) frontier.

Indeed, ever since the commencement of the space race with the launch of Sputnik 1 in 1957, there has been a breathtaking and seemingly endless development of space-related technologies. Humankind is now engaged in a multitude of space activities far beyond the contemplation of those involved at that time. The utilization of space technology now forms a crucial part of everyday society in all parts of the globe – irrespective of the (geo)political, economic and cultural characteristics of any one country. Our reliance on space technology is such that the world would cease to function in many respects without constant and unimpeded access, and this imperative is likely to become even more pronounced for future generations. This has primarily been driven by the increasing commercialization of outer space.

Yet, as is well known, there remains a vast gulf between the space capabilities of the relatively small number of space "powers" compared with the rest of the world. It has been estimated that approximately 60 States currently possess some form of direct space capability, although the extent that they are able to utilize space for their own development (and other) purposes varies quite significantly. Of course, this also means that perhaps up to 140 States thus far do not realistically have any independent capability to directly access space themselves, yet their reliance on the technology for many aspects of their functioning and development continues to grow. These countries are instead totally dependent on others for their space access, which therefore impacts upon their space "security" and may well also impede opportunities for creativity, innovation and progress among their citizens. The reality is that their access to satellite data and the ability to utilize vital space technology in a crisis would be largely dependent on and subject to the strength and enforceability of their existing contractual relationships and political ties.

It is in this context that the recent development and adaptation of so-called "small" satellite technology potentially represents a paradigm shift in the way humankind accesses space. These satellites are usually cheaper and less complex to develop, build and launch than conventional satellites. They therefore open the possibilities for a significantly greater degree of space access to a much larger range of space "actors". Already, groups such as university students and non-profit organizations in both developed and developing countries have increasingly been able to become involved in space through these means. The development of this technology may represent an important precursor to the establishment of indigenous and independent space programs in States that previously were not in a position to consider such activities. In effect, by eliminating some significant barriers to entry, small satellite technology may facilitate capacity building, broader collaborative opportunities and education/training programs, as well as bridging (some) technology gaps, for hitherto "non-space faring" States. It will also open up even more diverse commercial opportunities for a much broader range of potential service providers and, generally, "bring space to more people".

In addition, as the technology develops even further, it may open the door to "traditional" users of outer space – both States but also private commercial entities – to utilize it for existing as well as new purposes, thus expanding the scope of their capability at a significantly lower relative cost. Of course, this may require a mind-shift on the part of existing space actors, as they grapple with whether and how to adapt to this relatively new technology and adjust their activities to the challenges posed by the potential for new market entrants.

Consequently, the increasing advent of this technology could potentially redefine the landscape of many activities in space. This new space paradigm will not see the end of more traditional satellite technology since, naturally, small satellite technology will not quench our insatiable demand for all that space can provide. However, it does open up a plethora of possibilities, many of which we are simply not in a position to comprehend or even imagine at this point. In this regard, one might liken the potential of small satellites to the way that mobile phones have revolutionized terrestrial communications activities. We simply do not know where this technology might ultimately lead to and what it will allow us to do. However, we can confidently expect that it will open the door to an even more expansive array of commercial opportunities.

Thus, from a technological perspective at least, small satellite technology most likely represents a "win-win" possibility that enhances the momentum for change and further promotes commercial space activities. Indeed, in many respects, this has been the singular motivation for both developers and users thus far. As with many aspects related to the exploration and use of outer space, the technology continues to move forward at a rapid pace without sufficient attention being paid to the regulatory consequences and requirements. It is therefore important not to be too caught up in this wave of optimism and innovation, without at least also considering how these developments coexist with the current regulatory framework, which has largely been designed with "big" satellite technology in mind.

The purpose of this chapter is therefore to take pause and reflect on various regulatory requirements and challenges posed by the existing international legal regime in relation to the use of small satellite technology. While many of the users of this technology are no doubt cognizant of these requirements, it is the author's experience that many are not; or, put another way, they do not consider the regulatory issues with the same degree of attention as they do the technical factors.

What this discussion will highlight is the fact that the existing legal framework was not designed with small satellite technology specifically in mind. As a result, further regulation will be required – particularly at the national level – and this will necessitate a "balancing" of sometimes competing interests between protecting the present and future of State from potentially very significant liability on the one hand and encouraging innovation and research and development on the other. Although the discussion below focuses on the current regulatory requirements, it leads to the conclusion that the design of future legal regimes to deal specifically with small satellite technology will necessitate some fundamental policy decisions by national lawmakers and regulatory bodies.

13.2. The current international legal framework and regulatory requirements

The international regulation of the exploration and use of outer space is primarily based upon a series of five United Nations (UN) Space Treaties and several General Assembly (GA) Principles (see *supra*, Chapter 1). The Treaties in particular set out a number of fundamental rules, imposing various obligations on States Parties, some of which are also regarded as representing customary international law (Hobe et al. 2013). More and more States have come to recognize the need to promulgate national space laws to "transform" these international obligations into their respective domestic legal spheres. Given that the advent of small satellite technology presents opportunities for hitherto non-space faring States to engage in space activities, it may well be that the development of such technology in a particular country may "pre-date" any specific applicable national laws. Thus, the possibilities of greater access to this technology may be a driving force in the enactment of a further "wave" of national space law in various countries, for example, as was the case in Austria, which enacted its national space law in late 2011[2].

It should be noted that, in addition to these various instruments, and possibly also to supplement perceived "gaps" in them, there have recently been an increasing number of "soft-law" guidelines concluded that also relate to the conduct of particular activities in outer space. This has been for several reasons, partly related

2 The Austrian *Outer Space Act* of 2011 entered into force in December of that year and was introduced largely in response to the development of small satellite programs by a number of Austrian universities, and the impending launch of two Austrian-built small satellites, which were eventually launched from India into low earth orbit in February 2013.

to the strategic and political nature of space, which has made the finalization of internationally binding treaties more difficult to achieve[3].

This chapter will refer primarily to existing "hard-law" regulatory requirements that flow from the Space Treaties – although reference will be made to one important set of voluntary guidelines – from the perspective of how they may relate to the use of small satellites and seek to raise some pertinent questions that arise from their applicability. It is not intended in this chapter to be exhaustive in this regard, or comprehensive as to all precise details, but rather to raise the more significant issues and the challenges they pose. This will also serve to highlight the importance of properly addressing this issue by way of specifically directed regulation, given that the use of small satellite technology will most likely continue to grow exponentially into the future.

13.2.1. *International responsibility – authorization and supervision*

The regime for space activities is structured on the basis that States bear international responsibility for "national activities in outer space", including when such activities are carried out by non-governmental entities (OST, Art. VI). While there is no precise definition in the OST as to what constitutes a "national activity", the terms of the domestic space law of a particular State will clarify the scope of activities to which it refers – in essence, representing an interpretation by the drafters of that legislation of what they regard to be national activities in outer space, at least for the purposes of the specific domestic law.

A review of existing national space legislation indicates that, in most cases, States have legislated for the regulation of space activities based on the "territoriality" of the activity (i.e. where an activity, for example, a launch, involves the territory of that State) in accordance with general international law principles of jurisdiction. In addition, many States that have national space law also regulate space activities based on the nationality of the space actor (i.e. the person/entity engaged in the space activity). In these circumstances, the (international) responsibility under the OST may extend to extra-territorial activities, such as the launch of a satellite from overseas by a national of the relevant country.

Article VI of the OST goes on to require that the "appropriate State" – which is thought by most commentators to mean the State whose national activity it is – undertakes the "authorization and continuing supervision" of such activities.

3 For a discussion of the increasing trend toward the conclusion of non-binding instruments in the realm of outer space, see Marboe (2011) and Freeland (2011).

Typically, the authorization of space activities is implemented by way of a licensing regime established under national law, at least for those States with specific domestic space legislation (Freeland 2010b). This can be through the creation of a comprehensive "one size fits all" license regime or, more likely, via the establishment of different forms of license, depending upon the particular space activity for which authorization is being sought.

In relation to the use of small satellites, there is little conjecture that their launch and use does, indeed, constitute a space activity. Moreover, the satellite itself would in most circumstances be a space object for the purposes of international space law – including for the purposes of the Liability Convention (see below), as well as the domestic law of most countries. Activities involving small satellites therefore would typically fall within the scope of Article VI of the OST. This in itself is not surprising, what is surprising, is that this is not necessarily understood by the users of small satellites, particularly with respect to "experimental" projects. The reality is that those seeking to engage in small satellite activities, irrespective of where those satellites might be launched, should take careful note of the relevant national laws and apply for the requisite license (where applicable). As noted below, this might also have added consequences in terms of financial and liability concerns, as well as other aspects of conditionality.

Moreover, the requirement of continuing supervision on the part of the State may be quite complex. There is, for example, some conjecture as to how, in practice, the need for continuing supervision might be undertaken in circumstances where the relevant space activity is a cooperative venture between institutions in a number of States. Internal arrangements between the cooperating States should be put into place to allow for each State to, in some way, exercise a degree of supervision, at least in relation to those aspects of the activity (and over its nationals who may be involved in its ongoing operation) in which it has a specific interest. Yet, even this pre-supposes that the institutions or persons engaged in the small satellite activity have informed the relevant governmental agency of their involvement, and have provided specific details as to the scope of the program, design, issues of control, etc.

Adding to the complexity is the fact that most small satellites are not designed with control systems, and therefore cannot be maneuvered once they are launched and operative. As soon as they are placed in orbit, their position cannot be altered from Earth. This may also explain why this requirement may often have been disregarded, leaving the responsible State in a difficult position in terms of its obligations under the OST.

13.2.2. *International liability – national indemnity requirements*

The general international liability provisions found in the OST[4] and the more detailed regime specified in the Liability Convention impose liability on a "launching State" for certain damage caused by a space object[5]. There are no time limitations or caps on the amount of this liability under the Liability Convention, as long as it represents "damage"[6] by a "space object" as those terms are defined for the purposes of that Treaty[7]. In the absence of specific indemnities in relation to claims by third parties, or where the various exceptions and exonerations contained in the Liability Convention do not apply, a launching State will bear this international obligation of liability (Freeland 2001) even in circumstances where the space activity is undertaken by a non-governmental entity and perhaps also even where the State may not be aware of the activity at all[8].

This represents one compelling "incentive" for States to pass domestic space law. The enactment of national space law enables States to formalize domestic legal processes that would allow them to pass on financial responsibility and recover from their national non-governmental entities the full amount (or part thereof) of the damages for which the State may be liable at the international level. Of course, this does not remove the international obligation of liability of a launching State under

4 Article VII of the OST prescribes the general terms giving rise to international liability for damage caused by an object launched into outer space. The scope of international liability is then elaborated in the Liability Convention. However, even if it is not a State Party to the Liability Convention, a State would still be subject to the liability provisions in the OST, as well as any other potential claims based on the general public international law principles of State responsibility.

5 The identity of the relevant launching State(s) is determined at the time of launch, the Liability Convention defining a launching State as "(i) A State which launches or procures the launching of a space object; (ii) A State from whose territory or facility a space object is launched" (Liability Convention, Art. 1, c).

6 The Liability Convention defines "damage" as "... loss of life, personal injury or other impairment of health; or loss of or damage to property of States or of persons, natural or juridical, or property of international intergovernmental organizations" (Liability Convention, Art. I, a).

7 As noted, it would be difficult to argue that an operating small satellite was *not* a space object for the purposes of the Liability Convention, even if it is not manoeuvrable while in operation.

8 On this point, there may be an argument that where the only possible relevant mode by which a State could be a launching State in a specific case is by ''procuring'' the launch, there is a minimum ''threshold'' test to demonstrate such procuring, at least based on knowledge of the particular activity. However, it is unclear whether such an argument reflects the correct legal position.

the Liability Convention – this contingent liability remains in place in relation to any space object for which a particular State is deemed to be a launching State. However, it does enable the State to put in place a domestic mechanism by which it can transfer the financial "risk" associated with this potential international liability for third party claims. Indeed, this is precisely the practice that a number of States have followed in their national laws in relation to "traditional" satellite technology.

Consequently, national space legislation often attaches conditionality to the issue of a license to engage in a specific space activity the practical effect of which is to require the applicant to provide or somehow procure an indemnity to the Government for damage, although the amount may be subject to specific caps under a particular national law. Although it would be relatively straightforward to simply require the applicant in these circumstances to take out an appropriate commercial insurance against third party claims to the extent of the specified (maximum) damage, this would often be impractical (given the relative lack of depth of the international space insurance market) and, more specifically in the case of many small satellite operators, disproportionally costly[9]. Indeed, such a requirement might make the planned small satellite activity unaffordable, thus preventing it from going ahead at all.

This gives rise to difficult considerations that would require a "balancing" between the protection of the State from potential financial liability and the desirability of encouraging expertise, research and development, perhaps as a precursor to more profitable and commercial opportunities down the track. Such potentially conflicting interests between a need for regulation on the one hand and the provision of incentives for a new innovation on the other are not unique to the situation of small satellite operators – similar arguments have been raised in relation to the requirement for the "equitable sharing of benefits" derived from the exploitation of natural resources under the Moon Agreement. However, unlike the Moon Agreement, virtually every space-faring State is a party to both the Outer Space Treaty and the Liability Convention – and, in any event, the liability regime they establish arguably also reflects customary international law. It is therefore incumbent on all States with an (potential) involvement in space to somehow address this issue.

---

9 The Australian *Space Activities Act* of 1998 envisaged that in certain circumstances, rather than procure insurance, an applicant could instead demonstrate ''direct financial responsibility'' as an alternative (section 47). The Australian legislative framework was amended in 2018, and the relevant legislation is now titled the *Space (Launches and Returns) Act*, but this possibility has been retained (section 47, 2, b).

The ideal scenario would be for the small satellite operator to negotiate with the relevant launch service provider for the provision of insurance cover and/or an indemnity by that provider (and perhaps also the Government standing behind that provider) to the launching State and the payload owner (e.g. the University that has built the small satellite(s)), at least in relation to certain elements of potential third-party claims (again most likely subject to a cap). This is often the case in commercial launch service contract arrangements for large satellites. Some small satellite operators contend that the position is more complicated in the case of a collaborative small satellite program such as the QB50 project, involving Universities from several countries (and thus potentially a considerable number of launching States). However, the point remains that many such programs have proceeded without the issue even being raised with either the launch service provider or the intermediary arranging the launch.

Once again, this is something that should be negotiated coincidentally with the development of the technical aspects of such a program. A failure to do so potentially not only places the launching State in a difficult position but might also expose the institution supporting the small satellite operators to a real and unacceptable risk of liability. Obviously, this should be of practical concern to those involved.

13.2.3. *Registration – national and United Nations registers*

The Registration Convention creates a two-pronged regime of registers that are relevant in respect of space objects that are launched *inter alia* "into earth orbit" (Registration Convention, Art. II, 1). The State of Registry (as defined) maintains a national register in which such space objects are to be included and, in addition, shall provide certain specified information in relation to those objects to the UN, which itself maintains a central register (Registration Convention, Art. IV, 1). In accordance with the terms of the OST, the registration of a space object within a State's national register also has implications with regard to the "jurisdiction and control" of that object (OST, Art. VIII).

In situations where a State, has not, for instance, previously been involved in launching activities, it may not have in place a national register, nor a mechanism for the furnishing of the required information to the UN. There may be a time lag associated with the establishment of the national register, which, in most circumstances, could only be implemented under national space legislation. Once again, this will require consultation and information flows between the small satellite operator and the relevant Government agency (if indeed such an agency exists).

In addition, with widespread cooperative small satellite programs that may potentially involve institutions from many countries, there will be need for careful coordination between the various launching States as to who should be the State of Registry – it can only be one of the launching States (Registration Convention, Art. I, c). It may not, for example, make practical sense that each launching State would seek to be the State of Registry for its specific small satellites in the context of a joint program involving a large constellation of objects launched simultaneously from the one launch vehicle.

13.2.4. *Sustainability of the space environment – space debris mitigation*

One of the major challenges for the future exploration and use of outer space is the growing proliferation of space debris. Much has been written about the exponential growth of pollution in outer space and the hazards that it poses. There are many approaches on how the problems should be addressed, given that the whole issue of the environment of outer space is a complex one, with many interconnecting variables at play (Bohlmann and Freeland 2013, p. 375). As noted above, these variables, and the enormous financial implications that would arise from setting in motion binding requirements, have meant that to date only "soft-law" guidelines, rather than hard law treaty regulation, have been agreed to address this issue. Nonetheless, these guidelines[10], although voluntary and expressed in general terms, are significant in that they reflect existing practices as developed by a number of States and international organizations and set (minimum) standards toward which space-faring nations should strive.

The principles underpinning the debris mitigation guidelines are that care should be taken to minimize the risk of debris "creation" in the conduct of space activities[11].

10 See IADC (2020). IADC Space Debris Mitigation Guidelines [Online]. Available at: https://orbitaldebris.jsc.nasa.gov/library/iadc-space-debris-guidelines-revision-2.pdf. [Accessed 20 April 2022] (IADC Debris Mitigation Guidelines); COPUOS (2007), Space Debris Mitigation Guidelines of the Committee on the Peaceful Uses of Outer Space [Online]. Available at: http://www.unoosa.org/pdf/publications/st_space_49E.pdf [Accessed 20 April 2022] (UN Space Debris Mitigation Guidelines).

11 The UN Space Debris Mitigation Guidelines recognize two broad categories of space debris mitigation measures: those that curtail the generation of potentially harmful space debris in the near term – minimization of the production of mission-related space debris and the avoidance of break-ups; and those that limit their generation over the longer term – end-of-life procedures that remove decommissioned spacecraft and launch vehicle orbital stages from regions populated by operational spacecraft.

The importance of accessing and using space for all aspects of our lives necessitates a diligent adherence to these standards to the greatest extent possible. It is generally recognized that it is in the interest of all space-faring States to follow these guidelines, and this is, as noted, increasingly reflected in their practices. The long-term sustainability of outer space activities is a matter of interest and importance for the international community as a whole and is now one of the principal focal points for UN Committee on the Peaceful Uses of Outer Space (COPUOS).

There are some potentially significant environmental challenges that arise from the use of small satellite technology. Growing demand, the expanding range of functions and, ultimately, commercial services they can provide point to rapid increases in the numbers of small satellites that will be placed into Earth orbit. In order to utilize this technology to achieve "global" coverage, very large constellations of small satellites will be required, and are being planned (Zolfagharifard and Griffiths 2014)[12]. While these satellites will primarily be placed into a low Earth orbit, projects such as these will "populate" important orbits with a significant number of space objects and increasingly pose a potential collision risk.

Even with respect to less ambitious low-cost small satellite programs, the issue may still remain. Many small satellite programs to date have simply been "experimental" (although this is changing now). They have often utilized existing "off-the-shelf" components, and the expectations of mission success for any significant period of time have not necessarily been high. It is fair to say that such circumstances give rise to lower perceptions of risk and a higher tolerance towards failure. For many such programs, at least in the relatively early phases of small satellite development, the process has largely been about the journey (to space) rather than delivery of services – though of course this is now changing. Many of these programs have relied on "piggy-back" launches, which has meant that the satellites have been placed in orbits significantly higher than the very low orbits that would allow them to decay relatively quickly. For many small satellites, therefore, there is a potentially very long period (perhaps in excess of the 25 years cap suggested by the IADC Debris Mitigation Guidelines) before orbital decay, even though the satellite itself will have been functioning for only a short timeframe.

Moreover, as is well known, there are several variants of small satellite technology. While it is too simplistic to categorize them solely on the basis of their size and weight, some satellites may be too small to be picked up by conventional tracking systems. Yet, such low mass objects also can cause catastrophic damage in

12 This article reports that SpaceX plans to launch 700 satellites (each weighing 113 kilograms), and Google 180 satellites, both to provide internet services for the 4.8 billion people of the world who are still without online access.

certain circumstances. The potential consequences, and therefore the potential risks, would, of course, be greatly magnified should the development of a large-scale commercial human spaceflight industry, despite recent setbacks, ultimately come to fruition[13].

Of course, these issues are relevant to the question of potential liability raised above. They also point to the need to carefully consider how, and to what extent, the future implementation of small satellite programs can and will be undertaken, so as to as much as possible be consistent with the overarching goal of managing the long-term sustainability of outer space activities in such a way as to maximize the (commercial) benefits that can be derived, while maintaining appropriate and acceptable safety standards, particularly for missions involving humans.

In some senses, therefore, the environmental consequences relating to small satellite programs have not really been properly factored into the regulatory framework. This is also a question of education and awareness, but is a highly important factor to take into account when designing the future legal regime to apply to such programs.

13.2.5. *Other regulatory considerations – frequency allocation and traffic management*

As noted, these brief comments do not purport to be comprehensive as to the relevant regulatory factors associated with this new commercial space paradigm featuring large-scale small satellite programs. However, the primary regulatory issues that ultimately stem from the principal requirements under the UN Space Treaties have been raised. There are, of course, other equally relevant considerations that also arise. For example, as more such programs emerge, particularly offering commercial services, the issue of radio frequency usage becomes all important. The existing use of the "amateur band" frequencies will no longer be applicable and appropriate. The regulatory framework of the International Telecommunications Union (ITU) will become even more relevant. While the ITU operates effectively to manage the use of radio spectra, it is a large intergovernmental organization and can be highly bureaucratic. Decisions about allocations of valuable (commercial) frequencies need significant periods of time and are sometimes highly political. The coordination of frequencies to minimize harmful interference is complex. This lengthy process does not necessarily sit comfortably with the shorter timeframes associated with small satellite programs. Accordingly, procedures will need to be

13 For a discussion of the relevant legal issues related to the proposals to establish a commercial human spaceflight industry, see (Freeland 2010a).

established to accommodate this technology without compromising the important work of the ITU. This will not be an easy task.

In addition, the introduction of large numbers of small satellites will highlight even more the imperatives to consider the development of international traffic management systems involving space traffic, as well as its intersection with air traffic. Once again, there is much work to be done by all stakeholders.

13.3. Concluding remarks

This chapter has highlighted the fact that the current international legal framework continues to apply to new and developing technologies – such as small satellites – that will contribute to the further evolution of commercial space activities. Of course, the business case for those large programs that have been announced is yet to be proven and, while it is clear that small satellites will form a (significant) part of the future dimension of space, there may be some "false starts" along the way as to the most appropriate approach to be undertaken by those entities seeking to utilize the technology to develop very significant commercial businesses.

Consequently, the existing law and the technology, at least at the international level, do not represent a natural fit. The international regulatory framework was not designed specifically to deal with the advent of this technology, nor for the expansive range of new space actors. Moreover, these new actors in particular may not be completely aware of, or understand, the relevance and implications of the existing framework.

The UN is therefore conscious of the imperative to explore the potential dynamics of the small satellite industry and promote the need to address both the challenges and the opportunities posed by small satellites. It has, for example, established the Basic Space Technology Initiative (BSTI), which seeks to support capacity building in fundamental space technology and also to promote the use of space technology and its applications for sustainable development. This has partially been guided by the growth of small satellites technology and the increasing access to this technology by universities and smaller institutions, in countries along the full spectrum of economic development. Despite these initiatives, however, it seems unlikely that binding international frameworks will be put in place anytime soon to effectively deal with this technology.

Yet, even putting these initiatives aside, it is clear that such shifts in space technology require the development of appropriate regulatory standards in a relatively short timeframe. Small satellite entrepreneurs are anxious that any real (or perceived) barriers to entry posed by national regulatory requirements are removed. Many of these entities believe that if they are not able to develop and implement their plans in the short-term, then the opportunity will be lost, since someone else will do it instead, perhaps in a more "user friendly" domestic regulatory environment.

Whether or not these fears are justified in every case, what seems increasingly likely is that, in some respects, small satellite technology will become a mainstream means of utilizing space for commercial purposes in the future. Attempting to regulate this 21st century technology solely by reference to 20th century rules is therefore likely to create difficulties and uncertainties, and perhaps deter some who would otherwise consider engaging in the space industry.

In the meantime, however, there is no doubt that small satellite technology can offer great opportunities, but it also poses some significant challenges to the broader perspective of the exploration and use of outer space. The need for clear regulation to specifically address this technology is clear and it thus falls on national lawmakers to provide what is required within a more expedient timeframe (Freeland and Davis 2015).

In the end, therefore, clear domestic policies must be formulated. National legislatures have to come to grips with the ever-changing range of space technology, particularly if they wish to become increasingly involved in space activities. Some Governments are already attempting through their legislation to deal specifically with the issues that arise through the advent of small satellite technology (Marboe and Tranmuller 2012, p. 289)[14], but there is a long way to go. Whatever rules are put in place must find the right balance between, on the one hand, the need for regulation of the financial and technical elements, to minimize the risks to an acceptable level, and the facilitation of research and innovation to allow for greater and more efficient access to space, and the potential for commercial returns, on the other.

Public Policy questions arise as to whether, for example, to exempt small satellite operators from several of the existing national regulatory requirements that apply to their large satellite "brethren". Yet, to do so may have the ultimate effect of minimizing the "incentives" or motivation of these operators to engage in best practices, or to take simple, inexpensive steps to ensure that their local stakeholders are covered by existing protections. Naturally, this may not necessarily be the case

14 The authors describe how the national laws of Austria, Belgium and the Netherlands have been structured to deal with the possibility of future small satellite programs involving those countries.

when it comes to commercial small-satellite enterprises; however, it is suggested that the industry as a whole would not necessarily be unduly stifled by the requirement that in all circumstances, they take proper and appropriate risk management steps. Any relaxation of the rules for the users of this technology will bring with it added risks for the regulators and the relevant State, even though in many cases these might be quite small.

These are difficult choices and States will take differing paths, depending upon their specific circumstances. This will, unfortunately, mean that there is unlikely to be established a uniform international set of rules to address the complexities of small satellites, at least in the short-medium term. Perhaps we might see the emergence of a soft-law code of conduct at the international level, but this may not provide a sufficient base to determine the conduct of those new actors in the space paradigm.

This again points to the strong role that national law and lawmakers have to play, which will require close consultation between all stakeholders, and emphasizes the need for regulators, the scientific community, the entrepreneurs and the lawyers to all talk to each other to a far greater degree than has thus far been the case.

13.4. References

Bohlmann, U.M. and Freeland, S. (2013). The regulation of space activities and the space environment. In *Routledge Handbook of International Environmental Law*, Alam, S., Jahid, M., Tareq, H.B., Chowdhury, M.R., Techera, E.J. (eds). Routledge, London.

Coldeney, D. (2018). FCC fines Swarm Technologies $900K over unauthorized satellite launch [Online]. Available at: https://techcrunch.com/2018/12/20/fcc-fines-swarm-technologies-900k-over-unauthorized-satellite-launch/ [Accessed 24 December 2018].

Freeland S. (2001). There's a satellite in my backyard! – Mir and the Convention on International Liability for Damage Caused by Space Objects. *University of New South Wales Law Journal*, 24(2), 462–484.

Freeland, S. (2010a). Fly me to the moon: How will international law cope with commercial space tourism? *Melbourne Journal of International Law*, 11(1), 90–119.

Freeland S. (2010b). Matching detail with practice: The essential elements of national space legislation. In *Proceedings of the 53rd Colloquium on the Law of Outer Space*. American Institute of Aeronautics and Astronautics, Reston.

Freeland, S. (2011). For better or for worse? The use of "soft law" within the international legal regulation of outer space. *Annals of Air and Space Law*, 36, 409–445.

Freeland, S. and Davis, M. (2015). Space treaties are a challenge to launching small satellites in space [Online]. Available at: https://theconversation.com/space-treaties-are-a-challenge-to-launching-small-satellites-in-orbit-37971 [Accessed 20 April 2022].

Hobe, S., Schmidt-Tedd, B., Schrogl, K.-U. (eds) (2009). *Cologne Commentary on Space Law, Outer Space Treaty*, Vol. I. Carl Heymanns Verlag, Cologne.

Hobe, S., Schmidt-Tedd, B., Schrogl, K.-U. (eds) (2013). *Cologne Commentary on Space Law, Rescue Agreement, Liability Convention, Registration Convention, Moon Agreement*, Vol. II. Carl Heymanns Verlag, Cologne.

Marboe, I. (2012). *Soft Law in Outer Space: The Function of Non-binding Norms in International Space Law*. Bohlau Verlag, Vienna.

Marboe, I. and Traunmuller, K. (2012). Small satellites and small states: New incentives for national space legislation. *Journal of Space Law*, 38(2), 289.

Zolfagharifard, E. and Griffiths S. (2014). Elon Musk's new mission revealed: SpaceX founder confirms plans for tiny satellites that will provide cheap internet worldwide [Online]. Available at: http://www.dailymail.co.uk/sciencetech/article-2830263/Elon-Musk-s-new-mission-revealed-SpaceX-founder-confirms-plans-tiny-satellites-provide-cheap-internet-worldwide.html [Accessed 20 April 2022].

14

Exploitation of Material Resources of Celestial Bodies

Philip DE MAN
Centre for Global Governance Studies, University of Leuven, Belgium

14.1. State of play

At the time of writing, the exploitation of material resources from celestial bodies in outer space is in the early stages of planning, with various private initiatives competing for funding of their ambitious plans in a rapidly changing commercial landscape[1]. Future development of the exploitation of space resources will depend in no small part on a clarification of the existing legal regime on spacefaring activities as it applies to this novel activity. Similar to other activities that were developed in the wake of the negotiation and adoption of the current international space law framework, however, the regulation of material resource exploitation in space is subject to differing interpretations and applications of general principles discussing only in the broadest of terms the limits of lawful exploration and use of outer space, including the Moon and other celestial bodies.

1 The key private players in space mining as of 2022 are ispace, Moon Express, Asteroid Mining Corporation, TransAstra and Deltion Innovations. Two of the main companies that led the rush to space resources, Planetary Resources and Deep Space Industries, were either acquired by other companies or forced to reorganize and restructure following financial troubles; neither is still focused on space mining.

Space Law,
coordinated by Thomas LECLERC. © ISTE Ltd 2023.

Different States take different approaches in their attempts to provide the budding space resource industry with the necessary legal certainty. While some States have adopted national legislation on the presumption that this is the most efficient way of establishing legal certainty in order to attract players in the space mining industry, others have stressed the need for a multilateral clarification of these same principles, to ensure a common approach to the exclusive use of inclusive resources in a way that would avoid exacerbating existing inequalities. It should be noted in this respect that the States that have ratified the Agreement Governing the Activities of States on the Moon and Other Celestial Bodies (Moon Agreement) are bound by an obligation to undertake to establish an international regime, including appropriate procedures, to govern the exploitation of the natural resources of the celestial bodies of the solar system, other than Earth, as such exploitation is about to become feasible (Moon Agreement, Art. 11, 5)[2].

In 2016, the then 84 Member States of the Legal Subcommittee of the United Nations (UN) Committee on the Peaceful Uses of Outer Space (COPUOS) decided by consensus to add a single/issue item to its agenda, titled "General exchange of views on potential legal models for activities in the exploration, exploitation and utilisation of space resources"[3]. The annually renewed discussions under this item have mainly served to confirm the divergent views of States on space resource exploitation and, ultimately, the contentious nature of the regulation of this activity. In 2022, a Working Group on Legal Aspects of Space Resource Activities was established in the Legal Subcommittee, whose five-year workplan aims to finalize "a set of initial recommended principles" for space resource activities in 2027[4]. Notwithstanding, these international regulatory efforts, some States have already proceeded to adopt national legislation providing a legal basis for the authorization

2 Though the Moon Agreement is, at the time of writing, only binding on 18 States, its text was adopted by consensus by the COPUOS and the UNGA. Moreover, most of the 18 States Parties to the Moon Agreement are listed as States of Registry for multiple space objects in the UNOOSA Index of Objects Launched into Outer Space: Australia (23 entries), Austria (3), Belgium (32), Chile (4), Kazakhstan (9), Mexico (14), Morocco (3), Pakistan (6), Peru (4), Philippines (3), Saudi Arabia (13), Turkey (13), Uruguay (6) and Venezuela (3). These objects are governed by the laws of the States that have registered them, including their international obligations under the Moon Agreement. Moreover, Austria, Belgium and the Netherlands also have to respect their international obligations under the Moon Agreement when acting through the European Space Agency of which they are member.

3 UNGA (2016). Report of the Committee on the Peaceful Uses of Outer Space, Fifty-ninth session (8–17 June 2016), A/71/20.

4 COPUOS (2022). Report of the Legal Subcommittee on its sixty-first session, held in Vienna from 28 March to 8 April 2022, A/AC.105/1260, Annex II, Appendix.

and supervision of activities of space resource exploitation by their nationals[5]. We can also refer to the joint industry and academia initiative outside of the UN aimed at clarifying the need for and identifying possible building blocks of a future international legal regime for material space resource exploitation through the Hague International Space Resources Governance Working Group (The Hague International Space Resources Governance Working Group 2019).

14.2. Legal principles

Pending the further evolution and conclusion of the above legal developments, the activity of the exploitation of material resources from celestial bodies shall be governed by a good faith interpretation and application of the general principles of international space law as codified in the existing treaties on outer space, in particular the Treaty on Principles Governing the Activities of States in the Exploration and Use of Outer Space, including the Moon and Other Celestial Bodies (OST), in line with the principles of the Vienna Convention on the Law of Treaties (VCLT)[6]. As such, it is generally agreed that the exploitation of natural resources from celestial bodies is a form of exploration and use of outer space, which States are free to engage in as per Article I, paragraph 2 of the OST. This was indirectly confirmed by the discussions following the adoption of the Moon Agreement, when the matter of a possible moratorium on the exploitation of natural resources from celestial bodies pending the adoption of a future international regime was decisively rebutted by the delegate of the United States, without objection from any of the other States that participated in the negotiations of the Agreement.

Such freedom to exploit the resources from celestial bodies is not unfettered, however, as it is subject to the same limitations as should be applied to guarantee the legality of any other activity in spacefaring. Key among these limitations are restrictions flowing from the requirement that outer space should remain free for

5 See United States 2015 Commercial Space Launch Competitiveness Act (Public Law 114–90 Nov. 25, 2015); Luxembourg 2017 Law on the Exploration and Use of Space Resources (A674 Law of 20 July 2017); United Arab Emirates 2019 Law on the Regulation of the Space Sector (Federal Law No. 12 of 19 December 2019); Japan 2021 Act on the Promotion of Business Activities Related to the Exploration and Development of Space Resources (Law No. 83 of 15 June 2021). See also the Artemis Accords concluded as bilateral political agreements supporting the US Artemis Programme between the United States and 21 countries + 1 territory since 13 October 2020.

6 Though the VCLT entered into force after the conclusion of the Outer Space Treaty, it is accepted that its rules on treaty interpretation codify customary international law, and hence also apply to treaties concluded before the entry into force of the VCLT.

exploration and use by all States on a basis of equality (OST, Art. I, 2), that such exploration and use should be for the benefit and in the interests of all States and shall be the province of all humankind (OST, Art. I, 1), and that States shall conduct their exploitation activities with due regard to the corresponding interests of all States Parties to the Outer Space Treaty (OST, Art. IX). Though the general nature of these requirements means that their specific implications will be clarified primarily through developing State practice of resource exploitation, the consumable nature of material resources that are depleted by their use – unlike other space resources such as solar power and Earth orbits – requires that the interests of all other States, including in particular developing and non-spacefaring States, be taken into account through the establishment of a centralized management system to enforce certain restrictions agreed to by the international community. Such system would also facilitate the implementation of the obligation of States to undertake appropriate international consultations when they have reason to believe that their exploitation activities would potentially cause harmful interference with the activities of other States in the peaceful exploration and use of outer space (OST, Art. IX).

Other principles guiding the exploration and use of outer space, including the Moon and other celestial bodies, need to be applied indiscriminately to the exploitation activity as one of such forms of exploration and use. These include, among others, the obligation of States to register space objects engaged in mining on celestial bodies when launched into Earth orbit or beyond, even if current legislation and State practice are unclear as to whether such registration obligation can be extended to mining equipment that may be constructed on celestial bodies in the case of a more permanent mining settlement (see Convention on Registration of Objects Launched into Outer Space of 14 January 1975, Art. II). Likewise, launching States shall remain liable for any damages caused by their space objects used in mining activities, although the current rules of the Convention on International Liability for Damage Caused by Space Objects (Liability Convention) in this respect do not take into account possible damages to the space environment that such mining activities would likely cause[7]. However, to the extent that celestial bodies should be classified as an environment under international law – and there are no indications that they should not be – mining activities on these bodies are subject to any limitations that follow from the application of general principles of

7 Article I (a) of the Liability Convention defines ''damage'' as "loss of life, personal injury or other impairment of health; or loss of or damage to property of States or of persons, natural or juridical, or property of international intergovernmental organizations".

international environmental law, to the extent that their application does not depend on the existence of territorial sovereignty, which is not recognized in outer space[8].

A key provision of international space law contained in the OST requires that there shall be free access to all areas of celestial bodies (OST, Art. I, 2). Although the application of this principle determines in part on what a celestial body – an as of yet legally undefined notion – is, the more urgent question concerns the legal implications of the depletion and consumption of material resources found on celestial bodies to such an extent that they effectively make entire areas inaccessible to other States, either through the physical disappearance of an entire celestial body or an area on such body, or by rendering the conditions on such body or area ultimately inhospitable to any other activity. While this is commonly construed as a problem of *de facto* appropriation under Article II of the OST (see *supra*, Chapter 5), the inclusion of the rule on access to all areas of celestial bodies makes clear that it is rather a problem of interpretation of the limits to the free use of a part of outer space by competing actors. Whether these limits entail a *de facto* ban on exploitation activities that render unusable entire areas on celestial bodies remains to be determined. However, it is clear, at least, from other provisions in international space law that States are allowed to make exclusive uses of areas on celestial bodies, provided that reasonable efforts are made to preserve the accessibility of these areas for others. As such, States are allowed to establish stations, installations and other facilities on the surface of celestial bodies, on the condition that these shall be open to representatives of other States, on a basis of reciprocity (OST, Art. XII, Moon Agreement, Art. 9).

Exploitation activities on celestial bodies by private actors need to follow the same rules as govern the activities of States. This follows from the principle that all activities in the exploration and use of outer space by non-governmental actors, including mining companies, shall be considered national space activities for which States bear international responsibility (OST, Art. VI). Therefore, States are required to adopt appropriate regulations that allow them to authorize and continuously supervise potential mining activities performed by their nationals and/or using space objects launched for which they are a launching State. Such supervision is necessary to

8 Article III of the OST requires that "States Parties to the Treaty shall carry on activities in the exploration and use of outer space, including the moon and other celestial bodies, in accordance with international law, including the Charter of the United Nations, in the interest of maintaining international peace and security and promoting international co-operation and understanding". In international law, principles on State responsibility for transboundary harm are reflected in, among other instruments, Principle 21 of the 1972 Stockholm Declaration on the Human Environment (UNGA Resolutions 2994-2996/XXVII) and Principle 2 of the 1992 Rio Declaration on Environment and Development. We should also note that the environmental protection of celestial bodies is stronger in the Moon Agreement (see MA, Art. 7).

guarantee that private actors in space abide by the rules of international space law, in particular the above rules of the OST, such as access to all areas of celestial bodies and consultations in case of potential harmful interference with the activities of other States or their nationals. These rules also include certain obligations that are sometimes deemed difficult to reconcile with the commercial realities of private competition, such as the obligation for States to inform the Secretary-General of the UN, as well as the public and the international scientific community, of the nature, conduct, locations and results of their activities in the exploration and use of outer space (OST, Art. XI). Because an important aspect of space mining activities comprises the identification of commercially viable celestial bodies and areas on such bodies, the obligation to share information on the location of activities of exploitation may not be welcomed by mining companies. To be sure, this information sharing obligation only exists "to the greatest extent feasible and practicable", and commercial considerations may well be taken into account to determine the outcome of this test (OST, Art. XI). However, the ultimate nature and location of space exploitation activities may still be deduced from other information sharing provisions whose application leaves less room for discretion, such as those related to the registration in the UN index on objects launched into outer space, and the assignment of frequencies and associated orbits in the Master International Frequency Register of the International Telecommunication Union (see *infra*, section 14.4).

14.3. Terms: "exploitation" and "natural resources"

The application of the above and most other rules of space law to the novel activity of the exploitation of material resources from celestial bodies will in large part depend on the interpretation of a number of key terms and concepts employed to delineate the scope of these rules. As most of these notions are not exclusively limited in their application to the activities in the current entry, the following will only discuss those terms that are typical of the activity of space mining, namely natural resources and exploitation. The definition of related notions such as celestial bodies and non-appropriation, including the contentious issue of property rights over natural resources removed from celestial bodies, will not be covered here[9]. In short, while property rights over resources from celestial bodies may either be lawful or unlawful depending on the interpretation of the scope of the non-appropriation principle, it is clear that any use of such resources can never result in exclusive legal title over the celestial body from which they were recovered, or the mining area thereon, in accordance with the aforementioned principles that free access shall be

9 This part is only about the regulation of the exploitation of natural resources. For more on the appropriation of natural resources, see *supra*, Chapter 5.

retained over all parts of celestial bodies and that celestial bodies or parts thereof cannot be appropriated.

For the most part, the UN treaties on outer space avoid any reference to the terms: natural resources and exploitation. The one treaty that does contain references to these notions is the ill-ratified Moon Agreement, which in paragraph 5 of its preamble bears in mind "the benefits which may be derived from the exploitation of the natural resources of the Moon and other celestial bodies", and whose Article 11 contains the basic principles for a future regime governing the exploitation of natural resources from the Moon, principles applying to all other celestial bodies from our solar system, other than Earth (Moon Agreement, Art. 1, 1). These provisions are of course only binding on those States that have ratified the Moon Agreement, which excludes most current spacefaring nations. Regardless, none of the provisions in the Moon Agreement covering the activity of exploitation of natural resources contain any definitions of these terms.

As previously noted, the fact that the OST leaves out any reference to the activity of natural resource exploitation does not mean that we should not apply its principles to this novel activity when engaged in by spacefaring States that have ratified this instrument, nor does it imply that space resource exploitation should not be deemed lawful. The UN space treaties deliberately adopt a general phrasing of widely applicable principles, in keeping with the preference of their drafters to not unduly limit the application of these instruments in anticipation of future developments. It does mean, however, that in the absence of any indication that the drafters of the OST and the Moon Agreement intended to give a special meaning to the terms natural resources and exploitation, the treaties have to be interpreted in light of the ordinary meaning to be given to these terms when applying their principles to a novel activity such as space mining (VCLT, Art. 31, 1).

To determine the ordinary meaning of the term natural resource, we may refer to other relevant rules of international law that govern the relations between the States Parties to the OST and the Moon Agreement. Most conventions that contain rules on natural resources and which elaborate on the meaning of the term limit themselves to listing the different types of resources that are covered by the scope of the treaty, rather than defining the component that distinguishes a resource from other elements of the environment[10]. What few treaties that do define the characteristic trait of a

10 See, for example, the African convention on the conservation of nature and natural resources of September 15, 1968, entered into force June 16, 1969. The provision defines natural resources as "meaning, renewable resources, that is soil, water, flora and fauna" (conservation of nature and natural resources, Art. III, sub a). See also the defunct Convention on the regulation of Antarctic mineral resource activities of 2 June 1988, which defines

resource typically focus on the criterion of economic value as a result of human interaction. As such, the Convention on Biological Diversity of 5 June 1992, entered into force on 29 December 1993 (CBD), defines biological resources as those biotic components of ecosystems "with actual or potential use or value for humanity" (CBD, Art. 2). Likewise, genetic resources are described as genetic material "of actual or potential value". The same criterion is used to define the subject matter of the International treaty on plant genetic resources for food and agriculture of 3 November 2001, entered into force on June 29, 2004 (Art. 2) and developed by the Food and Agriculture Organisation (FAO)[11].

These definitions also correspond to the meaning attributed to natural resources outside the context of international law. Most legal dictionaries even omit a specific reference to natural resources, suggesting that the ordinary meaning of the notion should also inform our legal understanding, absent any indications to the contrary. As one of the only exceptions, the relevant entry of Black's Law Dictionary confirms our interpretation of natural resources as "any material from nature having potential economic value" (Garner 2009). It follows that all natural resources, including material resources on celestial bodies, are characterized by their susceptibility to exploitation, and that, in turn, exploitation can be defined with reference to the natural resource notion as the activity that realizes the economic potential of natural resources, thereby, for the purpose of the regulation of space mining, distinguishing them from other phenomena in space[12].

---

mineral resources as "all non-living natural non-renewable resources, including fossil fuels, metallic and non-metallic minerals" (Convention on the regulation of Antarctic mineral resource activities, Art. 1, 6). Likewise, the US Commercial Space Launch Competitiveness Act of 2015 merely defines a "space resource" as an abiotic resource in situ in outer space, including water and minerals.

11 Compare also the clear link with value production after human activity in Article 2 of the Convention on fishing and conservation of the living resources of the high seas of April 29, 1958, entered into force on March 20, 1966: "[a]s employed in this Convention, the expression 'conservation of the living resources of the high seas' means the aggregate of the measures rendering possible the optimum sustainable yield from those resources so as to secure a maximum supply of food and other marine products. Conservation programmes should be formulated with a view to securing in the first place a supply of food for human consumption".

12 See also the current online version of Black's Law dictionary [Online]. Available at: https://thelawdictionary.org [Accessed 22 April 2022]. The definition of natural resources does not appear to distinguish between this notion and the activity of exploitation. A natural resource is defined as any "[a]sset or material contributing to the nation's natural capital. To extract, process, or refine natural resources requires capital and human resources, as mental and physical labor. By these means, the nation obtains its true economic value".

While this cursory overview of the meaning of the notion "natural resource" and its related activity of exploitation may be insufficient to serve as a basis for a comprehensive regulation of mineral resources of celestial bodies, it does clarify a number of points that should be reflected in a future regime negotiated on the basis of the existing space treaties. First, as noted, the exploitation of natural resources from celestial bodies should be allowed as a specific form of exploration and use of these bodies, covered by the freedom to use outer space and subject to the same limitations. Second, the exploitation of natural resources from celestial bodies is nonetheless a specific type of use that is only completed when the economic potential of the resource concerned has been realized through a form of human activity. This means that any limited State practice concerning, for example, the removal and sale of rocks collected as samples from the Moon should not be taken into account as subsequent practice in the application of the OST (VCLT, Art. 31, 3, b), for these samples were never used as natural resources; they should rather be classified as parts of a celestial body[13]. Third, since the classification of parts of a celestial body as material natural resources depends on the realization of their economic potential, any rights granted over these resources upon removal should lapse in case of non-use.

14.4. Future international regime

The latter point is in line with the regime of the International Telecommunication Union (ITU) on the use of orbits associated with frequency assignments. This regime is based on the recognition in the ITU Constitution that frequencies and associated orbits are phenomena that should be classified as limited natural resources, and which are therefore to be used rationally, efficiently and economically, so that countries or groups of countries may have equitable access to them (ITU Constitution, Art. 44, 2). Though enshrined in a separate legal regime, the ITU principles guiding the use of orbits as limited resources can be seen as a

13 See in this regard the regulation of samples in Art. 6, 2 of the Moon Agreement, which even stops short of granting property rights over samples of minerals to those States that have collected them from celestial bodies: "In carrying out scientific investigations and in furtherance of the provisions of this Agreement, the States Parties shall have the right to collect on and remove from the moon samples of its mineral and other substances. Such samples shall remain at the disposal of those States Parties which caused them to be collected and may be used by them for scientific purposes. States Parties shall have regard to the desirability of making a portion of such samples available to other interested States Parties and the international scientific community for scientific investigation".

possible way of concretizing the general rules of the OST to regulate the use of natural resources. Indeed, the ITU regime on the use of orbits and frequencies confirms that all States have the freedom to use space resources, subject to an obligation of due regard for the corresponding interests of other States and the principle of non-appropriation. In this regard, it is instructive to note that the prolonged use of orbitals and orbital positions in accordance with ITU rules is regularly criticized as a form of appropriation in violation of Article II of the OST. However, these concerns do not take into account the legal nature of the rights granted through registration of a frequency assignment in the ITU Register, which only grants a form of international recognition and protection from harmful interference over the use of registered frequency assignments and their associated orbits (ITU Radio Regulations, Art. 8). Any rights over natural resources of orbits and frequencies are thus by no means property rights, but merely protected rights of use that exist only as long as actual and continued use is made of these resources. Importantly, this protected right of use lapses in case the frequency assignment is never actually brought into use or is no longer in regular operation after an initial period of actual use (ITU Radio Regulations, Art. 11).

The effective system of orbital and frequency management by the ITU is a useful example of how future needs for space resources by competing actors may be organized. The nearly universally accepted detailed and technical set of rules contained in regularly updated international instruments of the ITU that deals with day-to-day forms of use of a category of space resources surely provides a more legally relevant inspiration than terrestrial regimes such as the law of the sea and the mining of the deep seabed. What the ITU system most clearly demonstrates is the overarching need for a centralized system providing for the registration of rights, whatever their nature, over areas on celestial bodies and the material resources extracted from them. A workable regime that safeguards the interests of spacefaring countries and companies without sacrificing the legitimate rights and interests of others should primarily focus on setting up a procedure that registers assignments over areas on celestial bodies containing resources as soon as verifiable and advanced plans for their exploitation have been submitted, subject to definite temporal and geographical limitations to these activities and effective coordination with other users that may be affected by them. The purpose of this procedure should be twofold: guaranteeing access to material resources from celestial bodies for those users that have the capabilities to exploit them, both during the exploitation activity and the period preceding the actual mining operation; and ensuring continuous centralized checks on the feasibility of exploitation plans, both technologically speaking and in terms of the immediate usefulness of the recovered resources for meeting an actual demand, under fair conditions to ensure equitable access to the benefits from these resources by all interested States[14].

The above principles are based on a good faith interpretation of the provisions of the OST and the ITU regime concerning the use of orbits associated with frequency assignments. The principles also closely resemble the main purposes of the future international regime that the States Parties to the Moon Agreement undertake to establish to govern the exploitation of the natural resources of the Moon and all other celestial bodies of the solar system, as such exploitation is about to become feasible (Moon Agreement, Art. 11, 7 *jo.* 5). In particular, the Moon Agreement provides that celestial bodies and their natural resources are the common heritage of mankind, and that, therefore, a future international regime governing their exploitation should be based, *inter alia*, on the following principles:

a) the orderly and safe development of the natural resources of the Moon;

b) the rational management of those resources;

c) the expansion of opportunities in the use of those resources;

d) an equitable sharing by all States Parties in the benefits derived from those resources, whereby the interests and needs of the developing countries, as well as the efforts of those countries which have contributed either directly or indirectly to the exploration of the Moon, shall be given special consideration.

14.5. References

Barnes, R. (2009). *Property Rights and Natural Resources*. Hart, London.

Jakhu, R.S., Pelton, J.N., Nyampong, Y.O.M. (2016). *Space Mining and Its Regulation*. Springer, Berlin.

de Man, P. (2015). The exploitation of asteroids and the non-appropriation principle: Reflections on the nature of property rights in light of the US Space Resource Act of 2015. *J. Space L.*, 40(1), 1–55.

---

14 This is also in line with the requirements in the Moon Agreement to limit the use of areas on celestial bodies to the needs of the station and the use of minerals on celestial bodies in accordance with the requirements of the mission. See Moon Agreement, Art. 6, 2 of the: "States Parties may in the course of scientific investigations also use mineral and other substances of the moon in quantities appropriate for the support of their missions". See, also, Moon Agreement, Art. 9, 1: "A State Party establishing a station shall use only that area which is required for the needs of the station and shall immediately inform the Secretary-General of the United Nations of the location and purposes of that station".

McDougal, M.S., Lasswell, H.D., Vlasic, I.A., Smith, J.C. (1963). The enjoyment and acquisition of resources in outer space. *Univ. Pa. Law Rev.*, 111, 521–636.

Pritzsche, K.U. (1989). *Natürliche Ressourcen im Weltraum – das Recht ihrer wirtschaftlichen Nutzung*. Lang, Frankfurt am Main.

Smith, M.L. (1990). *International Regulation of Satellite Communication*. Martinus Nijhoff, Dordrecht.

The Hague International Space Resources Governance Working Group (2019). Building blocks for the development of an international framework on space resource activities [Online]. Available at: https://www.universiteitleiden.nl/binaries/content/assets/rechtsgeleerdheid/instituut-voor-publiekrecht/lucht–en-ruimterecht/space-resources/bb-thissrwg–cover.pdf [Accessed 19 August 2022].

Viikari, L. (2002). *From Manganese Nodules to Lunar Regolith: A Comparative Legal Study of the Utilization of Natural Resources in the Deep Seabed and Outer Space*. University of Lapland, Rovaniemi.

PART 4

Challenges and Issues Raised by the Development of Space Activities

15

Legal Issues Related to the Exploration of the Universe

Laetitia CESARI ZARKAN
Université du Luxembourg, Faculty of Law, Economics and Finance, Luxembourg

15.1. Introduction

The legal issues linked to the exploration of the universe are part of both the relations between States, which are changing by essence and the commercial, scientific and technological development of space systems, which is structural and evolves over the long-term. The exploration of the universe can be seen as a succession of experiments and attempts initiated by the launch of Sputnik-1 – the first artificial satellite – into orbit around Earth and then by sending the first human into space. To a certain extent, it is possible to consider that our understanding of the universe goes back to the invention of the first telescopes to observe the stars and the conception of sky maps representing the movement of the stars. Subsequently, the sending of probes around and to the surface of the other planets of the solar system constituted a real turning point for our scientific knowledge (Billings 2021). Each era corresponds to a certain political strategy which depends not only on the available technologies but also on the political will, closely linked to the state of relations between States, rivals or partners. On the basis of these different criteria, the development of space projects and the financing of space applications are also part of the relationship between government entities and national private actors. To develop sustainable and coherent exploration programs, space powers must take into account the state of scientific research and available operational technologies that define current and future technical possibilities and limitations.

This chapter is placed in the context of the exploration of the universe and proposes, as such, to analyze a non-exhaustive series of legal issues which could appear in the more or less near future. From a legal point of view, human beings plan the exploration of the universe from their own prism, the starting point being the knowledge and the technologies available to humanity at a given moment, then the envisaged developments based on the state of the research and the available financing. A large number of legal aspects must nevertheless be taken into account from the very beginning of an exploration project, whether it involves humans or robots: the responsibility of States, whether international or financial or of private actors, insurance, the rules of competition between companies, the intellectual property of onboard technologies, to name but a few. Inter-State relations, international cooperation and the opportunities given to the private sector constitute the basis of the investments that can be envisaged for the implementation of exploration projects and the development of launchers, onboard systems and space operations. It is thus possible to reduce expenses by avoiding duplication of costs while improving the quality of projects. These factors justify and condition the whole organization of the political decision process, the legal framework and the financing of such exploration programs with, at the end of the day, the success or failure of these undertakings. In the same way, from an ethical point of view, space exploration, such as it is currently taking shape, is anthropocentric. The scientific experiments carried out by human beings, their expansionist vocation, commercial profits and, in a sense, personal satisfaction are at the center of space exploration to the detriment of the protection of orbits and celestial bodies. From now on, it is not only a question of using space technologies for applications on Earth such as telecommunications, Earth observation, navigation or weather forecasting. Exploring the universe also allows us to increase our scientific knowledge beyond Earth's atmosphere and, to some extent, to envision a long-term human presence in space (Arnould 2018). However, space activities must be framed with caution. While a shortage of legal rules could create a lack of control over space actors, too strict a legal regime could slow down or hinder innovation and the development of future activities.

The approach taken in this chapter is thus based on an analysis of human exploration and the deployment of robotic systems. It also examines the development of communication links from Earth or in space and the status of space resources, as well as the rules applied to the infrastructures that will be built on and around celestial bodies. Finally, this chapter addresses the question of the long-term sustainability of space activities in terms of domain protection and management of resources generated by human activities.

While the Treaty on Principles Governing the Activities of States in the Exploration and Use of Outer Space, including the Moon and Other Celestial Bodies (1967 Treaty) establishes the freedoms to use and explore outer space (1967 Treaty, Art. I), the question is what are the limits to these freedoms and whether they should be framed, as each individual action in space has collective consequences (Degrange et al. 2021). The current generation is at a pivotal point in history, arriving too late to explore Earth but born too early to explore space truly. The applicable law of space, or *corpus juris spatialis*, relies primarily on States' sovereign power, despite the fact that no space power has exclusive rights (Couston 2022). The development of space activities, international cooperation and future permanent installations on celestial bodies, to name but a few factors, could thus push the international community to consider a framework for the respective national competencies. The aim here is to pave the way to a double reflection: the way to consider the legal norms that will apply once humans start to settle on other celestial bodies in a sustainable way and the rules applicable to the sending of remotely piloted systems, robotic or not, for specific missions.

15.2. Anthropocentrism through the ages: envoy of humanity, space tourist and Homo spatialis

The Agreement on the Rescue of Astronauts, the Return of Astronauts and the Return of Objects Launched into Outer Space, opened for signature on April 22, 1968 (1968 Agreement), currently provides the basis for the protection of spacecraft crew members. The 1968 Agreement provides for the protection of astronauts in the form of three obligations imposed on States Parties: information, rescue and repatriation (Kiss 1968). Considered "envoys of humanity" (1967 Treaty, Art. V), the assistance that States must provide extends to the territory under the jurisdiction of a State but is also applicable in the event of a ditching on the high seas, a landing or any other place not under the jurisdiction of a State (1968 Agreement, Art. 1). In this respect, two questions arise. The first pertains to the reclassification of the astronaut, who would no longer be only a member of the crew of a spacecraft. However, an astronaut should also be in charge of a precise mission in order to distinguish them from individuals travelling for recreational purposes and in return for a sum of money. The second question concerns the assistance and protection to be provided to individuals as a whole and regardless of their State of origin in the context of interplanetary facilities, particularly if the nationals of a State need to be rescued by another State party to the 1968 Agreement.

15.2.1. *Explore or consume: the evolution of the astronaut status*

The emergence of the space tourism market tends to call into question the notion of "[hu]mankind's envoys" as envisaged by the 1967 Treaty. Although astronauts are subject to strict selection and are sent on specific missions, the crew of a spacecraft can now be composed of individuals who "consume" space. They can sign up, for a fee, for a trip aboard a spacecraft, often for a suborbital flight, to a very high altitude, without putting into orbit (International Civil Aviation Organization 2015) or more rarely for a stay aboard the International Space Station (ISS).

It seems that these private actors, whose status could be more similar to that of a "consumer" rather than an explorer or scientist, will be more numerous in the future, especially when crewed flights reach other celestial bodies and settle there permanently. Nevertheless, the law applicable to space tourism differs from the classical conception of this activity, which is planned for leisure or travel purposes (World Tourism Organization 1999). Such a trip into space requires meticulous preparation that can range from 750 to 1000 h of preliminary training (Gohd 2022). Space exploration is also currently highly regulated. While individuals can participate in space flights, they are nevertheless subject to strict contractual conditions. Does this make them astronauts? With the normalization of this type of activity and the increase in the number of people able to travel in space without prior selection or a predetermined mission, space activities could eventually be perceived more through the prism of leisure rather than science or exploration as such. If individuals can join the crew of a spacecraft to explore the universe for purely recreational purposes, the risk would also be that they decide to go beyond the limits imposed and take reckless risks. In this case, the current responsibility of States to ensure that space activities are conducted in accordance with obligations under international law (1967 Treaty, Art. VI) would eventually have to be reduced to accommodate more corporate entities that send individuals into space, for payment, under contract. This could imply the creation of a new and more responsible status for those who wish to venture beyond our earth's atmosphere without being subject to a previously planned mission.

The subject of assistance to be provided to individuals by States when humanity begins to settle on other celestial bodies may also be the subject of reflection, particularly with regard to the parts occupied by certain States to carry out their space activities or to build facilities and the "safety zones" envisaged under the pretext of preventing any interference with activities carried out beyond the earth's orbit. For example, National Aeronautics and Space Administration (NASA) has provided emergency assistance to be provided by the signatories of the Artemis

Accords (NASA 2020) to personnel in distress in outer space, citing the 1968 Agreement (Artemis Accords, para. 6). The notion of a "safety zone" defined by the Agreements as "the area wherein this notification and coordination will be implemented to avoid harmful interference" may, however, prove problematic in the long run because of the perception of other States. The size and scope of these zones should reasonably reflect the nature of the operations being conducted and the environment in which they are conducted. However, a State's use of the "first come, first served" principle to conduct its activities in a delimited area over the medium to long-term could be interpreted as violating the principle of non-appropriation (1967 Treaty, Art. II). An agreement on the nature and existence of these safety zones will therefore have to be found between the space powers on the basis of a balance between this principle of non-appropriation and the prevention of a "potentially harmful interference with activities in the peaceful exploration and use of outer space" (1967 Treaty, Art. IX).

Other issues can be identified on the question of the limits and protection of human beings who will settle on a celestial body, thus the question of the survival of the human species in the long-term and the debates concerning self-determination and the right of peoples to self-determination. The rules governing space habitats could be based, on the one hand, on the supervision by States of space objects, installations and equipment that will be placed on other celestial bodies or in orbit and, on the other hand, on the biological needs of the human beings who will settle in space – human beings qualified as *Homo spatialis* (Robinson 1984) or *Homo alterios* (Fasan 1970) – if the human species succeeds in surviving in space and evolving in order to adapt to the hostile conditions that surround it. Even if the sending of human beings into space implies, at present, their protection at all times against radiation or the implementation of measures for their survival, when several generations of human beings will have lived on other celestial bodies and will have learned how to adapt, the link between the planet Earth and these individuals from elsewhere could attenuate. The question of governance in space and the legitimacy of the application of space law as it is currently envisaged will then necessarily arise.

15.2.2. *Rethinking Humanity away from Earth*

With the evolution of society on another celestial body, as *Homo spatialis* moves further away from Earth, certain principles could be called into question, notably that of the prohibition of national appropriation by proclamation of sovereignty (1967 Treaty, Art. II). In this context, will interplanetary travel and facilities require the drafting of new rules applicable to the use and exploration of space? Insofar as the fundamental rights and freedoms of human beings must be guaranteed, it is also a question of questioning the relevance of a constitutional text that would extend to all installations or of several texts that would be established according to the specificities of each celestial body on which human beings would settle. Concerning civil rights, the different questions related to participation in collective decisions, voting, expressing oneself and receiving information in a free and unlimited way also arise.

As such, the personal is always political, and it is impossible to separate societal problems from ethical considerations in the exploration of the universe. As soon as humanity projects itself into space, it brings the hopes and struggles that exist on Earth. The questions of social justice, fight against discriminations, inclusion of all the categories in the development of the human species but also the relations of force and the political and geopolitical questions are part of our luggage, whatever the distance that we think to put between our planet and the rest of the universe. Legally, these topics can be framed, especially with respect to gender, geographic and generational diversity, and social class. Rethinking space exploration from this perspective could provide a universal perspective for future exploration programs. For example, when designing a space mission, the physical, physiological and psychological performance and characteristics of male individuals are taken as a reference, which impacts the overall effectiveness of the mission (Thomas 2019). Whether it is the suits or equipment used, the electronic devices, the type of medication and the resulting bodily reactions, among other factors, their inventors and designers are often unaware of the importance of stepping outside of the pre-established standard, which reinforces gender stereotypes to the extent that all of these factors hinder the successful completion of some missions (Fortin and Zraick 2019; Criado Perez 2020). Addressing these inequalities could have a significant positive impact in the long run.

On the one hand, the exploration of the universe concerns humanity as a whole and it will not be possible to standardize all the elements necessary for a mission on the basis of a single gender. On the other hand, space exploration is inspiring. Contributing legally and ethically to the creation of an inclusive, stable and safe framework for space activities as a whole will thus be beneficial for the future of the industry but also for the whole of humanity in the very long-term.

Universal access to communications is also one of the points that would be useful to develop in order not to lose contact with Earth, as well as between celestial bodies, if the human species becomes interplanetary. The International Telecommunications Union (ITU) could have a significant role in the distribution of the different objects and activities that will take place in the future. In this sense, radio frequencies and associated orbits around Earth are limited natural resources (ITU Constitution, Art. 44). Considering this limitation is essential for the future of space activities. The ITU thus justifies equitable access and rational, efficient and economical use of frequencies and orbits. Therefore, sharing these resources is subject to certain administrative conditions that make their distribution rational and tend to avoid possible abuses by certain powers. This reference to equity is fundamental to guarantee a certain distributive justice in the long-term (Roth 1992). It will also be necessary in the future to work on the protection of different communications, especially because of the emergence of large satellite constellations, in order to prevent electronic interference and hostile operations in cyberspace and to prevent the interruption of services on Earth or the cutting-off of links between Earth and celestial bodies.

The question of the distribution of resources finally arises. As long as human beings are not self-sufficient, food and energy, sent from Earth or produced directly in space, oxygen, water, as well as all the substances necessary for life should not only be distributed among living beings, but their quality must also be subject to strict controls, especially if agricultural or breeding activities develop beyond our atmosphere. Some rules are already in place at the international level to prevent any detrimental effects of contamination of celestial bodies (1967 Treaty, Art. IX). In the future, it would also be useful to consider legal protection of the soil, subsoil and space above the surface of these bodies in order to preserve them to the maximum.

Suppose the survival of the human species depends on the factors mentioned above. In that case, its development seems to be conditioned to its capacity to adapt the space objects, their design, their uses and the related infrastructures to guarantee a viable exploration of the universe in the long-term.

15.3. Managing objects launched or built in space: communication links between infrastructures, protection and monitoring of systems, and management of space debris

The sustainable establishment of life beyond our Earth's atmosphere is closely tied to the use of robotic and computer systems. Projects related to sending communications infrastructure, robotic systems, or technological systems raise questions about the damage they could create and the debris that could result. The Convention on International Liability for Damage Caused by Space Objects, concluded on November 29, 1969 (1972 Convention) was adopted to settle disputes that might arise in the event of damage caused to a space object by a space object launched by another State. This type of incident is already the subject of a legal framework regarding relations between States (see *supra*, Chapter 7). However, new private actors have undertaken to launch a large number of space objects into orbit in recent years and will certainly have recourse, in the event of a dispute, to alternative means of resolution such as arbitration or mediation.

In the future, the proliferation of objects in orbit or on celestial bodies could also cause suspicions on the part of certain States, which could see in it the sign of a form of appropriation which would take the form of a large number of objects placed in the same place and the establishment of security zones. In this respect, the management of the devices that will move on the celestial bodies, notably by land or air, could be subject to a framework with an adapted registration system and rules of circulation and follow-up of the different systems (Osborn Frandsen 2022).

15.3.1. *Legal protection of space infrastructures*

Communication infrastructures and satellite positioning and navigation capabilities could evolve. These infrastructures that can serve as relays – around the Moon, for example (Jones 2022) – are currently being developed in parallel by different public and private entities. The emergence of similar projects could create duplication of resources but also force the States to coordinate the different systems in orbit and the services thus generated. The spacecraft and equipment that use these infrastructures will also need to be interoperable to ensure efficient use and resilience. The exploration of the universe relies on these technologies insofar as the different systems will have to transfer the data captured, receive instructions and allow for real-time communication. Space powers are working on exploration missions that use probes and robotic systems to collect information on the various celestial bodies. These technologies result from international cooperation in which different States contribute by investing in certain parts of the project. The mission orders, the design of the spacecraft, the integration of the components, the updating of the systems and their maintenance, etc., are the object of agreements and contracts between the various stakeholders who must plan a mission several years in advance. Depending on the case, the implementation and financing of the different projects may be the responsibility of the States and the competent space agencies.

Telescopes used to observe distant objects are also an integral part of scientific research aimed at exploring the universe. Placed on Earth or launched into space, these instruments allow us to collect essential information to increase our knowledge and understand, to some extent, the universe. These telescopes can also be placed in space to observe phenomena that are sometimes impossible to capture from Earth due to the Earth's atmosphere, thus allowing access to more astronomical data to increase our understanding of the universe. Once launched into space, telescopes are protected from light pollution, which is very present on Earth, but also from interference caused by the constellations of satellites that multiply in orbit and appear on the images captured. For this reason, the issue of dark and silent skies for the benefit of science and society is a subject debated within the Committee on the Peaceful Uses of Outer Space (COPUOS), with the scientific community on one side and the commercial industrial operators on the other side within the Scientific and Technical Subcommittee (STSC) of COPUOS (COPUOS 2022). Telescopes not only identify celestial bodies and observe their trajectories but also detect any astronomical event that could impact our planet or future space travels. In the context of the exploration of the universe, these telescopes can be considered as the eyes of the human species: they allow it to see beyond the planet Earth and to better anticipate, in the long-term, the scientific, commercial and strategic issues to come. The management of this data captured, stored and transferred is therefore essential to the development of our space activities.

Finally, the cybersecurity of spacecraft is essential for onboard systems, especially for communications technologies. Space missions increasingly depend on data circulating through computer systems that support "cyberspace". Spacecraft are connected to receive and transmit instructions and information through embedded software. Once these robotic systems are deployed for space activities, they are likely to be the target of external threats or be subject to inherent risks of malfunction. Software embedded in satellite systems should therefore benefit from increased protection in this area, with legal guidance from the industry on the minimum level of protection required for space systems and malicious actors conducting hostile cyber operations. To face this type of threat, protection must be extended not only to ground-based stations but also to space systems, particularly with regard to connected devices on celestial bodies and satellite constellations whose individual elements are built in series and may present identical vulnerabilities.

15.3.2. *The legal framework for space debris mitigation*

The management of debris and recycling of materials is one of the points that could be discussed in depth in the future, to protect the space domain by preventing contamination or accumulation of waste on the one hand and the safety of space objects by preventing collisions and possible damage on the other. These different aspects have been considered by the international community, particularly regarding the disintegration of objects and debris produced during the operation of the orbital stages of launchers and spacecraft (COPUOS 2007).

In the context of the exploration of the universe, it is useful to plan measures to reduce the production of debris in orbit and on celestial bodies to avoid collision or contamination. Indeed, space debris increases the risk of damage to spacecraft, which can cause space missions to fail or result in loss of life. Some materials used could also pollute certain celestial bodies by contamination or congestion. The solutions proposed by the international texts concern the reduction of debris from the design of systems, during the operational phases and after the end of a mission (COPUOS 2007). They also concern the collection, sharing and dissemination of space debris tracking data (COPUOS 2019). The issue of resource reduction regularly arises, particularly in the context of future space activities. Some space actors are working on new ways to dispose of debris currently orbiting Earth (ESA 2019), others are considering circular economy models to reduce, reuse, recycle and recover materials used in space exploration (Paladini et al. 2021).

The reduction of debris thus prevents the increase of accidental collisions that can occur between two objects and thus the legal and financial liability that could result from it. Given that debris are inactive objects or components that cannot be remotely piloted to avoid an impact, their multiplication has become a source of concern, especially for missions exploring the universe that rely on the use of advanced, rare and expensive technologies (Descalles 2019).

15.4. Finding harmony among space powers: from resource allocation to protecting the interests of the scientific community

The extraction, exploitation and sale of space resources are a source of debate, particularly with regard to the tensions that could arise from the moment these resources become the object of concrete appropriation, with the rise in tensions that this could create and the possible placement of defensive and offensive systems in space (see *supra*, Chapter 5). This question of the ownership of space resources is being debated at the international level within COPUOS.

The debates are particularly interested in the question of materials extracted from a celestial body and used to manufacture objects or supply energy to systems directly in outer space. The distribution and exploitation of these resources are considered differently in various States. The United States, Luxembourg, the United Arab Emirates and Japan have notably legislated on the subject by reaffirming the right to extract space resources and to appropriate them for sale or reuse. This question of resource distribution is not new. States have notably envisaged it in the Agreement Governing the Activities of States on the Moon and Other Celestial Bodies, concluded on 18 December 1979 (1979 Agreement).

Beginning with the first mission of the Apollo program, astronauts collected rocks and parts of lunar soil to be examined on Earth (Spudis 2017). This allowed scientists to analyze the various samples and gain invaluable information to improve our understanding of the universe. However, while using the resources for scientific reasons is accepted, making them available for profit is controversial. States are questioning the benefits derived from the extraction and use of space resources and whether these should be subject to "equitable distribution". In the event that the States proceed with such sharing, it would then be a matter of determining the manner of sharing these benefits by weighing up the interests and needs of the developing countries on the one hand and the efforts of the countries that have contributed to their extraction on the other (1979 Agreement, Art. 11). The major powers that invest in the development of means and technologies for the extraction of space resources and take risks in the conduct of their activities consider that they should benefit from a guaranteed return on their investment, while other States that have not developed this type of system consider that a delay in technological development should not prevent them from reaping the benefits of the use of the resources, with all the issues related to property rights that this implies (Hobe et al. 2009).

The Convention on Registration of Objects Launched into Outer Space, adopted by the General Assembly of the United Nations on November 12, 1974 (1974 Convention), finally imposes some rules on States, particularly with regard to the registration, position and function of space objects, which are attached to a particular State, whether they are launched into orbit or positioned on a celestial body. If, at present, the technical conditions to which operators are subjected differ from one State to another, the establishment of durable facilities on celestial bodies and of infrastructure and information systems should lead to the emergence of a set of common standards between States to allow interoperability between the different systems. The possibility of collaborating at the international level is important for robotic exploration, which relies on the quality and reliability of technologies sent into space, the security of activities against intentional threats, the safety measures taken to protect systems, their long-term sustainability and the standards applicable to systems to guarantee their sustainable use. Indeed, rapprochement operations could be misinterpreted by rival space powers, which could lead to an escalation of tensions. In the long run, one potential problem with the lack of rules of the road in outer space would be the difficulty of blaming an offending party for "aggressive" or "hostile" behavior if there is no objective standard by which to measure the dangerousness of space activities, whether they might damage other space objects or harm human life (Porras and Zarkan 2021).

15.5. Conclusion

The universe includes outer space and the matter, time and energy it contains. When discussing the universe, it is difficult to establish a precise territorial scope. Questions such as "how does the law apply?", "where does space begin?", "how big is it?" arise. In the absence of scientific and technical criteria for delimiting outer space, it is the responsibility of the States to work on the most relevant criteria for determining this vertical boundary, which is currently the subject of much debate, particularly within the Legal Subcommittee of COPUOS.

Limited, however, to an examination of human and robotic exploration of the universe, this chapter has outlined the legal framework for the exploration of the universe based on current knowledge. This chapter has also considered the long-term sustainability of space activities, both human and robotic, involved in the exploration of the universe. This chapter has undoubtedly omitted certain points which, in the near future, will seem unavoidable and will be the center of discussions within the largest international bodies. Without a doubt, as one moves away from planet Earth, the interest of the *corpus juris spatialis*, constituted of the law applicable to space matters, will diminish. But while these particular points are illustrations of a particular situation at a specific moment, the legal and ethical aspects will undoubtedly remain at the heart of the subject.

The point here is not to establish which activity is the most important for the exploration of the universe but rather to consider the way in which the human species – which will have developed far from planet Earth – will consider this exploration and will accept, or not, to be framed by rules, and if this is the case, these rules will be the fruit of the work of all of humanity or of a few States which see in the exploration of the universe an opportunity to pursue their own scientific, commercial and strategic interests. It is not a question of establishing a framework adopted collectively, during a given period, which would be limited to our earthly prism. It is a question of knowing if, for the future of the exploration of the universe, our species will be able to adapt and possibly reinvent itself. If the need to explore comes from an impulse, and if this impulse is constrained, it will necessarily be slowed down. On the contrary, if this momentum is left unsupervised, there is a risk that it will become corrupted. This is the difficulty. If the human exploration of the universe will inspire generations of human beings, it is currently mostly a future project that can only be prepared on legal and ethical levels.

15.6. References

Arnould, J. (2018). *Oublier la Terre ? La conquête spatiale 2.0.* Le Pommier, Paris.

Billings, L. (ed.) (2021). 50 years of solar system exploration: Historical perspectives, national aeronautics and space administration [Online]. Available at: https://www.nasa.gov/sites/default/files/atoms/files/50-years-of-solar-system-exploration_tagged.pdf [Accessed 5 June 2022].

COPUOS (2007). Space debris mitigation guidelines of the Committee on the Peaceful Uses of Outer Space [Online]. Available at: https://www.unoosa.org/pdf/publications/st_space_49E.pdf [Accessed 5 June 2022].

COPUOS (2019). Guidelines of the Committee on the Peaceful Uses of Outer Space for the Long-Term Sustainability of Space Activities [Online]. Available at: https://www.unoosa.org/res/oosadoc/data/documents/2021/stspace/stspace79_0_html/st_space79E.pdf [Accessed 5 June 2022].

COPUOS (2022). Report of the Scientific and Technical Subcommittee on its fifty-ninth session, A/AC.105/1258, Sixty-fifth session [Online]. Available at: https://www.unoosa.org/oosa/oosadoc/data/documents/2022/aac.105/aac.1051258_0.html [Accessed 5 June 2022].

Couston, M. (2022). *Fascicule 1501 : Droit spatial. – Principes directeurs du droit international public de l'espace.* JurisClasseur Transport, LexisNexis.

Criado Perez, C. (2021). *Invisible Women: Data Bias in a World Designed for Men.* Abrams Press, New York.

Degrange, V., Peter, H., Cesari Zarkan L. (2021). Entre force et diplomatie spatiale : le missile russe Nudol s'invite à la table des négociations, *Notes d'actualité de l'Institut d'Études de Stratégie et de Défense* [Online]. Available at: https://iesd.univ-lyon3.fr/wp-content/uploads/2021/12/Notes-dactualite%CC%81-CZ-D-P_2.5.pdf [Accessed 5 June 2022].

Descalles, J. (2019). Fatoumata Kebe, l'astrophysicienne qui chasse les débris spatiaux, *Usbek & Rica* [Online]. Available at: https://usbeketrica.com/fr/article/fatoumata-kebe-chasseuse-de-debris-spatiaux [Accessed 5 June 2022].

ESA (2019). ESA Commissions World's First Space Debris Removal [Online]. Available at: https://www.esa.int/Safety_Security/Clean_Space/ESA_commissions_world_s_first_space_debris_removal [Accessed 5 June 2022].

Fasan, E. (1970). *Relations with Alien Intelligence: The Scientific Basis of Metalaw.* Berlin Verlag, Berlin.

Fortin, J. and Zraick K. (2019). First all-female spacewalk canceled because NASA doesn't have two suits that fit. *The New York Times* [Online]. Available at: https://www.nytimes.com/2019/03/25/science/female-spacewalk-canceled.html [Accessed 5 June 2022].

Gohd, C. (2022). Don't call them "space tourists", says former NASA astronaut commanding private Ax-1 mission. *Space.com* [Online]. Available at: https://www.space.com/nasa-astronaut-not-space-tourists-spacex-ax1-mission [Accessed 5 June 2022].

Hobe, S., Schmidt-Tedd, B., Schrogl, K.-U., Goh, G. M. (2009). *Cologne Commentary on Space Law*, Vol. II. Carl Heymanns, Cologne.

International Civil Aviation Organization (2015). Commercial space flights, Appendix A LC/36-WP/3-2 [Online]. Available at: https://www.icao.int/Meetings/LC36/Working%20Papers/LC%2036%20-%20WP%203-2.en.pdf [Accessed 5 June 2022].

Jones, A. (2022). China to build a lunar communications and navigation constellation. *SpaceNews* [Online]. Available at: https://spacenews.com/china-to-build-a-lunar-communications-and-navigation-constellation/ [Accessed 5 June 2022].

Kiss, A.-C. (1968). L'accord sur le retour et le sauvetage des astronautes et la restitution des objets lancés dans l'espace extra-atmosphérique. *Annuaire français de droit international*, 14, 736–746.

NASA (2020). Artemis accords: The principles for cooperation in the exploration and civilian use of the Moon, Mars, Comets and Asteroids for peaceful purposes [Online]. Available at: https://www.nasa.gov/specials/artemis-accords/index.html [Accessed 5 June 2022].

Osborn Frandsen, H. (2022). Looking for the rules-of-the-road of outer space: A search for basic traffic rules in treaties, guidelines and standards. *Journal of Space Safety Engineering*, 9, 231–238.

Paladini, S., Saha, K., Pierron, X. (2021). Sustainable space for a sustainable Earth? Circular economy insights from the space sector. *Journal of Environmental Management*, 289, 112511.

Porras, D. and Zarkan, L. (2021). Safety norms for space security: How the development of STM norms can strengthen security in space. Advanced Maui Optical and Space Surveillance Technologies Conference (AMOS) [Online]. Available at: https://amostech.com/TechnicalPapers/2021/SSA-SDA/Porras.pdf [Accessed 5 June 2022].

Robinson, G. (1984). Astronauts and a unique jurisprudence: A treaty for spacekind. *Hastings International and Comparative Law Review*, 7(3), 483–499.

Roth, A.D. (1992). *La prohibition de l'appropriation et les régimes d'accès aux espaces extra-terrestres*. Graduate Institute Publications, Geneva.

Spudis, P. (2017). The robotic exploration of the moon. *NASA Science, Earth's Moon* [Online]. Available at: https://moon.nasa.gov/exploration/history/ [Accessed 5 June 2022].

Thomas, H. (2019). A place for women in space. *Foreign Policy* [Online]. Available at: https://foreignpolicy.com/2019/04/08/women-space-nasa-spacesuits-too-small-anne-mcclain/ [Accessed 5 June 2022].

World Tourism Organization (1999). Global code of ethics for tourism [Online]. Available at: https://www.unwto.org/global-code-of-ethics-for-tourism [Accessed 5 June 2022].

16

The COSPAR Planetary Protection Policy and International Law

Leslie I. Tennen
Law Offices of Sterns and Tennen, Phoenix, USA

16.1. Introduction

Spacecraft sent to explore our celestial neighbors could carry biological material that could contaminate the pristine natural environments and jeopardize the integrity of scientific experiments and investigations. Forward contamination of celestial bodies has particular ramifications in the search for present, past, or precursors of extraterrestrial life. Back contamination of the environment of Earth with potentially catastrophic results could occur from the return of samples of extraterrestrial matter. As mankind entered the space age and acquired the technological capability to send robotic probes to the Moon and planets, the international legal and scientific communities adopted rules, policies and procedures to prevent forward and back contamination. Legal regulations were developed through the United Nations (UN), and scientific standards were established through COSPAR, the Committee on Space Research.

COSPAR is an international, non-governmental organization composed of national scientific institutions and international scientific unions, created in 1958 by the International Council of Scientific Unions (ICSU), now known as the International Science Council (ISC). It has established the Planetary Protection Panel (PP Panel) to consider matters of forward and back contamination. The panel includes representatives of national and international authorities, and the COSPAR commissions on Space Studies of the Earth-Moon System, Planets and Small Bodies of the Solar Systems (Scientific Commission B), and Life Sciences as Related to Space (Commission F). One of the two Vice-Chair positions is reserved for a representative of the UN Office for Outer Space Affairs (UNOOSA) (Coustenis et al. 2019).

The focus of the COSPAR PP Panel is limited to policies to prevent biological and organic contamination of celestial bodies in the search for life and the protection of the environment of Earth from extraterrestrial matter. The PP Panel does not address preservation of historical sites, unique environments or protection of celestial environments other than from contamination by microbes and other biological or organic material. The planetary protection policy (PPP) articulated by COSPAR has evolved over time and consists of non-binding guidelines (although frequently referred to as requirements). The COSPAR PPP has been recognized and implemented by all space active States in their missions to celestial bodies.

16.2. Legal standards of protection of celestial environments

The UN Committee on the Peaceful Uses of Outer Space (COPUOS) has been the primary forum for the promulgation of legal protections for celestial environments. However, the initial impetus was on the preservation of celestial bodies in order to protect the right of States to conduct activities in space without obstacles created by other States, and not the protection of natural celestial environments. The UNGA Resolution 1962 (XVIII) (Legal Principles Declaration of 1963) reflected this primary concern for the rights of States and declared that States were to conduct activities in outer space with due regard for the corresponding interests of other States (Legal Principles Declaration, para. 6). In addition, States were directed to undertake appropriate international consultations prior to proceeding with any activity or experiment it had reason to believe would cause potentially harmful interference with activities of other States. A State could request consultations if it had reason to believe an activity or experiment planned by another State would cause potentially harmful interference with its activities.

COPUOS achieved a milestone when consensus was reached on the Treaty on Principles Governing the Activities of States in the Exploration and Use of Outer

Space, including the Moon and Other Celestial Bodies of 1967 (OST). Article IX incorporated the due regard and international consultation provisions of paragraph 6 of the Declaration of Principles and added prohibitions against forward and back contamination. States are to avoid harmful contamination of outer space, including the Moon and other celestial bodies, and adverse changes in the Earth's environment caused by the introduction of extraterrestrial matter (OST, Art. IX). The Agreement Governing the Activities of States on the Moon and Other Celestial Bodies of 1979 (Moon Agreement) goes further and specifies that States shall prevent the disruption of the existing balance of the lunar environment[1] and avoid harmfully affecting the Earth's environment through the introduction of extra-environmental matter (Moon Agreement, Art. 7). COPUOS has recognized the COSPAR PPP as the international reference standard for guiding compliance with the baseline legal requirement in article IX of the OST to avoid harmful contamination of celestial environments by terrestrial biological material[2].

16.3. Current COSPAR planetary protection policy

The COSPAR PPP began as a comprehensive policy that recognized pristine celestial environments to be scientific preserves, and in 1964 established Planetary Quarantine Requirements (PQR) applicable to all solar system bodies within our technological reach. These PQR set specific limits on the probability that interplanetary spacecraft could contaminate a celestial body. Landing spacecraft were subject to more stringent limits than were fly-by or orbiting craft. The probability of contamination was calculated by a complex formula which considered numerous factors, including the initial microbial burden on a spacecraft at launch, the probability of survival of such microbes and the probability of their growth in a celestial environment. Revisions to the policy have been made based on scientific explorations and results, and new perspectives and approaches to planetary protection. Initial modifications limited the application of the PQR to missions to celestial bodies deemed of biological interest in the search for life. Further revisions conditioned the obligation to utilize cleanliness and decontamination techniques on interplanetary probes on whether the spacecraft was equipped with instruments to search for evidence of past or present life (Tennen 2004).

---

1 The Moon Agreement specifies in Article 1.1 that references to the Moon include other celestial bodies within the solar system unless specific legal norms enter into force with respect to particular bodies.

2 COPUOS (2017). Report of the Committee on the Peaceful Uses of Outer Space, A/72/20.

The COSPAR PPP was restructured during the World Space Congress in 2002. Missions to celestial objects were divided into five categories based on the target body and the mission objectives. Four categories were directed to preventing forward contamination, with the lowest category, in terms of decontamination controls, being missions to target bodies which are "not of direct interest for understanding the process of biological or chemical evolution" (COSPAR 2005). The fifth category was designated to protect against back contamination by spacecraft intending to return samples of extraterrestrial matter to Earth, to which the most rigorous requirements are to apply. The 2002 revisions created the new category of "special region" for areas on Mars which may have water in the form of surface or subsurface ice and a temperature range conducive to sustaining life. All spacecraft sent to a special region, irrespective of whether it is intended to search for evidence of life, would be subject to the level of decontamination applicable to the Viking missions.

16.3.1. *Categories of mission objectives and target body combinations*

COSPAR continues to utilize the categorization system based on target body/mission objective combinations as the structure of the PPP. The current overarching policy as stated by COSPAR is as follows:

> The conduct of scientific investigations of possible extraterrestrial life forms, precursors, and remnants must not be jeopardized. In addition, the Earth must be protected from the potential hazard posed by extraterrestrial matter carried by a spacecraft returning from an interplanetary mission. Therefore, for certain space mission/target planet combinations, controls on contamination shall be imposed in accordance with issuances implementing this policy. (COSPAR 2021)

The policy retains the four categories of interplanetary missions regarding forward contamination originally established in 2002, although updates and revisions have been made as recently as June 2021.

16.3.1.1. *Category I*

Category I applies to fly-by, orbiter and landing missions to target bodies which are "not of direct interest for understanding the process of chemical evolution or the origin of life". Target bodies in this category include Io and undifferentiated, metamorphosed asteroids. The PPP states that "no protection of such bodies is warranted" (COSPAR 2021) and accordingly no specific decontamination controls are imposed and no specific documentation is required.

16.3.1.2. *Category II*

Category II consists of fly-by, orbiter and landing missions to target bodies of "significant interest relative to the process of chemical evolution and the origin of life, but where there is only a remote chance that contamination carried by a spacecraft could compromise future investigations" (COSPAR 2021). Target bodies in this category include most of the moons in the solar system, including the Moon of Earth. Also included are the planets other than Mars, but including Pluto, carbonaceous chondrite asteroids, comets and some Kuiper belt objects. The defining criterion of a "remote" chance of contamination of these target bodies "implies the absence of environments where terrestrial organisms could survive and replicate" (COSPAR 2021). Missions to these Category II target bodies are subject to documentary requirements of "a short planetary protection plan" (COSPAR 2021) that identifies possible impact sites and includes brief pre- and post-launch analyses of impact strategies. A post encounter or mission report is also to be prepared identifying any location of spacecraft impact.

In June 2021, COSPAR revised the policy by adding two new subdivisions for lander missions to the Moon based on the geographic region to be accessed. Category IIa applies to missions to most of the lunar surface and excludes only those areas described in Category IIb. Areas within Category IIb are the permanently shadowed regions (PSRs) and the lunar poles at latitudes south of 79° S and north of 86° N. Different standards of reporting the organic inventory onboard the spacecraft apply to missions within the two categories. Category IIa missions are to report organic products that may be released into the lunar environment by the propulsion system. The inventory report for Category IIb missions, however, is to include all organic material with a total mass greater than 1 kg.

16.3.1.3. *Category III*

Category III is composed of fly-by and orbiter missions to target bodies which are of significant interest in the search for evidence of life, and present a substantial risk that contamination could jeopardize future investigations. Target bodies in this category include Mars, Europa and Enceladus. Fly-by and orbiter spacecraft are not designed to land or intentionally impact the target body and can satisfy the planetary protection requirements by either orbital lifetime parameters or the reduction in the bioburden carried by the spacecraft. The orbital lifetime parameters standards are a probability of 99% or more that the spacecraft will not impact the target body within 20 years of launch, and 95% or more that no impact will occur within 50 years. If the spacecraft cannot meet these orbital lifetime parameters, then it must employ decontamination techniques to reduce the bioburden at launch to a total number of spores on and in the spacecraft of 5×10^5 or less[3].

16.3.1.4. *Category IV*

Category IV applies to landing craft to Category III target bodies of Mars, Europa and Enceladus, with specific requirements based on the mission objectives and landing area. Spacecraft which are not intended to conduct life detection experiments are designated as subcategory IV(a) and must satisfy specific bioburden limits of a total of 3×10^5 spores or less, with an average of 300 spores or less per square meter. Missions which carry life detection instruments are placed in subcategory IV(b) and must meet much more stringent controls of a total of 30 spores, or the "levels of bioburden reduction driven by the nature and sensitivity of the particular life detection experiments" (COSPAR 2021). However, the entire spacecraft is not required to meet this limit. If the life detection instruments can be cleaned and isolated from exposure to possible recontamination, the remainder of the spacecraft will be subject to the subcategory IV(a) limits. In June 2020, the COSPAR Bureau revised the PPP to specify that fly-by, orbiter and lander missions to Europa or Enceladus must meet a probability of 1×10^{-4} of contamination of subsurface liquid water, and that the period of exploration is 1,000 years commencing with the beginning of the 21st century.

The final subcategory, IV(c), applies to spacecraft sent to a special region, which are subject to the limits in subcategory IV(b) whether or not there are life detection experiments on board. Similar to the alternatives in standards in subcategory (b), the entire spacecraft must achieve the cleanliness limit unless the subsystems which will encounter the special region are designed to be cleaned and isolated from exposure to possible recontamination.

16.3.1.5. *Category V and sample return mission*

16.3.1.5.1. Restricted and non-restricted earth return

Category V addresses the issue of back contamination by sample return missions. There are two classifications within this category: restricted and non-restricted. The policy does not require any specific bioburden reduction measures for missions which are designated as Non-Restricted Earth Return, including to and from the Moon and Venus. At the other end of the spectrum are Restricted Earth Return missions to Mars or Europa, for which the highest levels of precautionary measures are imposed on both the outbound and return segments of the journey.

3 Spores are defined as aerobic microorganisms that survive a heat shock of 80°C for 15 min and cultured on Tryptic-Soy-Agar at 32°C for 72 h.

Spacecraft on a Restricted Earth Return mission are subject to the limits specified in Category IV(b) to prevent "false positives" in the search for extraterrestrial life. Samples collected must either be sterilized prior to return to Earth or stored in closed containers with a means provided to isolate the containers and "break the chain of contact" with the target body. The mission is to be reviewed and approved at several stages of the operation: prior to initial launch from Earth, during the period between the collection of samples and maneuvers to enter a biased Earth return trajectory, and before commitment to reentry into the Earth's environment. Samples of returned matter must be sterilized or tested for life detection and biohazards before any portion can be distributed.

The designation of a target as restricted or non-restricted sample return is to be made on a case-by-case basis for bodies which are not listed. This determination is to be made based on the following series of questions:

> "1. Does the preponderance of scientific evidence indicate that there was never liquid water in or on the target body?
>
> 2. Does the preponderance of scientific evidence indicate that metabolically useful energy sources were never present?
>
> 3. Does the preponderance of scientific evidence indicate that there was never sufficient organic matter (or CO_2 or carbonates and an appropriate source of reducing equivalents) in or on the target body to support life?
>
> 4. Does the preponderance of scientific evidence indicate that subsequent to the disappearance of liquid water, the target body has been subjected to extreme temperatures (i.e. >160°C)?
>
> 5. Does the preponderance of scientific evidence indicate that there is or was sufficient radiation for biological sterilization of terrestrial life forms?
>
> 6. Does the preponderance of scientific evidence indicate that there has been a natural influx to Earth, for example, via meteorites, of material equivalent to a sample returned from the target body?
>
> For containment procedures to be necessary ('Restricted Earth return'), an answer of 'no' or 'uncertain' needs to be returned to all six questions" (COSPAR 2021).

COSPAR was requested by the Japan Aerospace Exploration Agency (JAXA) to provide guidance regarding sample return missions to Phobos and Deimos. In 2019, pursuant to the above criteria, the PP Panel recommended that the outbound portion of such a mission be classified as Category II and that material returned from the two moons be designated as Category V Non-Restricted Earth Return (Coustenis et al. 2019).

16.3.1.5.2. Category V and sample return mission

Restricted Earth Return missions will be able to draw upon the experiences of the Apollo program and the Lunar Receiving Laboratory (LRL), which was built to hold lunar material and quarantine the returning astronauts until it was determined that they were not contaminated. The LRL was employed for the crews of Apollo 11, 12 and 14 until 1971, when the Interagency Committee on Back Contamination opined that further use was unnecessary and that lunar samples could be returned without restriction. The LRL experienced technical, operational and management issues that allowed the biological barrier to be breached, and the lessons learned provide valuable insights on how to avoid inadvertent release of extraterrestrial materials (Mangus and Larsen 2004).

The few sample return missions launched since Apollo have been non-restricted as their targets were comets, asteroids and Phobos, bodies deemed incapable of supporting life. A Restricted Earth Return mission is being conducted by the National Aeronautics and Space Administration (NASA), which started with the launch of Mars2020. The spacecraft successfully landed on the planet in 2021 and collected as many as 40 samples of Martian material. Although Mars2020 will hold the samples for return to Earth by a later mission, the samples collected are subject to Restricted Earth Return standards and protocols for containment, isolation, transportation and examination, some of which are yet to be developed and are not without controversy (MSR Science Planning Group 2019).

16.3.2. *Human missions to Mars*

The COSPAR PPP has been directed primarily to robotic missions, but also considers crewed missions to Mars. It is recognized that human explorers will not be able to operate in entirely closed systems and it is not possible to completely avoid exposure to Martian materials. Crew members "will carry microbial populations that will vary in both kind and quantity" (COSPAR 2021), rendering the bioburden limits for robotic spacecraft utterly inapposite and unsuitable for human missions.

The PPP recommends that before there are human missions to Mars, a better understanding is needed of the Martian environment, human associated contamination and the capabilities of terrestrial organisms. The PPP sets forth guidelines for crewed missions to Mars, including:

> "the development of a comprehensive protocol for addressing forward and backward contamination concerns during all phases of a mission;
>
> continuous monitoring and evaluation of microbes carried by a mission;
>
> the designation of a member of the crew with primary responsibility for implementing planetary protection provisions;
>
> a quarantine capability for the crew during and after the mission in case of potential contact with a Martian life form;
>
> robotic precursors to evaluate whether or not a site should be characterized as a special region prior to crew access;
>
> planetary protection requirements for human missions should be based on a conservative approach recognizing the lack of specific knowledge of the Martian environment, possible life and the utility and efficacy of human support systems;
>
> planetary protection requirements for subsequent missions should not be relaxed without scientific justification and consensus" (COSPAR 2021).

16.4. Implementation of the COSPAR PPP by space agencies and authorities

The European Space Agency (ESA), JAXA and NASA have adopted formal planetary protection policies to govern the management of missions, which are based on and implement the COSPAR PPP. China and India have conducted or are in the process of conducting missions to the Moon and Mars, the United Arab Emirates (UAE) has sent an orbiter to Mars, and Israel has sent a probe to the Moon, but none of these States have a publicly available formal planetary protection policy. Nevertheless, all States conducting missions to celestial bodies to date purportedly have complied with the COSPAR PPP.

	Category I	Category II	Category III	Category IV	Category V
Type of Mission	Any but Earth Return	Any but Earth Return	No direct contact (flyby, some orbiters)	Direct contact (lander, probe, some orbiters)	Earth Return
Target Body	See Category-specific listing	See Category-specific listing	See Category-specific listing	See Category-specific listing	See Category-specific listing
Degree of Concern	None	Record of planned impact probability and contamination control measures	Limit on impact probability Passive bioburden control	Limit on probability of non-nominal impact Limit on bioburden (active control)	If <u>restricted</u> Earth return: No impact on Earth or Moon; Returned hardware sterile; Containment of any sample.
Representative Range of Requirements	None	Documentation only (all brief): PP plan Pre-launch report Post-launch report Post-encounter report End-of-mission report	Documentation (Category II plus): Contamination control Organics inventory (as necessary) Implementing procedures such as:	Documentation (Category II plus): P_c analysis plan Microbial reduction plan Microbial assay plan Organics inventory	Outbound Same category as target body/ outbound mission Inbound If <u>restricted</u> Earth return:

(continued)					
			Trajectory biasing Cleanroom Bioburden reduction (as necessary)	Implementing procedures such as: Trajectory biasing Cleanroom Bioburden reduction Partial sterilization of contacting hardware (as necessary) Bioshield Monitoring of bioburden via bioassay	Documentation (Category II plus): P_C analysis plan Microbial reduction plan Microbial assay plan Implementing procedures such as: Trajectory biasing Sterile or contained returned hardware Continual monitoring of project activities Project advanced studies and research If unrestricted Earth return: None

Table 16.1. *COSPAR policy on planetary protection: categories for solar system bodies and types of missions*

Roscomos has not publicly issued any official documents incorporating the COSPAR PPP into Russian national policy (Dobrokhotsky et al. 2012). Article 4.1 of the Russian Federal Law "On space activities" specifies that space activities shall be conducted in conformity with the principle of environmental protection (Russian Federal Law 1993). Article 4.2 prohibits the creation of "harmful contamination of outer space which leads to unfavorable changes of the environment" in order to ensure "strategic and ecological security". Although Article 4 does not reference the Moon and other celestial bodies, Article 26.3 re-affirms Russia's duty to comply with the OST.

The planetary protection standards set forth in the ESA and JAXA policy documents apply to spaceflight missions conducted by the respective agency and to contributions to non-agency spaceflight missions conducted by third parties. NASA's participation in non-NASA missions is conditioned upon compliance by the third-party governmental or non-governmental entity with laws and procedures consistent with the COSPAR PPP. This was not always the policy of the US government. In late 2016, this provision was deleted from the applicable NASA directive, but it was restored 4 months later (Tennen 2020).

The Israel IL Beresheet lunar landing craft was launched on a private rocket from a NASA launch facility in Florida. Since NASA provided support for the launch, the payload was subject to the NASA planetary policy and requirements. The spacecraft crashed while attempting to land on the Moon, and several months later there were reports that a third party had secretly encased some tardigrades into components that were incorporated into the lander. The intentional concealment of the placement of living organisms in a spacecraft launched to a celestial body with NASA participation could constitute a violation of US law (Johnson et al. 2019).

In July 2020, NASA issued Interim Directives NID 8715.128, Planetary Protection Categorization for Robotic and Crewed Missions to the Earth's Moon (NASA 2020a), and NID 8715.129, Biological Planetary Protection for Human Missions to Mars (NASA 2020b). These documents were not formal policy statements but were to direct the activities of NASA on an interim basis pending future official agency policy promulgation. These NID's essentially were a form of recommendation to COSPAR for revisions to the PPP, which would lessen or remove planetary protection requirements for large areas of the Moon.

NASA Interim Directive 8715.128 would change the classification of most missions to the Moon from Category II to Category I, by dividing the Moon into two sections. The first section consists of areas with PSRs, which are the northern and southern polar zones, together with Apollo landing and other historic sites, and which are designated as II-L. The second section consists of all remaining areas of the Moon, and is labeled as I-L. The NID identifies different contents for organic inventory reports for category II-L robotic missions and crewed missions. Category I missions are not subject to any planetary protection constraints under the COSPAR policy, and the NID thereby exempts these missions from the PPP. This NID opens vast areas of the Moon for NASA and private companies to operate without any consideration of planetary protection concerns and is consistent with a recommendation of the 2019 Final Report of the NASA Planetary Protection Independent Review Board (PPIRB) (NASA 2019). The PPIRB also recommended that NASA should consider recategorizing portions of the Martian surface from Category IV to Category II. Although NID 8715.129 does not adopt this recommendation, it can be considered a harbinger to such action.

The June 2021 revisions to the COSPAR policy, discussed in section 16.3.1.2., adopted the conceptual structure of the NID by establishing different standards for different areas of the lunar surface, notably the poles and PSRs, compared to the remaining areas of the Moon. But COSPAR did not adopt the NID's reclassification of most of the Moon to Category I and thereby rejected a blanket exemption from any planetary protection considerations for lander missions to those areas. Nor did COSPAR include unspecified and undefined historical sites in the areas subject to organic inventory and other reporting requirements. The absence of reference to historical sites by COSPAR is not surprising as it is consistent with the limited object and purpose of the COSPAR PPP to protect against forward and back contamination from biological sources.

Following the COSPAR revisions to the PPP in 2021, NASA issued Procedural Requirements (NPR) 8715.24, Planetary Protection Provisions for Robotic Extraterrestrial Missions (NASA 2021), which superseded and cancelled NID 8715.128. However, NPR 8715.24 also cancelled the fundamental agency policies and directives which mirrored the COSPAR PPP and which had set forth detailed substantive and procedural provisions guiding NASA's implementation of planetary protection measures (NASA 2017). The cancellation of these fundamental agency documents appears to conflict with the longstanding NASA planetary protection policy directive still in effect (NASA 2008), which expressly refers to the cancelled policy document as continuing to be applicable. NASA is preparing a formal technical standard which should address this discrepancy. Potentially more significant, however, is that NPR 8715.24 reflects a shift in planetary protection from a blanket approach with detailed requirements to what is referred to as risk-informed decision-making.

16.5. Application of international law to the planetary protection policy

16.5.1. *Due regard*

Article IX of the OST addresses both forward and back contamination as well as the nexus between planetary protection and interference with the activities of States. The first sentence of this article emphasizes the protection of the rights of States and specifies:

> In the exploration and use of outer space, including the moon and other celestial bodies, States Parties to the Treaty shall be guided by the principle of co-operation and mutual assistance and shall conduct all their activities in outer space, including the moon and other celestial bodies, with due regard to the corresponding interests of all other States Parties to the Treaty. (OST, Art. IX)

The obligation of due regard requires that a State conduct its activities so as to not interfere with or hinder the exercise of rights by other States. The OST does not define "due regard" although the term has been utilized in international aviation and maritime instruments to refer "to the performance of an act with a certain standard of care, attention or observance" that is, a duty of due diligence (Marchisio 2017). The Convention on International Civil Aviation of 1944 (Chicago Convention) mentions due regard in Article 3, which specifies that State aircraft are exempt from the Treaty but are nevertheless required to operate with due regard for the safety of civil aviation. References to the obligation of due regard in the UN Convention on the Law of the Seas of 1982 (UNCLOS) include Article 87, paragraph 2, which requires States to exercise their rights to conduct activities on the high seas and in areas beyond the continental shelf with due regard to corresponding rights of other States, and Article 58(c) which specifies that ships operating in an exclusive economic zone shall give due regard to the coastal State.

In the South China Sea Arbitration under UNCLOS, it was recognized that the precise contours of the obligation to exercise due regard and diligence may be elusive, and the measures a State could have employed to avoid harm may not be readily apparent[4]. The exercise of due regard is not a guarantee that an activity will not impact other States. However, a State does not need be an active participant in an activity to violate its duty under international law, and tolerating offending conduct by an entity within the control of a State could contravene the obligation of due regard. This could be relevant in the context of harmful contamination to a

4 Permanent Court of Arbitration Award (2016). The South China Sea Arbitration (The Republic of Philippines v. The People's Republic of China), Case No 2013-19, para. 754.

celestial body caused by a non-governmental entity authorized by a State to conduct activities in space pursuant to Article VI of the OST (Tennen 2020).

Although the concept of due regard is set forth in Article IX of the OST, the obligation to consider the rights of other States is not restricted to planetary protection but applies broadly to all activities in the exploration and use of outer space and celestial bodies. The Moon Agreement reinforces this broad application by placing the principle of due regard in an article separate from the planetary protection provisions. Article 4.2 incorporates the principles of cooperation and mutual assistance and specifies that international cooperation "should be as wide as possible". Article 4.1 expands both the scope and the beneficiaries of due regard and specifies that "due regard shall be paid to the interests of present and future generations as well as the need to promote higher standards of living and conditions of economic and social progress and development in accordance with the Charter of the United Nations". The OST mandates that due regard be given to the corresponding interests of other States parties to the Treaty, but the Moon Agreement is not so limited.

16.5.2. *International consultation*

Article IX of the OST provides a procedure for international consultations for States to resolve disputes regarding potential interference with activities. A State which has reason to believe that its activities may cause potentially harmful interference with the activities of another State shall undertake consultations prior to engaging in the activity. Conversely, a State which believes that the activities of another State may interfere with its own activities may request consultations. However, Article IX does not mandate that States agree to participate in requested consultations.

The Moon Agreement expands upon this dispute resolution procedure in Article 15 and specifies that participation in requested consultations is compulsory. Pursuant to this article, consultations can be requested when a State believes that another State is interfering with its rights, or that the other State is not fulfilling its obligations under the Treaty. Moreover, the consultations are open to participation by other States. The Secretary-General (SG) shall be informed of the results of the consultations, and if they do not result in a successful resolution to the dispute, any State can request the SG to assist in resolving the matter. The consent of the other State to the controversy for the engagement of the SG is not required.

16.5.3. *Forward and back contamination*

Article IX of the OST contains express prohibitions against cross-contamination of Earth and celestial bodies and specifies that States:

> shall pursue studies of outer space, including the moon and other celestial bodies, and conduct exploration of them so as to avoid their harmful contamination also adverse changes in the environment of the Earth resulting from the introduction of extraterrestrial matter and, where necessary, shall adopt appropriate measures for this purpose. (OST, Art. IX)

As with the term "due regard", the OST does not define or clarify what is meant by "harmful contamination" or "adverse changes" to an environment. Nevertheless, in the context of biological contamination, the COSPAR PPP are founded on the recognition that at a minimum, the transportation of terrestrial organisms to celestial bodies such as Mars would potentially cause contamination that would be harmful to the integrity of scientific investigations in the search for life. Similarly, the return of samples from Mars and certain other bodies must be presumed as potentially hazardous to prevent back contamination.

After mandating that States are to avoid harmful contamination of space and adverse changes to the environment of Earth, Article IX directs that States "shall adopt appropriate measures for this purpose" where necessary. The OST does not identify what measures would be appropriate, and it could be argued that the reference to "this purpose" applies only to the avoidance of disruption of Earth's environment and not harmful contamination of space and celestial bodies. However, such an interpretation would be contrary to the due regard principle and the purposes underlying Article IX.

Article 7 of the Moon Agreement elaborates on the required standards of behavior regarding forward contamination and specifies that States "shall take measures to prevent the disruption of the existing balance of [the] environment, whether by introducing adverse changes in that environment, by its harmful contamination through the introduction of extra-environmental matter or otherwise". While the OST directs that States are to avoid harmful contamination, the Moon Agreement obligates States to take preventive measures to protect against disrupting the balance of the lunar environment. The formulation of Article 7 makes it clear that "harmful contamination" from the introduction of extra-environmental matter, such as terrestrial microbes or other biological material, is only one modality by which the balance of the environment can be disrupted. A restrictive reading of this provision would limit the definition of harmful contamination to the introduction of extra-environmental matter. However, notwithstanding such a narrow interpretation, the prohibitions in Article 7 apply to any activity that could disrupt the balance of the environment even without the presence of biological material (Sterns and Tennen 1988).

The Moon Agreement, according to the Preamble, was formulated in part, to "define and develop" the provisions of the space treaties, including the OST, in relation to the Moon and other celestial bodies. As such, the Moon Agreement clarifies the provisions of Article IX of the OST. The prohibitions against forward and back contamination in the Moon Agreement were not met with significant opposition when considered by COPUOS, and as such can be "taken to express the international will on such matters" (Hofmann et al. 2010).

The OST and the Moon Agreement prohibit activities that cause alterations in the environment of Earth, outer space or celestial bodies which are disruptive or adverse, or constitute harmful contamination. All activities on celestial bodies have impacted the pristine natural environment and thereby can to some extent be considered "adverse", "harmful" or "disruptive". The surface of the Moon has been altered by spacecraft which have crashed or soft landed, and by the experiments and medallions and other scattered debris as well as the footprints and vehicle tracks left behind. The law of outer space seeks to strike a balance between the right of States to conduct peaceful exploration and use of outer space and celestial bodies, and the protection of the natural environments. Thus, not all contamination is harmful, and not all alteration of a pristine celestial environment is disruptive. Activities expressly authorized by the treaties do not, in the abstract and without more information, contravene the environmental protection provisions of Article IX of the OST or Article 7 of the Moon Agreement.

Article I of the OST specifies that all States have the right to explore and use space on the basis of equality and in accordance with international law and shall have free access to all areas of celestial bodies. The Treaty describes rights and obligations of States conducting activities in space including the Moon and other celestial bodies but does not describe the specific activities themselves. Article XII of the OST alludes to allowable activities by providing for a right of visitation to "stations, installations, equipment and space vehicles" on the Moon and other celestial bodies, thereby recognizing the right of States to establish stations and installations and to locate equipment and space vehicles thereon.

The Moon Agreement expressly identifies several activities that States are permitted to conduct on the surface and subsurface of the Moon, including the right to collect and remove samples in Article 6, and to land objects on and launch them from the Moon, to establish facilities, stations and installations on or below the surface or subsurface, and to freely move them and their personnel, vehicles and equipment in Article 8. Even though the mere fact of such activities may not necessarily contravene the environmental protection provisions of the treaties, the Moon Agreement seeks to limit their impact both in regard to contamination as well as to changes to the environment resulting from the intruding activities. Article 9 specifies that States shall use only as much area as required when establishing stations on the surface or subsurface of the Moon. Although the admonition of Article 9 is directed only to "stations", there is no reason why the text should not also apply to "facilities" and "installations", none of which are defined in the Treaty.

16.5.4. *Disclosure and reporting requirements*

Article XI of the OST obligates States to disclose the nature, conduct, locations and results of their activities in space. These disclosures are to be made to the SG, the international scientific community and the public, to the greatest extent feasible and practicable. The OST does not contain any further guidance regarding the notifications, nor is there any requirement for disclosures to be made in advance of any activity. The Moon Agreement expands these reporting requirements and specifies that disclosure is to be made as soon as practicable after launch, and the scientific results released upon completion of the mission. For missions lasting more than 60 days, the disclosures are to be made at thirty-day intervals for the first 6 months, after which only "significant" additions need be reported.

The Moon Agreement specifically provides that States shall report to the SG the measures taken to prevent forward and back contamination pursuant to Article 7. However, these disclosures are not required to be made in advance of a mission. Advance notification is only required for the placement of radio-active materials on the Moon, and even in such circumstances is limited "to the maximum extent feasible" without further elaboration or clarification.

The OST contains specific requirements in Article V that States report to the SG, the international scientific community and the public, of the discovery in space of any phenomena that could constitute a danger to the life or health of astronauts. The Moon Agreement provides a similar reporting requirement in Article 5.3, which also includes disclosure of any phenomena that could endanger human life on Earth or elsewhere as well as astronauts in space. Although it is implicit in Article XI of the OST that the reporting requirements include the discovery of any indication of organic life, Article 5.3 of the Moon Agreement makes that reporting obligation explicit.

The planetary protection provisions of both the OST and the Moon Agreement would be reinforced if the reporting requirements mandated that the measures utilized by a mission to avoid harmful contamination or prevent disruption to the balance of the environment were disclosed prior to launch (Moore and Leaphart 1983). In addition, the requirement to report phenomena which may endanger human life should not be limited to natural phenomena in space or on a celestial body but should include phenomena caused by human activities which may pose such a risk. The State discovering potentially dangerous phenomena has significant discretion in whether, when and what to report (Sterns and Tennen 1988).

16.5.5. *International scientific preserves*

The final provision of Article 7 of the Moon Agreement introduced a new concept in space law that has profound implications for planetary protection. Paragraph 7.3 specifies that areas of special scientific interest can be considered for designation as "international scientific preserves for which special protective arrangements are to be agreed upon". The consideration of an area as an international scientific preserve is without prejudice to the rights of other States party to the Treaty, and the "competent bodies" of the UN are to be consulted in the promulgation of standards of environmental protection. The Moon Agreement does not mandate that international scientific preserves be designated, only that consideration be given to doing so. Furthermore, this paragraph specifies that any such consideration is without prejudice to the rights of other States parties, which indicates that neither a moratorium nor any particular prohibitions are imposed on States from commencing or continuing to conduct activities within geographic parameters of the proposed preserve area.

The concept of an international scientific preserve under the Moon Agreement is different than a special region as defined in the COSPAR PPP. The criteria for consideration as an international scientific preserve is that an area has "special scientific interest". The criteria for designation as a special region are expressions of this special scientific interest in relation to the search for evidence of life, specifically the possible presence of water and a temperature range conducive to sustaining life as we know it. An international scientific preserve could be considered for scientific interests other than or in addition to the search for evidence of life, as well as for the preservation and protection of historical sites, and areas of particular aesthetic or cultural significance.

16.5.6. *The non-appropriation principle*

Harmful contamination of a celestial environment by the introduction of terrestrial biological material would interfere with the rights of other States to explore and use the contaminated area. Such interference would restrict the ability of other States to conduct activities in the contaminated area and thereby contravene Article I of the OST. In addition, contamination that renders an area unavailable to other States could be considered a form of national appropriation by use, occupation or other means, in violation of the non-appropriation doctrine in Article II of the OST and Article 11.2 of the Moon Agreement (Sterns and Tennen 1998).

16.5.7. *International responsibility and liability*

A State which caused forward or back contamination would be internationally responsible pursuant to Article VI of the OST, and subject to claims for damages pursuant to Article VII of the Treaty. The liability provisions of the OST have been supplemented by the Convention on International Liability for Damage Caused by Space Objects of 1971 (Liability Convention), which establishes a regime to provide for compensation with reference to personal injuries and property damage caused by a space object to aircraft in flight, on the surface of Earth, or to another spacecraft. Neither the Liability Convention nor Article VII of the OST was directed to damages to a celestial environment caused by forward contamination. However, a spacecraft returning samples of extraterrestrial matter to Earth that causes damage may be the subject of a claim for compensation under the Liability Convention.

The Liability Convention imposes strict liability for damages caused by a space object to aircraft in flight or on the surface of Earth. Conversely, a fault-based system is applied to damages caused to another space object. The damages envisioned by the OST and the Liability Convention are premised upon the ramifications of physical impact by a space object with persons and property. It is open to interpretation whether the liability regime in space law extends to damages caused by exposure to, rather than physical contact with, harmful extraterrestrial material. Neither the OST nor the Liability Convention expressly conditions liability on direct physical contact, and in the event these space treaties were deemed to be inapplicable a claimant could resort to general principles of international law.

Whether pursued under Article VII of the OST, the Liability Convention or general principles of international law, the traditional compensatory regime would be unsuitable and fail to provide an effective remedy in the event of potential catastrophic injury from back contamination, especially in the event of adverse changes to the Earth's environment. A State responsible for the damage could well lack sufficient technological, economic and political resources to remedy the damage and compensate the whole of mankind[5]. Nevertheless, at a minimum, international responsibility and liability must impose the obligation on States to disclose all relevant details of the nature, scope, extent and location of any risk of harm, even if this necessitated the disclosure of proprietary data regarding the operations and design of spacecraft components or other sensitive information.

In this regard, Article XXI of the Liability Convention has particular relevance:

> If the damage caused by the space object presents a large scale danger to human life or seriously interferes with the living conditions of the population or the functioning of vital centers, the States Parties, and in particular the launching State, shall examine the possibility of rendering appropriate and rapid assistance to the State which has suffered the damage when it so requests. However, nothing in this article shall affect the rights or obligations of the States Parties under this Convention. (Liability Convention, Art. XXI)

This article provides important guidance but unfortunately is hampered by imprecision in terminology. The meaning of key terms is unclear, such as what is meant by "large scale", "danger" and "functioning". It is also unclear whether "vital centers" include military installations and facilities, and whether they also include centers located off-Earth (Sterns and Tennen 1988).

In the event of circumstances warranting the invocation of this article, definitional disputes over fine distinctions, real or imagined, must yield to humanitarian concerns. More significant is the limited scope of obligations of States to disclose information and cooperate with the international community. Article XXI neither imposes an affirmative duty on States to render any specific assistance nor provides any penalty for refusing to render aid. States are only required to examine the possibility of "rendering appropriate and rapid assistance" on request of a damaged State. Presumably "appropriate" assistance would include the disclosure of relevant information. The damaged State may request assistance from the launching State of the space object causing injury, as well as from any other State party to the Liability Convention. Again, if this article needs to be invoked, humanitarian considerations should prevail as appropriate regardless of the status of a State as a party *vel non* to the Liability Convention.

---

5 This unfortunately is illustrated all too well by the Covid-19 pandemic currently enveloping the globe.

Humanitarian concerns also should prevail in matters involving the rescue and return of any crew of a space object that becomes contaminated. Article VIII of the OST and the Article IV of the Agreement on the Rescue of Astronauts, the Return of Astronauts and the Return of Objects Launched into Outer Space of 1967 (Rescue and Return Agreement) specify that States shall rescue and return astronauts and space objects to the launching authority. States may be hesitant to engage in rescue and return operations where there is a possibility the craft or crew were exposed to or infected with harmful contaminants. Astronauts are considered to be envoys of mankind and are befitting of all possible aid and assistance especially during accident or distress. The return of space objects which may have been contaminated does not share these humanitarian considerations, and States may be reasonable in refusing or delaying rendering recovery and return activities until deemed safe to do so. The Convention on Registration of Objects Launched into Outer Space of 1974 (Registration Convention) specifies that where it is not possible to identify the State of registry of an object which caused damage or "which may be of a hazardous or deleterious nature", States party to the treaty shall render assistance, on request, to the greatest extent feasible (Sterns 2000).

16.6. Gaps in the application of the planetary protection policy

A variety of techniques are utilized by States to reduce the bioload on a spacecraft to reach the contamination limits set by COSPAR. These measures include the use of biologically controlled cleanrooms to prevent or limit the microbial burden during assembly and testing of spacecraft. Active decontamination is accomplished using various cleaning agents and solvents, dry heat, plasma or ionizing radiation. Spacecraft are protected against recontamination by barrier systems including purging, filters, and seals. However, and notwithstanding the use of state-of-the-art techniques, it is not considered possible to achieve absolute sterilization of a spacecraft (Meltzer 2012).

16.6.1. *Risks of contamination by failed missions*

A number of States have conducted interplanetary missions, especially to Mars, approximately half of which have ended in failure. Although all of these spacecraft were in compliance with the COSPAR planetary protection policy, contamination of the planet may nevertheless have occurred. Several orbiter and fly-missions were lost on arrival at the planet, including Mars Observer and Mars Climate Orbiter, and the Japanese Nozomi, or Planet B, orbiter. Failed landing craft include Mars Pathfinder, Mars Polar Lander, the European Space Agency Beagle 2 Lander and the Russian Phobos 2 lander. Each of these missions was classified as Category III or

IV-A, and each of the spacecraft carried significant terrestrial biological material to Mars. All are believed to have impacted the Martian surface, except for Nozomi, but the locations of these spacecraft are not known. Since the impact sites are not known, it cannot be ruled out that one or more spacecraft crashed in an area that could constitute a special region. NASA has acknowledged that biological contamination could have resulted from at least two of their failed missions. Whether and to what extent damage has been caused to the natural environment are open questions. The locations of successful landing craft, and rovers at the end of their missions, are known, as are inventories of terrestrial biological materials carried on the spacecraft, and future missions can factor such information into their mission plans. But spacecraft which are lost at undetermined impact sites on the planet pose a potential obstacle and hindrance that could interfere with future missions.

The missions that ended in failure demonstrate an inherent weakness with the classification of spacecraft based on the combination of mission objectives and target body. The policy tolerates the accidental and intentional deposition of significant microbial populations at multiple locations on the surface of a target body such as Mars. Even missions subject to the most stringent requirements in Category IV-B can limit the application of these requirements to systems and subsystems containing life detection equipment. The microbial burden limits and exemptions consider only the current mission, and not future explorations which might be conducted, and which will be impacted by the biological contaminants left behind. The dichotomy between life detection equipment and the remainder of a spacecraft has no importance to a future mission, and the different standards of cleanliness for each have no meaning for a later mission which must necessarily disregard the more restrictive life detection equipment standards and base all mission considerations on the higher bioburden allowed for the remainder of the spacecraft. For these same reasons, the intention to land in a special region should not be determinative of the standards of protection, as different areas of the entire planet could have been more hospitable to life at varying times in its history than now and the possibility that there could be evidence of the precursors or remnants of life cannot be summarily dismissed in the absence of scientific investigations (Sterns and Tennen 2007).

16.6.2. *Unintentional increase in risk to non-target bodies*

Another weakness of the mission type/target body combination structure is that it fails to adequately consider the possibility that the spacecraft could have unintentional contact and potentially contaminate other bodies, whether within the flight trajectory or in the relative vicinity of the target body of the mission. An example of this weakness was presented by the Phobos-Grunt mission, which was intended to land a spacecraft on Phobos and return to Earth with samples of surface dirt and rocks. Phobos orbits Mars at an altitude of 6,000 km, and the environment is considered to be inhospitable to life and present only a remote chance that terrestrial organisms could survive or replicate. As such, stringent decontamination of the spacecraft was not required to satisfy the COSPAR PPP.

The spacecraft carried a special "biomodule" canister which held 11 types of organisms representing all three domains of life, including extremophiles resistant to radiation, desiccation, salt and heat. This biomodule enclosure was designed to nurture the organisms and enhance their ability to survive and replicate. The mission, as with any sample return program, was fraught with the possibility of an off-nominal occurrence on or near Phobos during approach, landing, or return. The launch of a spacecraft to a target only 6,000 km from Mars, carrying extremophiles and other organic material in a canister designed to sustain survivability and promote replication, presented a substantial risk of unintended impact and contamination of the Martian environment, including locations that may qualify as special regions. This increased risk of contamination was unnecessary and avoidable and was averted only when the spacecraft failed to leave Earth orbit and burned up on re-entry into the atmosphere.

The Phobos-Grunt mission is especially disturbing from the perspective of planetary protection. It was reported by the sponsor of the biomodule experiment that serious deficiencies were present in management and execution of the program, leading to poor oversight and inadequate communications. Even more disturbing was a lack of quality control which allowed for the use of non-space-rated substandard parts. These defalcations in proper administration and operation are unacceptable for any space mission and take on increased importance in the context of a sample return mission like Phobos-Grunt.

The launch of the Phobos-Grunt spacecraft was neither a best practice nor a precedent that should be followed by other States. Although the mission purportedly was reviewed and approved as satisfying the COSPAR PPP, the reliability of such certification is open to question. Missions that intentionally transport terrestrial bacteria, extremophiles or other organisms in special survivability containers that could come into close proximity to Mars or another celestial body important for the search for life can be justified only if it can be established, *inter alia*, that the mission seeks to obtain critical and highly significant scientific data that cannot be obtained by any alternative means (Sterns and Tennen 2019).

16.6.3. *Proposed recategorization of areas of Mars*

The NASA PPIRB has recommended changes be considered to the categorization of portions of Mars from Category IV to Category II, which will reduce the standards of planetary protection for such areas to require only minimal documentation. These proposed changes in categorization highlight the inherent logical inconsistency that underlies the mission objective/target body classification structure of the PPP. The justification for a relaxation in forward contamination standards, from a scientific perspective, is the argument that there is an absence of data to support a conclusion that a terrestrial microbe or other organism could survive and replicate in the extraterrestrial environment. However, in the case of a sample return mission, even to and from an area of relaxed PPP controls, the extraterrestrial material is deemed to be potentially hazardous and the mission is subject to the most stringent back contamination as well as forward contamination standards.

16.7. Commercial new space ventures and planetary protection

16.7.1. *Application of the planetary protection policy to the private sector*

The COSPAR PPP was developed during an era when space activities were exclusively within the domain of governments. Legacy space contractors and innumerable smaller entities and start up New Space commercial ventures are assuming a greater role in space operations, which has important implications and ramifications for planetary protection. The OST anticipated that the private sector would conduct activities in space, and Article VI specifies that States are internationally responsible for the activities of their non-governmental entities in space. Article VI further specifies that the activities of non-governmental entities must be authorized and continuously supervised by the appropriate State. Several States have enacted or are in the process of drafting national licensing or other authorization systems to regulate the private sector (Jakhu and Pelton 2017, pp. 87–112), but to date none have included an express requirement that private entities comply with the COSPAR PPP (Babb, Erb and Howard 2018).

This gap is partially filled for missions which have government involvement or support, in which event the State is required to follow their internal planetary protection policies such as adopted by ESA, NASA and JAXA. However, neither ESA nor NASA nor JAXA are regulatory agencies that authorize private sector activities, and commercial missions without such government participation are not subject to their internal policy requirement documents. There is no reason why private entities should not be required to comply with the COSPAR PPP or similar strictures. The interests to be served by imposition of the PPP on activities conducted by governments to prevent forward and back contamination apply with equal force to activities conducted by non-governmental entities. The contamination of Mars with organic matter from Earth by a commercial mission will jeopardize future missions by both the public and the private sectors. The international scientific community and government agencies have expended considerable resources and engaged in concerted efforts to avoid the harmful contamination of celestial environments to protect the integrity of future explorations and investigations. If private sector activities are not required to comply with, at a minimum, the same standards of planetary protection as States, all of the measures taken in the past 60 years will have been for naught and rendered a nullity (Tennen 2020).

16.7.2. *Duties of disclosure and preservation of the discovery of extraterrestrial life*

The COSPAR PPP is directed to the preparation and operation of experiments and instruments but does not address the results of those activities in the event evidence of past or present extraterrestrial life is detected. It can be expected that the scientific community would act with the utmost care and take appropriate measures to protect and preserve evidence of alien life. A commercial entity that discovers evidence of life would not have the same interests as the scientific community. Consideration must be given to the formulation of appropriate rules and procedures to protect and preserve any discoveries of scientific interest, especially evidence of extraterrestrial life, made by non-governmental entities.

These policies should include procedures and requirements for the prompt, full and complete disclosure of the discovery of evidence of any indication of living alien organisms, or past life or the precursors thereof, which would be consistent with the reporting obligations in Article 5.3 of the Moon Agreement. In addition to immediate notification, the discovery should be fully documented, and all commercial activities that could disturb or destroy the discovery site and surrounding areas in proximity should be temporarily suspended.

Without a mandatory disclosure requirement, a commercial enterprise may be reluctant to give notification of a discovery without first conducting its own detailed examination of the evidence and determining whether a financial opportunity and advantage can be gained from exploiting potential intellectual property rights derived from a non-terrestrial microbe (Long 2017). However, the United States has already declared as a matter of policy that a commercial entity is not able to assert exclusive proprietary rights in extraterrestrial life that may be discovered. The *US Commercial Space Launch Competitiveness Act* of 2015 granted recognition to certain property rights in a "space resource" which is specifically defined to be abiotic, thereby excluding claims of ownership of any living organisms that may be discovered in ice or other materials extracted or mined on a celestial body (Jakhu and Pelton 2017 p. 406).

16.7.3. *Restrictions on areas of operations*

As it will be important for a commercial venture to protect and preserve the area after a discovery of evidence of life is made, it also may be questioned whether it will be important to prohibit in advance private sector activities from extracting resources and conducting other activities in areas such as special regions on Mars and other places that have special scientific interest pending scientific investigation. This question has implications beyond planetary protection and the search for life, as for historic and other unique sites of special interest. This question also has implications for the OST, which specifies in Article I that States have the right to explore and use all areas of celestial bodies, which could be effectuated through non-governmental entities. This right of free access, however, is subject to limitations set forth in the Treaty, including the environmental protection provisions in Article IX. Nevertheless, Article IX does not authorize a State to unilaterally declare an area off limits to all commercial activity, nor to bar other States from exploring and using the area, as such a declaration would violate the non-appropriation principle in Article II (Sterns and Tennen 2019).

A State could adopt a policy as part of its national authorization and supervision regime that restricted or prohibited access by its licensees to specific areas. COSPAR could play a role at the international level by considering guidelines that could be recommended to States for inclusion in their national licensing regimes. A more formal instrument, if deemed necessary, could be prepared by the international community in an appropriate forum such as COPUOS.

16.7.4. *Disclosure of locations and operations*

The consideration of New Space ventures vis-à-vis planetary protection should include the need for some method of advance notice of the areas to be impacted by commercial activities prior to the issuance of authorizations. This is necessary to provide the scientific community with information and allow an appropriate opportunity for any concerns about proposed missions to be communicated to the regulating agency prior to the issuance of any authorizations.

The interests of scientific integrity would be served by a requirement for non-governmental entities to prepare and disclose an inventory of biological matter and hazardous materials on-board a spacecraft. Commercial ventures also should be required to provide detailed pre- and post-mission environmental impact assessments for activities conducted on celestial bodies. Further in this regard, the amount of disturbance to a celestial environment that can be tolerated from New Space, and other commercial operations will need to be determined, as well as whether private entities should be required to remove their space objects and equipment at the end of the venture, and if remediation and restoration should be required of any celestial body surface and subsurface which was altered or disturbed (Tennen 2020).

16.7.5. *Limitations on otiose and vanity payloads*

One final aspect of New Space and planetary protection is whether there should be limits and prohibitions on the uses that can be made of, and the materials that can be transported and deposited on, celestial bodies, including the Moon. This issue involves scientific as well as cultural, philosophical, religious, ethical and aesthetic concerns (Rummel et al. 2012) and is exemplified by transport "services" to the Moon which currently are being marketed. Two companies are peddling the transportation of human remains to the Moon for lunar "burial". One of these companies also wants to take the remains of pets for lunar interment, while a third company is selling canisters to send to the Moon that customers can fill with almost any items they desire. Canisters of cremated human remains carried by one of the companies are to be put together in a large group and placed inside a container with commemorative markers for each individual in the group, to become a permanent monument on the surface of the Moon.

Erecting a monument to vanity on the Moon filled with cremated human or pet remains and establishing a site as a dedicated repository for a space object filled with what may be considered as treasured mementoes or useless detritus present obvious planetary protection concerns. If the Moon can be utilized for the purpose of a shrine to conceit, the only limit on what can be sent for permanent emplacement, and the monuments to mark their locations, will be the financial ability of the sender, and if someone can afford it, there likely will be a commercial company willing to transport anything it can to the Moon for a profit. Unfortunately, a precedent for these kinds of activities was set with the Lunar Prospector mission in 1999. The spacecraft, which intentionally crashed into the ice deposits at the south pole of the Moon, carried a canister containing cremated remains of Dr. E. Shoemaker for "burial". Although this payload did not violate COSPAR policy, and with all respect for Dr. Shoemaker and the intent of the Lunar Prospector mission planners to honor his memory, this set a terrible precedent and was ill advised (Sterns 2000).

As a matter of policy, the interest of the global public demands that the Moon must not become a preferred venue for narcissistic displays of vanity to the detriment of humanity. Not only does Article I of the OST declare that the exploration and use of outer space, including the Moon and other celestial bodies, is the province of all mankind, and that all States have the right to explore and use space on a basis of equality and without discrimination of any kind, the *corpus juris spatialis* also recognizes that there is an obligation to respect the rights of future generations (Sterns and Tennen 2019). The deposit of otiose and useless material on the surface of the Moon for no purpose other than private profit is inherently provocative and threatens to destabilize the security of peaceful exploration and use of space (Tennen 2020). Commercial ventures of this kind raise issues regarding violation of the non-appropriation principle in Article II of the OST and create obstacles that can hinder the activities of others in contravention of Article IX. In recognition of these concerns, the Board of Directors of the International Institute of Space Law issued a formal Statement in July 2021 to provide guidance to regulatory authorities in the authorization and continuing supervision of commercial activities in space (International Institute of Space Law 2021). The Statement exhorts and implores States to consider the interests of the public, the scientific community and all other stakeholders, together with the consequences of authorizing certain activities for the future exploration and peaceful use of outer space. The protection of celestial environments is essential for the success of both scientific exploration as well as the sustainable utilization and economic development of the resources of space for the benefit of all mankind.

16.8. References

Babb, R.J., Erb, H., Howard, D. (2018). Cost reduction solutions in regard to PP for commercial companies. *IAF Paper*, No. IAC-18-F1.2.3.

Coustenis, A., Kminek, G., Hedman, N., Ammannito, E., Deshevaya, E., Doran, P.T., Grasset, O., Hayes, A., Lei, L., Nakamura, A. et al. (2019). The COSPAR Panel on Planetary Protection Role, Structure and Activities [Online]. Available at: https://cosparhq.cnes.fr/assets/uploads/2019/07/PPP_SRT-Article_Role-Structure_Aug-2019.pdf [Accessed 19 June 2022].

COSPAR (2005). Planetary protection policy October 2002 (as amended 2005) [Online]. Available at: https://w.astro.berkeley.edu/~kalas/ethics/documents/environment/COSPAR%20Planetary%20Protection%20Policy.pdf [Accessed 19 June 2022].

COSPAR (2021). Planetary protection policy [Online]. Available at: https://cosparhq.cnes.fr/assets/uploads/2021/07/PPPolicy_2021_3-June.pdf [Accessed 19 June 2022].

Dobrokhotsky, O.N., Dyatlov, I.A., Orlov, O.I., Novikova, N.D., Hamidullina, N.M., Deshevaya, E.A. (2012). Ensuring biosafety in the study of samples of extraterrestrial origin by an example of preparation for the "Phobos-Grunt" Mission [Online]. Available at: https://cyberleninka.ru/article/n/obespechenie-biologicheskoy-bezopasnosti-pri-issledovanii-materialov-vnezemnogo-proishozhdeniya-na-primere-podgotovki-ekspeditsii/viewer (in Russian) [Accessed 19 June 2022].

Hofmann, M., Retberg, P., Williamson, M. (eds) (2011). *Protecting the Environment of Celestial Bodies: The Need for Policy and Guidelines*. International Academy of Astronautics, Paris.

International Institute of Space Law (2021). Statement of the Board of Directors on the consideration of the interests of the public and other stakeholders in the authorization and continuing supervision of commercial space activities [Online]. Available at: https://iislweb.space/wp-content/uploads/2021/07/IISL_Statement_Authorization_and_continuing_supervision_2021.pdf [Accessed 19 June 2022].

Jakhu, R.S. and Pelton, J.N. (eds) (2017). *Global Space Governance: An International Study*. Springer International Publishing AG, Cham.

Johnson, C.D., Porras, D., Hearsay, C.M., O'Sullivan, S. (2019). The curious case of the transgressing tardigrades (part 2) [Online]. Available at: https://www.thespacereview.com/article/3786/1 [Accessed 19 September 2022].

Long, G.A. (2016). The meaning of life and close encounters of the commercial kind. In *Proceedings of the International Institute of Space Law 2015*, Jorgenson, C.M. (ed.). Eleven International Publishing, The Hague.

Mangus, S. and Larsen, W. (2004). Lunar receiving laboratory project history [Online]. Available at: https://www.lpi.usra.edu/lunar/documents/lunarReceivingLabCr2004_208938.pdf [Accessed 19 June 2022].

Marchisio, S. (2017). Article IX. In *Cologne Commentary on Space Law: In Three Volumes*, Vol. 1, Hobe, S., Schmidt-Tedd, B., Schrogl, K.-U. (eds). Carl Heymanns Verlag, Cologne.

Meltzer, M. (2012). When biospheres collide: A history of NASA's planetary protection programs [Online]. Available at: http://www.nasa.gov/connect/ebooks/when_biospheres_collide_detail.html [Accessed 19 June 2022].

Moore, A.L. and Leaphart, J.V. (1983). Manipulation and modification of the outer space environment: International legal considerations. In *Proceedings of the 25th Colloquium on the Law of Outer Space*. American Institute of Aeronautics and Astronuatics, Reston.

MSR Science Planning Group (2019). Science-driven contamination control issues associated with the receiving and initial processing of the MSR samples [Online]. Available at: https://mepag.jpl.nasa.gov/reports.cfm [Accessed 19 June 2022].

NASA (2008). Biological Contamination Control for Outbound and Inbound Planetary Spacecraft, NASA Policy Directive (NPD) 8020.7G [Online]. Available at: https://nodis3.gsfc.nasa.gov/npg_img/N_PD_8020_007G_/N_PD_8020_007G__main.pdf [Accessed 19 June 2022].

NASA (2017). Planetary Protection Provisions for Robotic Extraterrestrial Missions, NASA Procedural Requirements (NPR) 8020.12D / NASA Interim Directive (NID) 8020.109A [Online]. Available at: https://nodis3.gsfc.nasa.gov/OPD_Docs/NID_8020_109A_.pdf [Accessed 19 June 2022].

NASA (2019). Report of the Planetary Protection Independent Review Board [Online]. Available at: https://www.nasa.gov/sites/default/files/atoms/files/planetary_protection_board_report_20191018.pdf [Accessed 19 June 2022].

NASA (2020a). Planetary protection categorization for robotic and crewed missions to the Earth's Moon, NASA Interim Directive (NID) 8715.128 [Online]. Available at: https://nodis3.gsfc.nasa.gov/OPD_docs/NID_8715_128_.pdf [Accessed 19 June 2022].

NASA (2020b). Biological planetary protection for human missions to Mars, NASA Interim Directive (NID) 8715.129 [Online]. Available at: https://nodis3.gsfc.nasa.gov/OPD_docs/NID_8715_129_.pdf [Accessed 19 June 2022].

NASA (2021). Planetary Protection Provisions for Robotic Extraterrestrial Missions, NASA Procedural Requirements (NPR) 8715.24 [Online]. Available at: https://nodis3.gsfc.nasa.gov/npg_img/N_PR_8715_0024_/N_PR_8715_0024_.pdf [Accessed 19 June 2022].

Rummel, J.D., Race, M.S., Horneck, G. (2012). Ethical considerations for planetary protection in space exploration: A workshop. *Astrobiology*, 12(11), 1017–1023.

Russian Federal Law (n.d.). "On space activities", Decree No. 5663-1 [Online]. Available at: http://www.unoosa.org/oosa/en/ourwork/spacelaw/nationalspacelaw/russian_federation/decree_5663-1_E.html [Accessed 19 June 2022].

Sterns, P.M. (2000). The scientific/legal implications of planetary protection and exobiology. In *Proceedings of the 42nd Colloquium on the Law of Outer Space*. American Institute of Aeronautics and Astronuatics, Reston.

Sterns, P.M. and Tennen, L.I. (1988). Principles of protection of the outer space environment in the corpus juris spatialis. In *Proceedings of the 30th Colloquium on the Law of Outer Space*. American Institute of Aeronautics and Astronuatics, Reston.

Sterns, P.M. and Tennen, L.I. (1998). Exobiology and the outer space treaty: From planetary protection to the search for extraterrestrial life. In *Proceedings of the 40th Colloquium on the Law of Outer Space*. American Institute of Aeronautics and Astronuatics, Reston.

Sterns, P.M. and Tennen, L.I. (2007). The future of planetary protection: Is there reason for optimism? In *Proceedings of the 49th Colloquium on the Law of Outer Space*. American Institute of Aeronautics and Astronuatics, Reston.

Sterns, P.M. and Tennen, L.I. (2019). Lacuna in the updated planetary protection policy and international law. *Life Sciences in Space Research*, 23(10), 10–21 [Online]. Available at: https://www.sciencedirect.com/science/article/pii/S2214552418301007 [Accessed 27 September 2021].

Tennen, L.I. (2004). Evolution of the planetary protection policy: Conflict of science and jurisprudence? *Advances in Space Research*, 34, 2354–2362.

Tennen, L.I. (2020). The role of COSPAR for space security and planetary protection. In *Handbook of Space Security. Policies, Applications and Programmes*, Schrogl, K.-U. (ed), 2 Volumes, 2nd edition. Springer, New York.

17

Military Activities in Outer Space: Legal Aspects

Georges D. KYRIAKOPOULOS
National and Kapodistrian University of Athens, Greece

17.1. Introduction

Inter-state relations in the context of outer space are part of a broader framework of international relations and, in many cases, can affect States' policies on Earth. According to Article III of the Treaty on Principles Governing the Activities of States in the Exploration and Use of Outer Space, including the Moon and Other Celestial Bodies of 19 December 1966 (OST) – which provides for the application, in outer space, of international law "including the Charter of the United Nations, in the interest of maintaining international peace and security and promoting international cooperation and understanding" (OST, Art. III) – international security in outer space must be approached in the context of the fundamental obligation to "maintain international peace and security" (UN Charter, Preamble, Art. 1 and Art. 2, paras. 3 and 4). It is thus obvious that in the light of the applicability of the UN Charter in space affairs, fundamental principles of international law such as the prohibition of the use of force in international relations (UN Charter, Art. 2, para. 4) or the "inherent" right of States to self-defense (UN Charter, Art. 51) also apply with respect to the activities of States in outer space.

With regard to the military uses of outer space, the fundamental provision is that of Article IV OST:

Space Law,
coordinated by Thomas LECLERC. © ISTE Ltd 2023.

> States Parties to the Treaty undertake not to place in orbit around the Earth any objects carrying nuclear weapons or any other kinds of weapons of mass destruction, install such weapons on celestial bodies, or station such weapons in outer space in any other manner.
>
> The Moon and other celestial bodies shall be used by all States Parties to the Treaty exclusively for peaceful purposes. The establishment of military bases, installations and fortifications, the testing of any type of weapons and the conduct of military manoeuvres on celestial bodies shall be forbidden. The use of military personnel for scientific research or for any other peaceful purposes shall not be prohibited. The use of any equipment or facility necessary for peaceful exploration of the Moon and other celestial bodies shall also not be prohibited (OST, Art. IV).

The first paragraph of Article IV of the OST prohibits the placement in orbit around the earth of nuclear weapons or weapons of mass destruction (WMDs) and the installation of such weapons in the celestial bodies. However, this provision does not extend to the placement of conventional weapons. On the contrary, the demilitarization regime that is being established for the Moon and the other celestial bodies is complete, similar to that of Antarctica. Indeed, the second paragraph of Article IV requires States to use the Moon and other celestial bodies exclusively for peaceful purposes. The installation of military bases, the testing of all types of weapons and the conduct of military exercises are prohibited.

The provision of paragraph 2 of Article IV reflects previous resolutions of the United Nations General Assembly (UNGA): UNGA resolution 1348 (XIII) of 1958, UNGA resolution 1472 (XIV) of 1959 and UNGA resolution 1962 (XVIII) of 1963, the latter being a "Declaration of Legal Principles Governing the Activities of States in the Exploration and Use of Outer Space". However, none of the aforementioned "Principles" refers explicitly to the "peaceful uses" of outer space, which is mentioned only in one paragraph of the preamble: "recognizing the common interest of all mankind in the progress of the exploration and use of outer space for peaceful purposes".

It is clear that Article IV of the OST does not prohibit any military use of outer space and the celestial bodies, the "peaceful" character being open to multiple interpretations. As noted, Article IV, paragraph 1 of the OST clearly prohibits the placement of nuclear weapons or WMDs in orbit around the Earth, their installation on the celestial bodies or their stationing in outer space, in any other means. However, this provision does not appear to cover the use of "direct-ascent" missiles, launched from the surface of the Earth and carrying nuclear weapons or WMDs in outer space. Furthermore, the wording of Article IV, paragraph 2 of the OST does not exclude the military uses of outer space in toto. Under said provision, States are prohibited from deploying nuclear weapons or any kind of WMDs on the celestial bodies. In addition, celestial bodies must be "used exclusively for peaceful purposes". In this regard, the creation of military bases, installations and fortifications, the testing of any kind of weapons and the conduct of military exercises on the celestial bodies are prohibited. States are entitled to use equipment or facilities "necessary for peaceful exploration of the Moon and other celestial bodies" (OST, Art. IV, para. 2 and Moon agreement, Art. 3, paras. 1, 2 and 4). According to the prevailing interpretation and State practice, "peaceful" has the meaning of "non-aggressive". This approach has the merit of considering the (broad) practice to put reconnaissance satellites for military purposes into orbit as compatible with international law.

17.2. Weaponization of outer space

Despite Article IV of the OST, the issue of weaponization of outer space is of particular concern to the international community, in the context of which initiatives have been repeatedly taken to prevent an arms race in outer space (PAROS).

Although the use of celestial bodies for non-peaceful purposes is prohibited (OST, Art. IV), the issue is far from being settled. It is true that States in possession of the necessary technical infrastructure and the means to develop space military activities – such as the United States (US), Russia or China – have so far avoided putting weapons in outer space. However, the international legal framework in force does not prevent States from deploying conventional weapons in space, and related scenarios have developed in military circles. Therefore, the issue of placing weapons in outer space always remains an open question. In this respect, reference should be made to a recent US legislative initiative to create a Space Force.

The US Space Policy Directive-4 of 2019 justifies this initiative as follows:

> It is imperative that the United States adapt its national security organizations, policies, doctrine, and capabilities to deter aggression and protect our interests. Toward that end, the Department of Defense shall take actions under existing authority to marshal its space resources to deter and counter threats in space, and to develop a legislative proposal to establish a United States Space Force as a sixth branch of the United States Armed Forces within the Department of the Air Force (Trump White House 2019).

Further, the French Space Command (CDE), created in 2019, in order to "better protect [French] satellites, including in an active way" (President Emmanuel Macron), launched its first military space exercise (AsterX) in March 2021, with the participation of the US Space Force and the German Space Situational Awareness Centre.

On the other hand, the North Atlantic Treaty Organization's (NATO) overarching Space Policy constitutes an illustrative example on how space security can be understood at the regional level (NATO 2022). The foundations of this policy were laid at the 2018 NATO Brussels Summit, when NATO leaders recognized that "space is a highly dynamic and rapidly evolving area, which is essential for the Alliance's security". NATO's Space Policy was adopted at the June 2019 Defence Ministers' meeting. At the December 2019 NATO Summit in London, NATO Member States declared space "a fifth operational domain, alongside air, land, sea and cyberspace, recognising its importance in keeping [NATO States members] safe and tackling security challenges, while upholding international law".

In any case, it is reiterated that military activities in outer space that would be permissible under space law must also be compatible with Article 2, paragraph 4 of the UN Charter. In other words, they must not be of an aggressive character. In addition, the "inherent" right to self-defense, in accordance with Article 51 of the UN Charter and customary international law, also applies to space activities, despite the restrictions imposed by Article IV of the OST.

Military, non-aggressive, activities of States in outer space have existed since the beginning of the space adventure of humankind, mainly through the use of reconnaissance and early warning satellites. However, the term "weaponization of outer space" refers to the placement of offensive weapons in space, including the development of weapons systems on Earth whose mission is to destroy targets in space. Anti-satellite weapons (ASAT) is a case in point, as they have already been used to destroy (friendly) satellites in orbit as part of military tests.

International practice records significant incidents of use of ASAT for test purposes. In January 2007, China destroyed an old meteorological satellite (Feng Yun 1C) with a ballistic missile. The destruction of the Chinese satellite took place in low earth orbit, at an altitude of 850 km above the Earth's surface and created a significant amount of space debris. On February 21, 2008, USA-193, a US military reconnaissance satellite, which was launched on December 14, 2006 and malfunctioned, was destroyed by a modified ballistic missile fired from a US warship. It should be noted that prior to the destruction of the satellite, the US Department of Defense (DoD) had justified the operation by stating that the destruction of USA-193 was necessary due to the fact that the satellite contained the hazardous material hydrazine, which would have survived re-entry. On March 27, 2019, India destroyed a (Indian) low-orbit satellite with a ground-to-space missile. According to the National Aeronautics and Space Administration (NASA) estimates, the Indian anti-satellite test created about 400 fragments of space debris, capable of endangering the International Space Station. On April 5 and December 5, 2020, the US Space Command announced that Russia had tested direct ascent anti-satellite missiles, with no apparent target. Finally, on November 15, 2021, another Russian anti-satellite test took place, targeting a Russian satellite and generating a considerable amount of space debris, which alarmed the crew of the International Space Station (ISS).

In addition, according to existing scenarios, space devices can be used to destroy targets on Earth (e.g. to intercept missiles), while cyberattacks against satellites in orbit are also a serious concern for States. The situation is further complicated by the fact that most space systems are of "dual-use" (civil/military).

17.3. International action for the "prevention of an arms race in outer space"

This issue has been on the agenda of the Conference on Disarmament (CD) since the early 1980s. In 1985, a committee was set up within the Conference in order to address the key issues related to the PAROS, such as the protection of satellite systems, the use of nuclear power sources in space activities as well as the adoption of relevant confidence-building measures. It is noteworthy that the United States openly opposed the Commission's mandate, choosing the path of direct talks with the (then) Soviet Union.

In 2008, China and Russia jointly proposed to the CD a draft "Treaty on Prevention of the Placement of Weapons in Outer Space and of the Threat or Use of Force against Outer Space Objects" (PPWT). According to Article II of this draft, States would undertake the obligation "…not to place in orbit around the Earth any objects carrying any kinds of weapons, not to install such weapons on celestial bodies and not to place such weapons in outer space in any other manner" (CD 2008). A revised draft was submitted to the CD by Russia on June 10, 2014.

17.4. Transparency and confidence-building measures in outer space activities

In 1990, the UNGA requested the Secretary General (SG) to carry out a study on potential confidence-building measures in outer space. In this respect, a group of governmental experts (GGE) delivered a report in 1993[1]. Further, the UNGA resolution 68/68 of 2011 asked again the SG to set up a GGE to conduct a Transparency and Confidence-Building Measures (TCBMs) survey. The new GGE submitted its report on July 29, 2013[2]. The report contains a set of TCBMs in outer space, including the proposal to establish coordination between the UN Office for Disarmament Affairs (UNODA), the UN Office for Outer Space affairs (UNOOSA) and other appropriate UN entities. Among the measures proposed, particular reference is made to (a) exchanges of information on the principles and goals of a State's outer space policy, on major military outer space expenditure and other national security space activities or on the forecast of natural hazards in outer space; (b) expert visits to space launch sites; (c) invitation of international observers to launch sites, flight command and control centers; (d) consultative mechanisms; and (e) measures of coordination.

The UNGA further encouraged the implementation of the TCBMs through UNGA resolution 68/50 of 2013, which urged Member States to review and implement, "to the greatest extent practicable", the proposed TCBMs, "through relevant national mechanisms, on a voluntary basis and in a manner consistent with the national interests of Member States".

Obviously, the proposed TCBMs constitute non-legally binding voluntary measures. However, as mentioned in the GGE report, "such measures can augment the safety, sustainability and security of day-to-day space operations and can contribute both to the development of mutual understanding and to the strengthening of friendly relations between States and peoples".

In 2017, a new GGE was established on the same matter, pursuant UNGA resolution 72/250 of 2017, in order to "consider and make recommendations on substantial elements of an international legally binding instrument on the prevention of an arms race in outer space, including, inter alia, on the prevention of the placement of weapons in outer space". The GGE met in two sessions, the first from 6 to 17 August 2018 and the second from 18 to 29 March 2019. However, no

1 UNGA (1993). Study on the application of confidence-building measures in outer space, A/48/305.

2 UNGA (2013). Group of Governmental Experts on Transparency and Confidence-Building Measures in Outer Space Activities, A/RES/68/189.

consensus was reached on a substantive report[3]. On December 7th, 2020, the UN General Assembly adopted resolution A/RES/75/36 on “Reducing space threats through norms, rules and principles of responsible behaviours”. Said resolution encouraged Member States, inter alia, “to... characterize actions and activities that could be considered responsible, irresponsible or threatening and their potential impact on international security, and share their ideas on the further development and implementation of norms, rules and principles of responsible behaviours and on the reduction of the risks of misunderstanding and miscalculations with respect to outer space”.

In line with the aforementioned resolutions of the UNGA, a European Union initiative, a Draft International Code of Conduct for Outer Space Activities, has also highlighted the importance of adopting TCBMs in outer space in order to “prevent confrontation and foster national, regional and global security and stability” (EU 2014). In particular, general Principle 28 of the Code emphasized “the responsibility of States, in the conduct of scientific, civil, commercial and military activities, to promote the peaceful exploration and use of outer space for the benefit, and in the interest, of humankind and to take all appropriate measures to prevent outer space from becoming an arena of conflict”.

Through repeated resolutions, the UNGA, recognizing the importance of TCBMs in outer space, constantly urges Member States, “in particular those with major space capabilities, to contribute actively to the objective of the peaceful use of outer space and of the prevention of an arms race in outer space and to refrain from actions contrary to that objective and to the relevant existing treaties in the interest of maintaining international peace and security and promoting international cooperation” (see UNGA resolution 74/32 of 2019, UNGA resolution 75/35 of 2020 and UNGA resolution 76/22 of 2021).

17.5. References

Blount, P.J. (2012). Targeting in outer space: Legal aspects of operational military actions in space. *Harvard National Security Journal Features*, 1–22.

3 For relevant positions expressed by States, see Annex II to the report of the Group of Governmental Experts on further practical measures for the prevention of an arms race in outer space, Report by the Chair of the Group of Governmental Experts on further practical measures for the prevention of an arms race in outer space, United Nations General Assembly, 9 April 2019, A/74/77.

Bourbonnière, M. and Lee, R.J. (2008). Legality of the deployment of conventional weapons in earth orbit: Balancing space law and the law of armed conflict. *European Journal of International Law*, 18(5), 873–901.

CD (2008). Draft "Treaty on Prevention of the Placement of Weapons in Outer Space and of the Threat or Use of Force against Outer Space Objects (PPWT)" [Online]. Available at: https://digitallibrary.un.org/record/633470?ln=en [Accessed 28 June 2022].

Cheng, B. (2000). Military uses of outer space: Article IV of the 1967 Space Treaty Revisited. In *The Utilization of the World's Air Space and Free Outer Space in the 21st Century*, Cheng, C.-J. and Kim, D.H. (eds). Kluwer, The Hague.

EU (2014). Draft International Code of Conduct for Outer Space Activities [Online]. Available at: https://www.eeas.europa.eu/sites/default/files/space_code_conduct_draft_vers_31-march-2014_en.pdf [Accessed 28 June 2022].

Goh, G.M. (2004). Keeping the peace in outer space: A legal framework for the prohibition of the use of force. *Space Policy*, 20(4), 259–278.

Gunasekara, S.G. (2012). Mutually assured destruction: Space weapons, orbital debris, and the deterrence theory for environmental sustainability. *Air & Space Law*, 37, 141–164.

Koplow, D.A. (2008–2009). ASAT-isfaction: Customary international law and the regulation of anti-satellite weapons. *Mich. J. Int'l. L.*, 30, 1187–1272.

Koplow, D.A. (2018). The fault is not in our stars: Avoiding an arms race in outer space. *Harvard International Law Journal*, 59(2), 331–388.

Kyriakopoulos, G.D. (2015). The current legal regulation on the use of ASAT weapons under international law and space law. In *Global Space Governance*, Jakhu, R., Chen, K.-W., Nyampong, Y. (eds). McGill Centre for Research in Air and Space Law Monograph Series III, Quebec.

Maogoto, J.N. and Freeland, S. (2008). From Star Wars to space wars – The next strategic frontier: Paradigms to anchor space security. *Air & Space Law*, 33(1), 10–37.

Mehdi, M. and Su, J. (2020). Prevention of an arms race in outer space and developing countries. *The Korean Journal of Defense Analysis*, 32(2), 253–270.

NATO (2022). NATO's overarching Space Policy [Online]. Available at: https://www.nato.int/cps/en/natohq/official_texts_190862.htm?selectedLocale=en [Accessed 28 June 2022].

Petras, M. (2002). The use of force in response to cyber-attack on commercial space systems – Reexamining "self-defense" in outer space in light of the convergence of U.S. military and commercial space activities. *J. Air L. & Com.*, 67, 1213–1268.

Schrogl, K.-U. and Neumann, J. (2009). Article IV. In *Cologne Commentary on Space Law*, Vol. I, Hobe, S., Schmidt-Tedd, B., Schrogl, K.-U. (eds). Carl Heymanns Verlag, Cologne.

Trump White House (2019). Space Policy Directive-4: Establishment of the United States Space Force [Online]. Available at: https://trumpwhitehouse.archives.gov/presidential-actions/text-space-policy-directive-4-establishment-united-states-space-force/ [Accessed 28 June 2022].

18

Outer Space in the Face of Contemporary Conflicts: Limits and Impasses of International Law

Julien ANCELIN[1] and Chloé DUFFORT[2]

[1] *Laboratoire de droit international et européen, Institut de la Paix et du Développement, Université Côte d'Azur, Nice*

[2] *Institut d'études politiques, Bordeaux, France*

18.1. Introduction

Has the principle of the peaceful use of outer space gone out the window (Imhof 2019)? Inscribed in the first recitals of the Treaty on Principles Governing the Activities of States in the Exploration and Use of Outer Space, including the Moon and other celestial bodies (1967 Treaty), according to which States acknowledge "the common interest of all mankind in the progress of the exploration and use of outer space for peaceful purposes" (1967 Treaty, Preamble), this principle is today confronted with State practices that directly challenge it.

Several events that have occurred in this geographical area in recent months seem to attest to an acceleration of belligerent practices. In November 2021, the International Space Station (ISS) was forced to modify its trajectory, due to a risk of collision with debris coming from the destruction of a satellite, caused by the firing into outer space of a Russian Direct-Ascent Anti-Satellite Missile (DA ASAT), and

Space Law,

this movement was repeated in June 2022[1]. In March 2022, during the first maneuvers of the armed conflict in Ukraine, the authorities of the Roscosmos Agency of the Russian Federation threatened the ISS with an "uncontrolled deorbit" in case of difficulties in cooperation with the other parties. Prior to this, in the summer of 2021, a Chinese military exercise to test the effectiveness of a "fractional orbital bombardment system" from outer space, employing a Long March 2C rocket launched from a hypersonic glider capable of carrying a nuclear payload, was widely documented (Davis 2020; Lagneau 2021). These recent examples demonstrate that space, and the area between space and the airspace (high-altitude

---

* The authors wish to inform the readers that this article was submitted to the editor at the beginning of the year 2022, which explains why the research has not been updated since. For this reason, it is important to place this article in the context in which it was written in order to obtain an objective reading. The examples mentioned here should not be seen as guidance for the reading of the article, but as proven if not exhaustive facts.

The geopolitical context has, as everyone knows, greatly evolved since 2022, and more and more links can be made between the provision of satellite services and the conduct of hostilities in theaters of military operations on land, in the context of armed conflicts.

Furthermore, in an effort to preserve international peace and security in space, following an anti-ASAT moratorium initiated by the U.S. government, a United Nations resolution for the prohibition of direct-ascent anti-satellite missile tests with destructive intent called for "all States to continue to use the relevant forums to discuss, establish and develop new practical measures to reduce risk and prevent both conflicts and a potential arms race in outer space. These could include transparency and confidence-building measures and additional moratoria, contributing to the development of legally binding instruments for the prevention of an arms race in outer space in all its aspects" (7 December 2022).

At the UNODA in Geneva, an open-ended working group is already seeking consensus on the creation of norms of responsible behavior to reduce the threat of misunderstandings in Outer Space and avoid international conflict.

1 See, in this sense, the declaration, on behalf of the European Union, of the High Representative of the Union for Foreign Affairs and Security Policy concerning the anti-satellite missile test launch carried out by Russia on November 15 and 19, 2021. According to the document: "The European Union strongly condemns the Russian Federation's conduct of a kinetic direct-ascent anti-satellite (ASAT) weapon its own satellite, COSMOS 1408, resulting in its destruction by a missile, as a clear act of irresponsible behavior in outer space". Council of the EU (2021) [Online]. Available at: https://www.consilium.europa.eu/en/press/press-releases/2021/11/19/statement-by-the-high-representative-of-the-union-for-foreign-affairs-and-security-policy-on-behalf-of-the-eu-on-the-russian-anti-satellite-test-on-15-november-2021/ [Accessed 1 July 2022]. See, also, Lesage, N. (2022). L'ISS est toujours menacée par les débris du tir de missile russe, 7 mois après. *Numerama*, [Online]. Available at: https://www.numerama.com/sciences/1009984-liss-est-toujours-menacee-par-les-debris-du-tir-de-missile-russe-7-mois-apres.html [Accessed 1 July 2022].

zone), has become an operational military theater for some States, but also for some international organizations, in addition to being a major strategic issue[2].

From now on, armed conflict between contemporary Powers, whatever its intensity, is no longer confined to the geographical areas that are contiguous to us (land, sea, air). It tends to arise more permanently in areas that were not, until now, theaters of major opposition between nations. We already know that these “new theaters of military operations” (Gaillard-Sborowski et al. 2016) have singular characteristics, they employ aerospace devices – understood as objects capable of lifting and flying in airspace (during ascent and/or descent) and moving in outer space – and they are deployed in places hostile to any human presence. After having had massive recourse to aerial means throughout the 20th century, the main military powers are now turning to the operational opportunities offered by high altitude areas and outer space, which is becoming a decisive marker of influence on the international scene. The main powers are therefore constantly developing their space capabilities, notably by putting innovative combat means into orbit and by installing, on Earth, command structures specifically dedicated to the new space component of their military organizations (Laurent 2021). This change is not without consequences, and it brings many challenges for the branches of international law that apply to conflicts[3] and that were conceived under the influence of eminently earthly considerations. The “terrestrial” spectrum (Couston 2004) that has accompanied the axiological construction of this law thus generates multiple questions when it comes to analyzing spatial conflicts from a legal perspective.

In practice, this translates into great ambiguity in the interpretation of international law by certain space powers. Among the worrying findings, the weaponization of outer space[4] by some States comes to the fore (Steer and Hersch 2021). Despite the prohibition on placing certain categories of weapons in orbit – Article IV of the 1967 Treaty stipulates that the following should not be undertaken: “any objects carrying nuclear weapons or any other kinds of weapons of mass destruction” – and despite the prohibition on the use of outer space for hostile

2 NATO (2021). NATO’s approach to space [Online]. Available at: https://www.nato.int/cps/en/natohq/topics_175419.htm [Accessed 1 July 2022].

3 The contemporary international law that applies to armed conflicts provides a fundamental distinction between the rules that prevent its occurrence – the *Jus ad bellum* – and those that regulate its conduct – the *Jus in bello* – also referring to international humanitarian law (Kolb 1997; Sassoli 2007).

4 The weaponization of outer space refers to the placement in orbit of military installations or conventional armaments. This practice is not contrary to the prescriptions of the Space Treaty (and its subsequent practice by the space powers) which imposes a peaceful use understood as a non-aggressive activity and not as a non-military activity (Gorove 2008).

purposes, most of the major powers pursue strategic goals in outer space and sometimes play directly on these prohibitions (Achilleas and Hobe 2020, p. 35). These practices have been presented as a set of behaviors that violate international law by a part of the doctrine (Cumin 2019, p. 82). This tendency tends to demonstrate the flaws or inadequacies of the existing normative framework. In other words, the fact that the applicable law is no longer sufficient to prevent the conduct of real space operations with strong unfriendly potentialities, meaning that a certain contrary practice tends to take hold in outer space[5]. Thus, this tendency to variable interpretations of the 1967 Treaty seriously confronts the guarantee of international peace and security promoted by international law, without being able to recognize the illicit, since the threshold of conflictuality required to legally qualify an armed conflict has not been reached. On the other hand, we may ask whether the application of *Jus ad bellum* – mainly the law of collective security intended to prevent threats to international peace and security – remains effective in the face of contemporary threats of conflict, or whether it is now necessary to reclassify these threats as hostile acts in order to apply *Jus in bello* – which is specifically intended to regulate hostile acts between belligerent parties to an armed conflict, once the latter can be legally qualified[6].

Without this legal reclassification, positive international law states that as long as there is no armed conflict in outer space, the management of conflicts that do not reach this threshold of hostility remains the responsibility of the United Nations (UN) Charter. However, the introduction of a climate of exacerbation of conflictualities (Schu 2020) in contemporary international relations casts doubt on the capacity of collective security law to prevent the risks of escalation.

The analysis of the ambiguity of this situation makes it possible to identify two explanatory cases: either the conduct of space operations with hostile potentialities in a context of peace is only permitted because of the lack of political credibility that

5 Hitchens, T. (2021). Biden administration to propose new global norms for military space [Online]. Available at: https://breakingdefense.com/2021/12/biden-administration-to-propose-new-global-norms-for-military-space/ [Accessed 1 July 2022]. See, also, NASA (2020). The Artemis Accords - Principles for cooperation in the civil exploration and use of the moon, mars, comets, and asteroids for peaceful purposes [Online]. Available at: https://www.nasa.gov/specials/artemis-accords/img/Artemis-Accords-signed-13Oct2020.pdf [Accessed 1 July 2022].

6 NATO confirmed at the Madrid summit on 30 June 2022 that the existence of hostile operations in relation to space could reach the level of an armed attack and lead the North Atlantic Council to invoke Article 5 of the North Atlantic Treaty. NATO (2022). NATO Strategic Concept 2022 [Online]. Available at: https://www.nato.int/strategic-concept/ [Accessed 1 July 2022].

certain space powers give to the existing prerogatives in *Jus ad bellum* or the contemporary space situation suffers from a lack of legal qualification concerning certain operations. If the first case reveals the importance of the geopolitical weight of the question, the second case reveals the existing difficulties concerning the qualification in public international law of space operations with ambiguous character and effects.

In the absence of an appropriate legal qualification allowing the application of *Jus in bello* in outer space[7], any conflict situation that might develop there is likely to raise important difficulties. Nevertheless, as things stand, it is the law of the non-use of force that continues to be the sole means of preventing armed conflict.

18.2. Spatial conflict subject to the law of collective security

Space conflict can enter into resonance with the categories of the law of collective security, as it is contained in the UN Charter. The applicability of *Jus ad bellum* to events related to the conflict in space does not seem to be in doubt, and the mobilization of this body of rules which envisages the construction of peace through law may appear useful. This observation remains, however, dependent on a projection: no behavior in space has, so far, been qualified as a threat to international peace and security, likely to engage the action of the UN Security Council (SC) (Perez 2019)[8]. In the case of the United States, it is not possible to determine whether the use of the term "threat to international peace and security" in the context of space conflict is justified by the fact that the term "threat to international peace and security" is not defined in the UN Charter but rather by the resolutions of the Security Council, whose power to determine such a threat is largely discretionary.

Moreover, several elements of space law attest to the applicability of the law of collective security to the conflictual events that could take place there. The preamble of the 1967 Treaty, referring explicitly to Resolution 110 (II) of the UN General Assembly (UNGA) of November 3, 1947, recalls that the prohibition of acts

---

7 The term "international armed conflict" is a generic notion, notably envisaged by Article 2, paragraph 1 of the Geneva Convention (I) for the Amelioration of the Condition of the Wounded and Sick in Armed Forces in the Field, signed on 12 August 1949. This notion is thought of in a terrestrial context (wherever operations are deployed on land).

8 According to the author, "it should be pointed out that, for the time being, no space behaviour could be considered as a use of armed force or armed aggression, in the sense of the *jus ad bellum*, and that there has never yet been an armed conflict in space. While the above-described behavior of the Russian satellite appears to be unfriendly, it does not constitute a use of force".

intended or likely to "provoke or encourage any threat to the peace, breach of the peace or act of aggression" (1967 Treaty, Preamble) is applicable to outer space, without it being necessary to modify the general scheme. The same is true of Article II of the same Treaty, which provides for an obligation of conformity of acts performed by States with the UN Charter, with a view to "maintaining international peace and security and promoting international cooperation and understanding" (1967 Treaty, Art. II).

The law of collective security also considers space as a territory subject to its prescriptions. The UNGA thus recalled, in its Resolution 1348 (XIII) on the question of the peaceful use of outer space, dated December 13, 1958, the need "to avoid the extension of present national rivalries into this new field" and, consequently, to subject outer space to the normative mechanism of the UN Charter. Space does not therefore constitute a zone that is an exception to the rules classically applicable in matters of collective security, providing, if necessary, for the setting up of international police operations. Thus, if an armed aggression were to be characterized in outer space, in violation of Article 2§4 of the UN Charter, the SC could be led to exercise its office and authorize a State or a group of States to resort to force. This aggression could take many forms: an armed attack, whether its origin is terrestrial or space based; kinetic or not; committed using offensive means such as anti-satellite weapons, directed energy weapons, dazzling, jamming or usurpation techniques; or offensive cyber weapons. It should be noted, incidentally, that if the aggression that triggered the action also constituted a violation of Article IV of the 1967 Treaty, which requires the use of the Moon and other celestial bodies "exclusively for peaceful purposes" (1967 Treaty, Art. IV), the SC's reaction could be facilitated, since the act committed would generate cumulative violations.

The applicability of the law of collective security to space conflict implies, moreover, questioning the particularities of the perimeter of the applicable norms. Indeed, the mechanisms conceived in reaction to armed aggression for terrestrial application (principally the use of force decided by the UN Security Council by virtue of Chapter VII and self-defense governed by Article 51 of the UN Charter) could rapidly encounter important limits in a context of operation in outer space. Indeed, the outer space theater in which hostilities take place raises very specific characteristics, in particular with regard to the mainly dual nature of the objects that gravitate there or to the unpredictable effects that could be generated by the creation of space debris (on the Kessler effect, see Vigné 2021). In the event that the threat to international peace and security originates from outer space, the SC could order a series of coercive measures aimed at preserving or re-establishing international peace and security, which would have to be carefully articulated with the principle of freedom in outer space, an axiom of the functioning of the medium (Martin

2022). It would therefore be necessary to limit the mandate of the peace operation, decided by the SC by virtue of Chapter VII of the UN Charter, so that it does not exceed the limits imposed by the shared use of the zone, between peaceful activities and military operations. In terms of individual or collective self-defense, it is noted that the applicability of the principle does not avoid the occurrence of practical difficulties, arguing in favor of clarifications. The regime of self-defense could thus be largely modulated: the criteria of necessity and proportionality, if they are valid for outer space, do not cover the same aspects on Earth. With regard to necessity, we can quickly see that there is a difficulty that is tricky to overcome in terms of qualifying a legitimate military objective. In this context, the existence of installations, whose utility/use is mainly dual in outer space, risks preventing a clear determination of what the notion of legitimate military space target can cover. It is thus possible to question whether only the use of the space object for military purposes, in connection with the aggression, would be necessary to justify the military objective. The law of collective security does not hold a definitive position and could suffer from a perilous variable geometry application (Corten and Dubuisson 2002, p. 73). It is thus reasonable to consider that the assessment of the criterion of necessity should be increased in outer space. By virtue of the principle of peaceful use, a corollary of freedom of exploration, the neutralization of a military target in space, whose nature is essentially dual, could produce harmful effects for civilians on Earth. With regard to proportionality, other difficulties appear and demonstrate that the applicability of the device is not sufficient to dispel the difficulties. Indeed, the use of force by the aggressed State, in outer space, could have serious consequences, both terrestrial and spatial, notably because of the debris caused by the destruction of a legitimate target. Thus, targeting a satellite used for data transfers could produce much more damage if it is linked to the operation of critical installations[9].

9 At the European level, communication satellites and, more generally, the Space and Research sector appear respectively in 11th and 36th place of an indicative list of critical sectors of activity, drawn up by the Commission of the European Communities in 2005 in a "Green Paper" which has not had much resonance since then, perhaps because of the "sensitivity [of the subject] for the public". See European Commission (2005). Green Paper on a European Programme for Critical Infrastructure Protection, COM 576 final. In France, the regulatory authority defined "operators of vital importance" as well as "sectors of activity of vital importance". France (2007). Decree No. 2007-585 on certain regulatory provisions of the first part of the Defense Code. This stated that a sector of activities of vital importance to the nation consists of activities that "relate to the production and distribution of essential goods or services", "or [that] may present a serious danger to the population". France (2006). Decree No. 2006-212 on the security of activities of vital importance.

Finally, it should be noted that if the law of collective security postulates its applicability, the UN bodies remain aware of the need to adapt the normative framework to the risks induced by space conflicts. In this sense, the UNGA Resolution aimed at "reducing space threats" by building new norms was adopted on November 12, 2021[10]. Furthermore, a report by the UN Secretary General (SG), dated July 13, 2021, offered confirmation that several "actions and activities in space [...] could be considered as having "a potential impact on international security" and, more specifically, that the "growing military dependence [on outer space] exacerbates the risk of armed conflict ever being triggered in space or spreading there"[11].

On July 13, 2021, the expression "armed conflict" was thus decided on by the UN SG. This phrase is full of meaning for international law and initiates, through the recognition of a new interest, the prospect of a paradigm shift. By using the term "armed conflict", which falls exclusively within the scope of the *Jus in bello*, the UN SG lends credibility to the thesis of a space war and establishes it as a strategic reality[12]. Moreover, the UN SG "has the competence, according to the Charter, [...] to draw the attention of States or organs of the United Nations to any matter which may endanger international peace and security (Article 99)" (Cumin 2015, p. 213). In doing so, international armed conflict in space – such as armed conflict on Earth – is no longer a mere doctrinal hypothesis, nor a science fiction scenario[13]. The international community, through this report, is becoming aware that outer space can be considered as the new theater of military operations, in the context of a future international armed conflict (Gaillard-Sborowski et al. 2016). The law of collective security thus offers means to capture space conflictuality and prevents it through the possibility of actions directed toward peacekeeping. It is now appropriate to consider these new conflict spaces from the perspective of the rules likely to frame their conduct in order to highlight the challenges that the law of armed conflict will face.

10 UNGA (2021). Prevention of an arms race in outer space, A/76/442.

11 UNGA (2021). Reducing space threats through norms, rules and principles of responsible behaviors, A/76/77.

12 ASAT technology could easily be used for hostile purposes against foreign satellites, as evidenced by the ASAT launch by the Russian Federation. See Le Monde (2021). La Russie reconnaît avoir détruit un satellite avec un missile [Online]. Available at: https://www.lemonde.fr/sciences/article/2021/11/15/apres-un-tir-antisatellite-russe-les-etats-unis-fustigent-un-comportement-dangereux_6102181_1650684.html.

13 On the criteria expected by the international judge to qualify an armed conflict, see, ICTY, 1995, Prosecutor v. Dusko Tadic, *Arrêt relatif à l'appel de la défense concernant l'exception préjudicielle d'incompétence*, IT-94-1-A, para. 70.

18.3. Spatial conflict challenges the application of the law of armed conflict

It is now rare for the use of armed force between states to be accompanied by a formal declaration of war. This is why the four Geneva Conventions states that it shall be applicable "to all cases of declared war or of any other armed conflict which may arise between two or more of the High Contracting Parties" (Geneva Convention (I) of 1949, Art. 2, 1). In theory, the so-called "international" armed conflict is supposed to be the exception in a world where the state of peace is generally taught as the norm (Cumin 2015, p. 573). However, in practice, when situations disturbing the state of international peace could not be curbed through diplomatic means, many international armed conflicts are identified on Earth. An international armed conflict is legally qualified as soon as the use of armed force is observed between States, which are then considered by Geneva Conventions as belligerent parties. Several clarifications can be made at this stage. First, it is the exercise of sovereignty – and incidentally the protection of sovereignty – which is targeted by the use of force, in the absence of effective diplomatic means. Second, the use of force can take various forms and international law cannot anticipate them all. This is why international law and its commentators have indicated that with regard to the use of force, all acts of hostility against a State could be retained to qualify the international armed conflict[14] . It is therefore on the qualification of actions disturbing international peace as "hostile acts" that the application of international humanitarian law (IHL) depends.

Thus, in doctrine, there are several definitions of the notion of hostility. When viewed broadly, these definitions allow many military operations to be considered hostile acts when they harm the enemy, that is, the full exercise of its territorial sovereignty. Indeed, the doctrine has clarified that the use of a weapon is not necessary and that acts of violence are sufficient to characterize hostilities constituting an international armed conflict within the meaning of international

---

14 ICRC (2016). Commentary on the First Geneva Convention (I) for the Amelioration of the Condition of the Wounded and Sick in Armed Forces in the Field, 12 August 1949, Art. 2, 1, §§210-218. The commentators cite the work of the International Law Association in support of their observation. According to the document: "the factual existence of an armed conflict is sufficient for international humanitarian law to apply". They add that "Indeed, Article 2(1) [of the Conventions] emphasizes the primacy of the factual existence of an armed conflict over the formal status of war. Consequently, the demonstration of the existence of an armed conflict within the meaning of Article 2, paragraph 1, must be based solely on facts proving the *de facto* existence of *hostilities* between the belligerents, even without a declaration of war" (emphasis added). ILA (2010). Final Report on the Meaning of Armed Conflict in International Law, p. 2 (translation from French).

humanitarian law[15]. Although conceived in an era when conflict took place mainly on Earth (or in its immediate environment), these rules do not suffer from exclusions *ratione loci* that would prevent them from being projected outside the geographical framework for which they were theorized.

Thus, we can envisage questioning the space operations that would constitute a possible international armed conflict in outer space, and if necessary, the adaptability of international humanitarian law to this hitherto uncharted theater of military operations. Can it be said today, with regard to the concepts used by international law, that space operations could be the cause of an armed conflict? As we have seen, the Geneva Conventions are silent on the geographical limits of the field of application of their law[16]. On the other hand, several elements argue in favor of an extended applicability. Article 1, common to the four Geneva Conventions, mentions that the Parties undertake to ensure respect for these instruments "in all circumstances"; Article 2, on the other hand, states that the provisions apply in "any [...] conflict", which offers real possibilities for international humanitarian law, such as that of envisaging the application of its provisions in outer space. This solution, while a useful palliative, may nevertheless prove to be limited in terms of verifying compliance with the principles of international humanitarian law, given the distinct characteristics of ground and space operations.

15 According to the ICRC, the concept of hostilities refers to "the (collective) use by parties to a conflict of *methods and means to harm the enemy*". ICRC (1988). Dictionary of the International Law of Armed Conflict, Geneva, p. 57 (emphasis added). Or, according to the Commentaries on the First Protocol Additional to the Geneva Conventions, the term "hostilities" "covers not only the time when a civilian uses a weapon, but also, for example, the time when he carries it, as well as situations where the civilian engages in hostile acts *without using a weapon*". In the Dictionary of the International Law of Armed Conflict, hostilities are defined as "*acts of violence* committed by a belligerent against an enemy in order to break down his resistance and impose obedience". Finally, in Professor Jean Salmon's *Dictionnaire de droit international public*, hostilities are "all offensive or defensive acts and military operations carried out by a belligerent in the context of an armed conflict" (emphasis added) (translation from French).

16 The only reference to outer space as a medium in international law relating to the conduct of hostilities can be found in Article 2 of the Convention on the Prohibition of Military or Any Other Hostile Use of Environmental Modification Techniques, which entered into force on 5 October 5, 1978 (ENMOD Convention). It should also be noted that "this assertion [the applicability of international humanitarian law to outer space] is reinforced by an opinion of the International Court of Justice, which states that the principles governing armed conflict apply to all forms of warfare". ICJ, Opinion of 8 July 1996, *Legality of the Threat or Use of Nuclear Weapons*, § 86 (translation from French).

International humanitarian law is, first and foremost, a law that promotes a principle of humanity that must remain even during armed conflict. To this end, several rules must be observed by belligerents taking part in hostilities. In particular, the rules emanating from articles 35, 36 and 37 of the Protocol Additional to the Geneva Conventions of 12 August 1949, and relating to the Protection of Victims of International Armed Conflicts, adopted on June 8, 1977 (Protocol I), which relate to the prevention of superfluous evils, imposing that the choice of methods and means of warfare by the Parties should not be unlimited, the control of new weapons and the prohibition of perfidy. The same applies to the rules imposing respect for the cardinal principle of distinction between the civilian population, combatants, civilian objects and, finally, military objectives, which must be the sole object of an attack, if any (Protocol I, Art. 48). Belligerents must also respect the principles of proportionality and necessity in the conduct of the attack, which translates into a balance between the estimated loss of life, damage to civilian objects and the concrete and direct military advantage anticipated (Protocol I, Art. 51, 5, b and Art. 52). Moreover, international humanitarian law imposes on the parties the obligation to take precautions in the attack and its effects (Protocol I, Art. 57 and 58). Finally, IHL provides a protective framework for the natural environment in which conflicts take place, in order to minimize widespread, long-term and severe damage induced by battle (Protocol I, Art. 55).

Is it possible to transpose these principles, which were conceived for terrestrial armed conflicts, to outer space, given its physical particularities? Indeed, from the point of view of physics, outer space is different from terrestrial environments and obeys the "extreme" laws (Verrger et al. 2002, p. 7) of celestial mechanics that are specific to it. If, as we have already noted, there is nothing to prevent the applicability *ratione loci* of international humanitarian law to the "outer space" zone, on the other hand, the verification of the principles that it intends to defend proves, once again, to be complex in practice. Since humans have little or no control over space mechanics, how can one know if an operation, creating a multitude of space debris, will not do "widespread, long-term and severe" (Protocol I, Art. 35, paras. 2 and 3) damage to the environment in a few months, or even a few years? How can we assess the "study, development, acquisition or adoption of a new weapon, means or method of warfare" (Protocol I, Art. 36), when we are not able to legally define what a weapon in outer space is at this time? Furthermore, how can we recognize perfidy in space operations, when in outer space, civilian and military needs are so intimately linked by the dual use of goods, as nowhere else (Protocol I, Art. 37)? In this sense, does the 2017 approach of the Russian satellite *Louch-Olymp* to the Franco-Italian satellite *Athena-Fidus* constitute an act of perfidy, a "ruse of war" within the meaning of international humanitarian law (Protocol I, Art. 37, para. 2) or a simple false move? This kind of event also raises the question of the applicability of the principle of distinction between civilian, military and dual-use space assets. Incidentally, the questions of verification of the principles of proportionality, necessity and precaution in the attack of these assets also arise. We are confronted here with real difficulties for an applicability of the law that is coherent and in adequacy with the reality of the facts that we face in space. On the one hand, satellites are recognized as "critical installations"[17] on which the vital interests of the international community depend. This could lead one to believe that they are assets protected by international law, even if, to date, they are not subject to special protection under international humanitarian law (Protocol I, Art. 54). On the other hand, according to international humanitarian law, "military objectives are limited to those objects which by their nature, location, purpose or use make an effective contribution to military action and whose total or partial destruction, capture or neutralization, in the circumstances, offers a definite military advantage" (Protocol I, Art. 52, para. 2).

Therefore, how can the joint application of the principles of distinction, proportionality and military necessity be verified when targeting a satellite, while being assured that the attack will not cause "incidental loss of civilian life, injury to civilians, damage to civilian objects, or a combination thereof, which would be excessive in relation to the concrete and direct military advantage anticipated" (Protocol I, Art. 51, 5, b)?

18.4. Conclusion

In the 20th century, outer space was, as commonly described, a sanctuary space in which peace and security should reign, in accordance with the principles of the 1967 Treaty. However, certain political actors were already addressing the subject of space war (Baumel 1984). Such a projection aroused varied reactions, ranging from sarcastic rejection to moderate interest, considering the analysis as belonging exclusively to an ominous prospective.

Recently, we have seen a shift in both policy and legal analysis. This change may be due to at least two factors that are not necessarily mutually exclusive. On the one hand, it may be linked to the observation of exponential technological evolution in the space field and to the new possibilities offered by the exploration and trade of space resources. On the other hand, it may be linked to the desire to act in anticipation of the harmful effects of this evolution, by making the law correspond to the reality of the political and military ambitions of the space powers, in order to

17 European Commission (2005). Green Paper on a European Programme for Critical Infrastructure Protection, COM(2005) 576 final.

prevent its perilous obsolescence. The International Committee of the Red Cross, for example, referred in a 2019 report to the "humanitarian consequences and constraints under IHL related to the *potential* use of weapons in outer space"[18]. This effort to anticipate already seems far away as the potential situation, which was mentioned in 2019, has seen its likelihood increase sharply in recent months[19].

However, despite this acceleration, the elements developed in this chapter show that several questions remain. The choice of the applicable legal framework arises, in particular in the case where an ambiguous space operation does not strictly speaking consist of a use of armed force – in the classic "kinetic" sense of the term (Libicki 2014)[20] – but aims to achieve, in spite of everything, by concealed hostile means, a military objective[21].

Thus, positive law offers a variable geometry protection in outer space, linked to the interpretation that can be made of it. This perspective is eminently dangerous, especially for States that depend largely on the space vector on the geostrategic

18 ICRC (2019). International Humanitarian Law and the Challenges of Contemporary Armed Conflicts, 33IC/19/9.7, pp. 30ff (emphasis added).

19 UNGA (2021). Reducing space threats through norms, rules and principles of responsible behaviors, A/76/77, p. 5.

20 This refers to cyber operations, or any other usual neutralization attempts, such as space weapons by destination represented by debris, or the obstruction of the view of military satellites by civilian space assets.

21 See, in this sense, ICRC (2010). Customary International Humanitarian Law Database. Available at: https://shop.icrc.org/customary-international-humanitarian-law-database-flyer-print-en.html [accessed 1 July 2022]. According to Principle 8, "military objectives are limited to those objects which by their nature, location, purpose or use make an effective contribution to military action and whose total or partial destruction, capture or neutralization, in the circumstances ruling at the time, offers a definite military advantage". Notwithstanding the difficulty of the issue, the ICRC has taken a position and considers, in its comments published in 2020 to the Geneva Convention (I) for the Amelioration of the Condition of the Wounded and Sick in Armed Forces in the Field, August 12, 1949, that "it is generally accepted that when cyber operations have effects similar to conventional kinetic operations, they constitute an international armed conflict. Indeed, if such operations result in the destruction of civilian or military property or in the killing or wounding of soldiers or civilians, there is no reason to treat the situation differently from equivalent attacks conducted by more traditional means and methods of warfare" (§255). The commentary continues that "where, for example, a situation objectively reveals that a State is actually engaged in military operations or other hostile action against another State, neutralizing enemy military personnel or objects, impeding its military operations or using/controlling its territory, it is an armed conflict. [And that] the existence of an armed conflict must be inferred from the facts" (§ 241).

level. If in the 20th century they were considered as "space powers", in the 21st century, this asset could turn into a vulnerability. In other words, only those states that are less dependent than others on the space theater, but that have the technological capacity to act kinetically in outer space, could have an interest in triggering a conflict there. However, since countermeasures can be taken on Earth, it would not seem appropriate for anyone to start an armed conflict in space.

18.5. References

Achilleas, P. and Hobe, S. (eds) (2020). *Cinquante ans de droit de l'espace, Fifty Years of Space Law*. Travaux du Centre de recherche de l'Académie de droit international de la Haye, Brill Nijhoff, Leiden.

Baumel, J. (1984). Guerre spatiale, science-fiction ou réalité de demain. *Revue des deux mondes*, 70–79.

Bories, C. and Rapp, L. (eds) (2021). *L'espace extra atmosphérique et le droit international*. Pedone, Paris.

Bories, C., Eudes, M., Rapp, L., Rass-Masson, L. (eds) (2021). *Droit de l'espace extra-atmosphérique : questions d'actualité*. Presses de l'Université Toulouse 1 Capitole, Toulouse.

Corten, O. and Dubuisson, F. (2002). Opération "liberté immuable" : une extension abusive du concept de légitime défense. *Revue générale de droit international public*, 106(1), 51–77.

Couston, M. (2022). Droit spatial – Principes directeurs du droit international public de l'espace. Report, JurisClasseur Transport, LexisNexis.

Cumin, D. (2015). *Le droit de la guerre. Traité sur l'emploi de la force armée en droit international*, Vol. 1. L'Harmattan, Paris.

Cumin, D. (2019). Militarisation et arsenalisation de l'espace extraatmosphérique : perspectives stratégiques et éthico juridiques. *Journal international de bioéthique et d'éthique des sciences*, 30(3), 77–101.

Davis, S. (2020). Les armes hypersoniques : un défi technologique pour l'OTAN et ses pays membres ? Report, 039 STC 20 F rév. 2 fin.

Gaillard-Sborowski, F. et al. (2016). Sécuriser l'espace extra-atmosphérique : éléments pour une diplomatie spatiale à l'horizon 2030. Fondation pour la recherche stratégique. Final report No. 152/FRE/SEEA.

Gorove, S. (2008). International space law in perspective – Some major issues, trends and alternatives. *Collected Courses of the Hague Academy of International Law*, Vol. 181. Martinus Nijhoff, Leiden.

Hamann, A. (2016). Le statut juridique de la déclaration de guerre. *Jus Politicum Revue de droit politique* [Online]. Available at: http://juspoliticum.com/article/Le-statut-juridique-de-la-declaration-de-guerre-1086.html [Accessed 1 July 2022].

Imhof, P. (2019). L'utilisation militaire de l'espace extra-atmosphérique est-elle licite selon le droit international public ? *Revue québécoise de droit international*, 32(2), 203–241.

Kolb, R. (1997). Sur l'origine du couple terminologique *ius ad bellum/ius in bello*. *Revue internationale de la croix rouge*, 79, 593–602.

Lagneau, L. (2021). Particulièrement sophistiquée, la dernière arme hypersonique testée par la Chine intrigue le Pentagone. *Zone militaire* [Online]. Available at: http://www.opex360.com/2021/11/23/particulierement-sophistiquee-la-derniere-arme-hypersonique-testee-par-la-chine-intrigue-le-pentagone/ [Accessed 1 July 2022].

Laurent, J.-M. (2021). La frénésie du spatial de défense : prémonition d'un conflit spatial imminent ou intuition visionnaire d'une transformation du paradigme aérospatial ? *La revue diplomatique*, 13, 164–171.

Libicki, M. (2014). De Tallinn à Las Vegas, une cyberattaque d'importance justifie-t-elle une réponse cinétique ? *Hérodote*, 152–153(1), 221–239.

Martin, J.-C. (2022). De l'applicabilité et de l'adéquation du jus ad bellum à l'espace extra-atmosphérique. In *L'espace extra atmosphérique et le droit international*, Bories, C. and Rapp, L. (eds). Pedone, Paris.

Perez, L. (2019). L'application du droit des conflits armés à l'espace extra-atmosphérique. *Note de recherche IRSEM* [Online]. Available at: https://www.irsem.fr/data/files/irsem/documents/document/file/2966/NR_IRSEM_n69_2019.pdf [Accessed 1 July 2022].

Perez, L. (2021). Le régime juridique du conflit armé dans l'espace extra-atmosphérique : une consolidation du droit international humanitaire par le droit de l'espace. In *Droit de l'espace extra-atmosphérique : questions d'actualité*, Bories, C., Eudes, M., Rapp, L., Rass-Masson, L. (eds). Presses de l'Université Toulouse 1 Capitole, Toulouse.

Sassoli, M. (2007). *Ius ad Bellum* and *Ius in Bello* – The separation between the legality of the use of force and humanitarian rules to be respected in warfare: Crucial or outdated? In *International Law and Armed Conflict: Exploring the Faultlines, Essays in Honour of Yoram Dinstein*, Schmitt, M. and Pejic, J. (eds). Martinus Nijhoff, Leiden.

Schu, A. and Laurent, J.-M. (2020). *Demain la guerre ? Étude sur le risque de guerre entre les États-Unis, la Chine et la Russie*. Bord de l'eau, Bordeaux.

Steer, C. and Hersch, M. (2021). *War and Peace in Outer Space, Law, Policy and Ethics*. Oxford University Press, Oxford.

Verrger, F., Ghirardi, R., Sourbes-Verger, I. (2002). *L'espace, nouveau territoire. Atlas des satellites et des politiques spatiales*. Belin, Paris.

Vigné, M. (2021). La protection de l'environnement en cas de conflit armé dans l'espace extra-atmosphérique. In *Droit de l'espace extra-atmosphérique : questions d'actualité*, Bories, C., Eudes, M., Rapp, L., Rass-Masson, L. (eds). Presses de l'Université Toulouse 1 Capitole, Toulouse.

List of Authors

Philippe ACHILLEAS
Université Paris-Saclay
France

Julien ANCELIN
Laboratoire de droit international et européen
Institut de la Paix et du Développement
Université Côte d'Azur
Nice
France

Ntorina ANTONI
Eindhoven University of Technology
The Netherlands

Setsuko AOKI
Keio University Law School
Tokyo
Japan

Olavo DE O. BITTENCOURT NETO
Catholic University of Santos
Brazil

Laetitia CESARI ZARKAN
Université du Luxembourg
Faculty of Law, Economics and Finance
Luxembourg

Magda COCCO
Vieira de Almeida & Associados
Lisbon
Portugal

Helena CORREIA MENDONÇA
Vieira de Almeida & Associados
Lisbon
Portugal

Philip DE MAN
Centre for Global Governance Studies
University of Leuven
Belgium

Chloé DUFFORT
Institut d'études politiques
Bordeaux
France

Frans VON DER DUNK
Harvey and Susan Perlman Alumni
Lincoln
USA
and
Lazarski University
Warsaw
Poland

Steven FREELAND
Western Sydney University
and
Bond University
Robina
Australia

Daniel FREIRE E ALMEIDA
Catholic University of Santos
Brazil

Christina GIANNOPAPA
European Union Agency for
the Space Programme
Prague
Czech Republic

Yvon HENRI
ITU Radio Regulations Board
Échenevex
France

Armel KERREST
Université de Bretagne Occidentale
Brest
France

Georges D. KYRIAKOPOULOS
National and Kapodistrian
University of Athens
Greece

Thomas LECLERC
Université de Bretagne Occidentale
Brest
France

Thierry LEMAIRE
General Counsel
Marseille
France

Sergio MARCHISIO
Sapienza University of Rome
Italy

Attila MATAS
Senior Satellite Regulatory Consultant
Geneva
Switzerland

Kai-Uwe SCHROGL
German Federal Ministry for
Economic Affairs and Energy
Berlin
Germany

Jenni TAPIO
Ministry of Economic Affairs
and Employment
and
University of Helsinki
Finland

Leslie I. TENNEN
Law Offices of Sterns and Tennen
Phoenix
USA

Index

D

E, F

G, H, I

J, L

M, N

O, P

T

U, V, W